AF877834

U.S.
CARRIERS

A COMPLETE ILLUSTRATED HISTORY
[FROM CV-1 to CV-81]

Edited By
Marc Schulman

U.S. CARRIERS
A COMPLETE ILLUSTRATED HISTORY
[FROM CV-1 to CV-81]

Edited by Marc Schulman

Published by HistoryCentral.com Press, New York
177A East Main Street • New Rochelle, NY 10801
Email: info@historycentral.com

© 2024 HistoryCentral.com Press
ISBN # 978-1-885881-85-4

© 2024 All rights reserved by the author Marc Schulman

The right of Marc Schulman to be identified as editor of this work has been asserted (with exceptions in the case of the reading excerpts that are reprinted here with permission), in accordance with the US 1976 Copyright 2007 Act. No part of this book may be reproduced or utilized in any form or by any means, electronic or mechanical, or by any information storage and retrieval system without the prior permission of the publisher. The only exception to this prohibition is "fair use" as defined by U.S. copyright law.

Table of Contents

CANCELLED/UNCOMPLETED CARRIERS

Preface

U.S. CARRIERS • A COMPLETE ILLUSTRATED HISTORY provides a detailed examination of the evolution of the United States Navy from the era of battleships to the era of aircraft carriers. We begin with the first carrier, the USS Langley CV-1 and discuss both the pre-war carriers and the World War II carriers that were pivotal in the Pacific theater, some of which remained in service through the Vietnam War, followed by a look at the postwar carriers such as the FDR and the Constellation. The nuclear age of the Naval surface ships is introduced with the USS Enterprise CVN 65.

The book details the newest carriers, including the George Washington. The histories are meticulously sourced from the US Navy's authoritative "History of US Fighting Ships," with minor edits made to enhance readability and ensure a seamless narrative that spans over a century of naval innovation and valor. Since the last volume of the US Navy's series was published in the 1980s, we have conducted additional research to bring the history of these ships up to date.

Explore the evolution of American carrier classes, a key focus of this book. It tracks the technological progress and strategic doctrines that shaped their design and use, highlighting how each class responded to the expanding needs of naval aviation and showcased American ingenuity and foresight.

Step into the past with the vividly colorized images within these pages, all sourced from the esteemed collections of the National Archives, the Library of Congress, and the US Navy Collection. The photographs, originally in black and white until the 1960s, have been carefully colorized to breathe life into the scenes, allowing readers to appreciate the grandeur of these vessels in their full glory. The more recent photos are presented in their original color, providing a visually stunning journey through the history of the US Navy.

From the original Langley, constructed atop the hull of a transport ship as an experimental platform, to the latest USS Ford, which heralds a new class of supercarriers, this book offers an in-depth analysis of the development of American carriers. During the Second World War, the carrier became the key capital ship. With planes that could strike hundreds of miles in the distance it was the American carriers that won the Pacific war at battles such as Midway.

This book only covers the attack carriers and light carriers but does not cover the escort carrier, which played a crucial role in escorting ships across the Atlantic and defeating the U-Boat. Today, there are those who question whether the our super carriers are vulnerable, however, they still remain the backbone of the US Fleet.

CV-1 • USS LANGLEY

(AC-3: dp. 19,360; l. 542'; b. 65'; dr. 27'8"; s. 15 k.; cpl. 163; a. 4 4").

Jupiter (AC-3) was laid down on 18 October 1911 by Mare Island Navy Yard, Vallejo, Calif.; launched on 14 August 1912; sponsored by Mrs. Thomas F. Ruhm; and commissioned 7 April 1913, Comdr. Joseph M. Reeves in command.

After successfully passing her trials, Jupiter, the first electrically propelled ship of the U.S. Navy, embarked a Marine detachment at San Francisco and reported to the Pacific Fleet at Mazatlan, Mexico, on 27 April 1914, bolstering U.S. naval strength on the Mexican Pacific coast during the tense days of the Vera Cruz crisis. She remained on the Pacific coast until she departed for Philadelphia on 10 October. En route, the collier steamed through the Panama Canal on Columbus Day—the first vessel to transit it from west to east.

Prior to America's entry into World War I, she cruised the Atlantic and Gulf of Mexico attached to the Atlantic Fleet Auxiliary Division. The ship arrived at Norfolk on 6 April 1917 and, assigned to NOTS, interrupted her coaling operations by two cargo voyages to France in June 1917 and November 1918. She was back in Norfolk on 23 January 1919, whence she sailed for Brest, France, on 8 March for coaling duty in European waters to expedite the return of victorious veterans to the United States. Upon reaching Norfolk on 17 August, the ship was transferred to the west coast. Her conversion to an aircraft carrier was authorized on 11 July 1919, and she sailed to Hampton Roads, Va., on 12 December, where she decommissioned on 24 March 1920.

Jupiter was converted into the first U.S. aircraft carrier at the Navy Yard, Norfolk, Va., to conduct experiments in the new idea of seaborne aviation, a field of unlimited possibilities. Her name was changed to Langley on 11 April 1920; she was reclassified CV-1 and recommissioned on 20 March 1922, Comdr. Kenneth Whiting in command. Her statistics as an aircraft carrier are: CV-1: displacement. 11,500; length. 542'; beam. 65'; draft. 18'11"; speed. 15 k.; cpl. 468; armor. 4 5", a/c 55 (max.), class. Langley.

Langley was the scene of numerous momentous events as the first Navy carrier. On 17 October 1922, Lt. Virgil C. Griffin piloted the first plane, a V-E-7-SF, launched from her decks. Though this was not the first time an airplane had taken off from a ship, and though Langley was not the first ship with an installed flight deck, this one launching was of monumental importance to the modern

U.S. Navy. The era of the aircraft carrier was born, introducing into the Navy what was to become the vanguard of its forces in the future. With Langley underway nine days later, Lt. Comdr. G. DeC. Chevalier made the first landing in an Aeromarine. On 18 November, Commander Whiting, at the controls of a PT, was the first aviator to be catapulted from a carrier's deck.

By 15 January 1923, Langley had begun flight operations and tests in the Caribbean for carrier landings. In June, she steamed to Washington, D.C., to give a demonstration at a flying exhibition before civil and military dignitaries. She arrived at Norfolk on 13 June and commenced training along the Atlantic coast and Caribbean, which carried her through the end of the year. In 1924, Langley participated in more maneuvers and exhibitions and spent the summer at Norfolk for repairs and alterations. She departed for the west coast late in the year and arrived at San Diego on 29 November to join the Pacific Battle Fleet. For the next 12 years, she operated off the California coast and Hawaii, engaged in training fleet units, experimentation, pilot training, and tactical-fleet problems. On 25 October 1936, she put into Mare Island Navy Yard, Calif., for overhaul and conversion to a seaplane tender. Though her career as a carrier had ended, her well-trained pilots proved invaluable to the next two carriers, Lexington (CV-2) and Saratoga (CV-3).

Langley completed conversion on 26 February 1937 and was reclassified AV-3 on 11 April. She was assigned to the Aircraft Scouting Force and commenced tending operations out of Seattle, Sitka, Pearl Harbor, and San Diego. She departed for a brief deployment with the Atlantic Fleet from 1 February to 10 July 1939 and then steamed to assume her duties with the Pacific fleet at Manila, arriving on 24 September.

At the outbreak of World War II, Langley was anchored off Cavite, Philippine Islands. She departed on 8 December and proceeded to Balikpapan, Borneo, and Darwin, Australia, where she arrived on 1 January 1942. Until 11 January, Langley assisted the RAAF in running antisubmarine patrols out of Darwin. She was then assigned to American-British-Dutch-Australian forces assembling in Indonesia to challenge the Japanese thrust in that direction. She departed Fremantle, Australia, on 22 February in convoy and left the convoy five days later to deliver 32 P-40's to Tjilatjap, Java.

Early in the morning of 27 February, Langley rendezvoused with her antisubmarine screen, destroyers Whipple (DD-217) and Edsall (DD-219). At 1140, nine twin-engine enemy bombers attacked her. The first and second Japanese strikes were unsuccessful, but during the third, Langley took five hits. Aircraft topside burst into flames, steering was impaired, and the ship took a 10-degree list to port. Unable to negotiate the narrow mouth of Tjilatjap Harbor, Langley went dead in the water as in rushing water flooded her main motors. At 1332, the order to abandon ship was passed. The escorting destroyers fired nine 4-inch shells and two torpedoes into the old tender to ensure her sinking. She went down about 75 miles south of Tjilatjap with a loss of 16.

CV-2 • USS LEXINGTON

(The fourth Lexington (CV-2) was originally designated CC-1, laid down as a battlecruiser 8 January 1921 by Fore River Shipbuilding Co., Quincy, Mass., authorized to be completed as an aircraft carrier 1 July 1922; launched 3 October 1925; sponsored by Mrs. Theodore Douglas Robinson, wife of the Assistant Secretary of the Navy, and commissioned 14 December 1927, Capt. Albert W. Marshall in command.)

After fitting out and shakedown, Lexington joined the battle fleet at San Pedro, Calif., on 7 April 1928. Based there, she operated on the west coast with Aircraft Squadrons, Battle Fleet, in-flight training, tactical exercises, and battle problems. Each year, she participated in fleet maneuvers in the Hawaiians, in the Caribbean, off the Panama Canal Zone, and in the eastern Pacific. In the fall of 1941, she sailed with the battle force to the Hawaiians for tactical exercises.

On 7 December 1941, Lexington was at sea with TF 12 carrying marine aircraft from Pearl Harbor to reinforce Midway when word of the Japanese attack on Pearl Harbor was received. She immediately launched search planes to hunt for the Japanese fleet and, at midmorning, headed south to rendezvous with Indianapolis and Enterprise task forces to conduct a search southwest of Oahu until returning to Pearl Harbor on 13 December.

Lexington sailed the next day to raid Japanese forces on Jaluit to relieve pressure on Wake; these orders were canceled 20 December, and she was directed to cover the Saratoga force in reinforcing Wake. When the island fell 23 December, the two carrier forces were recalled to Pearl Harbor, arriving 27 December.

Lexington patrolled to block enemy raids in the Oahu Johnston-Palmyra triangle until 11 January 1942, when she sailed from Pearl Harbor as flagship for Vice Adm. Wilson Brown, commanding TF 11. On 16 February, the force headed for an attack on Rabaul, New Britain, scheduled for 21 February. While approaching the day previous, Lexington was attacked by two waves of enemy aircraft, nine planes to a wave. The carrier's own combat air patrol and anti-aircraft fire splashed 17 of the attackers. During a single sortie, Lt. E. H. (Butch) O'Hare won the Medal of Honor by downing five planes.

Her offensive patrols in the Coral Sea continued until 6 March, when she rendezvoused with Yorktown's TF 17 for a thoroughly successful surprise attack flown over the Owen

Stanley mountains of New Guinea to inflict heavy damage on shipping and installations at Salamaua and Lae on 10 March. She now returned to Pearl Harbor arriving 26 March.

Lexington's task force sortied from Pearl Harbor on 15 April, rejoining TF 17 on 1 May. As Japanese fleet concentrations threatening the Coral Sea were observed, Lexington and Yorktown moved into the sea to search for the enemy's force, covering a projected troop movement. The Japanese must now be blocked in their southward expansion, or sea communication with Australia and New Zealand would be cut, and the dominions threatened with invasion.

On 7 May, search planes reported contact with an enemy carrier task force, and Lexington's air group flew an eminently successful mission against it, sinking light carrier Shoho. Later that day, 12 bombers and 15 torpedo planes from still unlocated heavy carriers Shokaku and Zuikaku were intercepted by fighter groups from Lexington and Yorktown, who splashed nine enemy aircraft.

On the morning of the 8th, a Lexington plane located Shokaku's group; a strike was immediately launched from the American carriers, and the Japanese ship was heavily damaged.

The enemy penetrated the American carriers at 1100, and 20 minutes later, Lexington was struck by a torpedo to port. Seconds later, a second torpedo hit the port directly abreast of the bridge. At the same time, she took three bomb hits from enemy dive bombers, producing a 7-degree list to port and several raging fires. By 1300, her skilled damage control parties had brought the fires under control and returned the ship to an even keel, making 25 knots, and she was ready to recover her air group. Then suddenly, Lexington was shaken by a tremendous explosion caused by the ignition of gasoline vapors below, and again, the fire raged out of control. At 1508, Capt. Frederick C. Sherman, fearing for the safety of men working below, secured salvage operations and ordered all hands to the flight deck. In 1707, he ordered, "abandon ship!" and the orderly disembarkation began, men going over the side into the warm water, almost immediately to be picked up by nearby cruisers and destroyers. Admiral Fitch and his staff transferred to cruiser Minneapolis. Captain Sherman and his executive officer, Comdr. M. T. Seligman ensured all their men were safe, and they were the last to leave their ship.

Lexington blazed on, flames shooting hundreds of feet into the air. Destroyer Phelps closed to 1500 yards and fired two torpedoes into her hull. With one last heavy explosion, the gallant Lexington sank at 1956, in 15°20'S, 120°30'E. She was part of the price that was paid to halt the Japanese overseas empire and safeguard Australia and New Zealand, but perhaps an equally great contribution had been her pioneer role in developing the naval aviators and the techniques which played so vital a role in ultimate victory in the Pacific.

Lexington received two battle stars for World War II service.

CV-3 • USS SARATOGA

(Displacement. 33,000; length. 888'0"; beam. 106'0"; draft. 24'11.5"; speed. 33.91 k.; complement. 2,111; armor. 8 8", 12 5", 4 6 pdrs., 81 dct; class. Lexington.)

The fifth Saratoga (CV-3) was laid down on 25 September 1920 as Battle Cruiser #3 by the New York Shipbuilding Co., Camden, N.J.; ordered converted to an aircraft carrier and reclassified CV-3 on 1 July 1922 in accordance with the Washington Treaty limiting naval armaments, launched on 7 April 1925, sponsored by Mrs. Curtis D. Wilbur, wife of the Secretary of the Navy, and commissioned on 16 November 1927, Capt. Harry E. Yarnell in command.

Saratoga, the first fast carrier in the United States Navy, quickly proved the value of her type. She sailed from Philadelphia on 6 January 1928 for shakedown; and, on 11 January, her air officer, the future World War II hero, Marc A. Mitscher, landed the first aircraft on board. In an experiment on 27 January, the rigid airship Los Angeles (ZR-3) moored to Saratoga's stern and took on fuel and stores.

Saratoga sailed for the Pacific via the Panama Canal. She was diverted briefly between 14 and 16 February to carry marines to Corinto, Nicaragua, and finally joined the Battle Fleet at San Pedro, California, on 21 February. The rest of the year was spent in training and final machinery shakedown.

On 15 January 1929, Saratoga sailed from San Diego with the Battle Fleet to participate in her first fleet exercise, Fleet Problem IX. In a daring move, Saratoga was detached from the fleet with only a single cruiser as escort to make a wide sweep to the south and "attack" the Panama Canal, which was defended by the Scouting Fleet and Saratoga's sister ship, Lexington. She successfully launched her strike on 26 January, and despite being "sunk" three times later in the day, proved the versatility of a fast task force centered around a carrier. The idea was incorporated into fleet doctrine and reused the following year in Fleet Problem X in the Caribbean. This time, however, Saratoga and carrier Langley were "disabled" by a surprise attack from Lexington, showing how quickly air power could swing the balance in a naval action.

Following the fleet concentration in the Caribbean, Saratoga took part in the Presidential Review at Norfolk in May and returned to San Pedro on 21 June 1930.

During the remaining decade before World War II, Saratoga exercised in the San Diego-San Pedro area, except for the annual fleet problems and regular overhauls at the Bremerton Navy Yard. In the fleet problems, Saratoga continued to assist in the development of fast carrier tactics, and her importance was recognized by the fact that she was always a high priority target for the opposing forces. The fleet problem for 1932 was planned for Hawaii, and, by coincidence, occurred during the peak of the furor following the "Manchurian incident" in which Japan started on the road to World War II. Saratoga exercised in the Hawaii area from 31 January to 19 March and returned to Hawaii for fleet exercises the following year between 23 January and 28 February 1933. On the return trip to the west coast, she launched a successful air "attack" on the Long Beach area.

Exercises in 1934 took Saratoga to the Caribbean and the Atlantic for an extended period, from 9 April to 9 November, and were followed by equally extensive operations with the United States Fleet in the Pacific the following year. Between 27 April and 6 June 1936, she participated in a fleet problem in the Canal Zone, and she then returned with the fleet to Hawaii for exercises from 16 April to 28 May 1937. On 15 March 1938, Saratoga sailed from San Diego for Fleet Problem XIX, again conducted off Hawaii. During the second phase of the problem, Saratoga launched a surprise air attack on Pearl Harbor from a point 100 miles off Oahu, setting a pattern that the Japanese copied in December 1941. During the return to the west coast, Saratoga and Lexington followed this feat with "strikes" on Mare Island and Alameda. Saratoga was under overhaul during the 1939 fleet concentration, but, between 2 April and 21 June 1940, she participated in Fleet Problem XXI, the last to be held due to the deepening world crisis.

Between 14 and 29 October 1940, Saratoga transported a draft of military personnel from San Pedro to Hawaii, and, on 6 January 1941, she entered the Bremerton Navy Yard for a long-deferred modernization, including widening her flight deck forward and fitting a blister on her starboard side and additional small anti-aircraft guns. Departing Bremerton on 28 April 1941, the carrier participated in a landing force exercise in May and made two trips to Hawaii between June and October as the diplomatic crisis with Japan came to a head.

When the Japanese struck at Pearl Harbor on 7 December 1941, Saratoga was just entering San Diego after an interim drydocking at Bremerton. She hurriedly got underway the following day as the nucleus of a third carrier force (Lexington and Enterprise were already at sea), carrying Marine aircraft intended to reinforce the vulnerable garrison on Wake Island. Presence of

these aircraft on board made Saratoga the logical choice for the actual relief effort. She reached Pearl Harbor on 15 December and stopped only long enough to fuel. She then rendezvoused with Tangier (AV-8), which had relief troops and supplies on board, while Lexington and Enterprise provided distant cover for the operation. However, the Saratoga force was delayed by the low speed of its oiler and by a decision to refuel destroyers on 21 December. After receiving reports of Japanese carrier aircraft over the island and Japanese landings on it, the relief force was recalled on 22 December. Wake fell the next day.

Saratoga continued operations in the Hawaiian Island region; but, on 11 January 1942, when heading towards a rendezvous with Enterprise, 500 miles southwest of Oahu, she was hit without warning by a deep-running torpedo fired by Japanese submarine, I-16. Although six men were killed and three firerooms were flooded, the carrier reached Oahu under her own power. There, her 8-inch guns, useless against aircraft, were removed for installation in shore defenses, and the carrier proceeded to the Bremerton Navy Yard for permanent repairs and installation of a modern anti-aircraft battery.

Saratoga departed Puget Sound on 22 May for San Diego. She arrived there on 25 May and was training her air group when intelligence was received of an impending Japanese assault on Midway. Due to the need to load planes and stores and to collect escorts, the carrier was unable to sail until 1 June and arrived at Pearl Harbor on the 6th after the Battle of Midway had ended. She departed Pearl Harbor on 7 June after fueling; and, on 11 June, transferred 34 aircraft to Hornet and Enterprise to replenish their depleted air groups. The three carriers then turned north to counter Japanese activity reported in the Aleutians, but the operation was canceled and Saratoga returned to Pearl Harbor on 13 June.

Between 22 and 29 June, Saratoga ferried Marine and Army aircraft to the garrison on Midway. On 7 July, she sailed for the southwest Pacific; and, from 28 to 30 July, she provided air cover for landing rehearsals in the Fiji Islands in preparation for landings on Guadalcanal. As flagship of Rear Admiral F. J. Fletcher, Saratoga opened the Guadalcanal assault early on 7 August when she turned into the wind to launch aircraft. She provided air cover for the landings for the next two days. On the first day, a Japanese air attack was repelled before it reached the carriers, but since further attacks were expected, the carrier force withdrew on the afternoon of 8 August towards a fueling rendezvous. As a result, it was too far away to retaliate after four Allied cruisers were sunk that night in the Battle of Savo Island. The carrier force continued to operate east of the Solomons, protecting the sealanes to the beachhead and awaiting a Japanese naval counterattack.

The counterattack began to materialize when a Japanese transport force was detected on 23 August, and Saratoga launched a strike against it. The aircraft were unable to find the enemy, however, and spent the night on Guadalcanal. As they were returning on board the next day, the first contact report on enemy carriers was received. Two hours later, Saratoga launched a strike which sent Japanese carrier Ryujo to the bottom. Later in the afternoon, as an enemy strike from other carriers was detected, Saratoga hastily launched the aircraft on her deck, and these found and damaged seaplane tender Chitose. Meanwhile, due to cloud cover, Saratoga escaped detection by the Japanese aircraft, which concentrated their attack on, and damaged, Enterprise. The American force fought back fiercely and weakened enemy air strength so severely that the Japanese recalled their transports before they reached Guadalcanal.

After landing her returning aircraft at night on 24 August, Saratoga refueled on the 25th and resumed her patrols east of the Solomons. A week later, a destroyer reported torpedo wakes heading toward the carrier, but the 888-foot flattop could not turn quickly enough. A minute later, a torpedo from I-26 slammed into the blister on her starboard side. The torpedo killed no one and only flooded one fireroom, but the impact caused short circuits which damaged Saratoga's turbo-electric propulsion system and left her dead in the water. Cruiser Minneapolis took the carrier under tow while she flew her aircraft off to shore bases. By [341] early afternoon, Saratoga's engineers had improvised a circuit out of the burned wreckage of her main control board and had given her a speed of 10 knots. After repairs at Tongatabu from 6 to 12 September, Saratoga arrived at Pearl Harbor on 21 September for permanent repairs.

Saratoga sailed from Pearl Harbor on 10 November and proceeded, via Fiji, to Noumea which she reached on 5 December. She operated in the vicinity of Noumea for the next twelve months, providing air cover for minor operations and protecting American forces in the Eastern Solomons. Between 17 May and 31 July 1943, she was reinforced by the British carrier, Victorious, and, on 20 October, she was joined by Princeton (CVL-23). As troops stormed ashore on Bougainville on 1 November, Saratoga's aircraft neutralized nearby Japanese airfields on Buka. Then, on 5 November, in response to reports of Japanese cruisers concentrating at Rabaul to counterattack the Allied landing forces, Saratoga conducted perhaps her most brilliant strike of the war. Her aircraft penetrated the heavily defended port and disabled most of the Japanese cruisers, ending the surface threat to Bougainville. Saratoga, herself, escaped unscathed and returned to raid Rabaul again on 11 November.

Saratoga and Princeton were then designated the Relief Carrier Group for the offensive in the Gilberts; and, after striking Nauru on 19 November, they rendezvoused on 23 November with the transports carrying garrison troops to Makin and Tarawa. The carriers provided air cover until the transports reached their destinations, and then maintained air patrols over Tarawa. By this time, Saratoga had steamed over a year without repairs, and she was detached on 30 November to return to the United States. She underwent overhaul at San Francisco from 9 December 1943 to 3 January 1944, and had her antiaircraft battery augmented for the last time, receiving 60 40-millimeter guns in place of 36 20-millimeter guns.

The carrier arrived at Pearl Harbor on 7 January, and, after a brief period of training, sailed from Pearl Harbor on 19 January with light carriers, Langley and Princeton, to support the drive

in the Marshalls. Her aircraft struck Wotje and Taroa for three days, from 29 to 31 January, and then pounded Engebi, the main island at Eniwetok, the 3d to the 6th and from the 10th to the 12th of February. Her planes delivered final blows to Japanese defenses on the 16th, the day before the landings, and provided close air support and CAP over the island until 28 February.

Saratoga then took leave of the main theaters of the Pacific war for almost a year, to carry out important but less spectacular assignments elsewhere. Her first task was to help the British initiate their carrier offensive in the Far East. On 4 March, Saratoga departed Majuro with an escort of three destroyers, and sailed via Espiritu Santo; Hobart, Tasmania; and Fremantle, Australia, to join the British Eastern Fleet in the Indian Ocean. She rendezvoused at sea on 27 March with the British force, composed of carrier, Illustrious, and four battleships with escorts, and arrived with them at Trincomalee, Ceylon, on 31 March. On 12 April, the French battleship, Richelieu, arrived, adding to the international flavor of the force. During the next two days, the carriers conducted intensive training at sea during which Saratoga's fliers tried to impart some of their experience to the British pilots. On 16 April, the Eastern Fleet, with Saratoga, sailed from Trincomalee, and, on the 19th, the aircraft from the two carriers struck the port of Sabang, off the northwest tip of Sumatra. The Japanese were caught by surprise by the new offensive, and much damage was done to port facilities and oil reserves. The raid was so successful that Saratoga delayed her departure in order to carry out a second. Sailing again from Ceylon on 6 May, the force struck at Soerabaja, Java, on 17 May with equally successful results. Saratoga was detached the following day, and passed down the columns of the Eastern Fleet as the Allied ships rendered honors to and cheered each other.

Saratoga arrived at Bremerton, Washington, on 10 June 1944 and was under repair there through the summer. On 24 September, she arrived at Pearl Harbor and commenced her second special assignment, training night fighter squadrons. Saratoga had experimented with night flying as early as 1931, and many carriers had been forced to land returning aircraft at night during the war; but, only in August 1944, did a carrier, Independence, receive an air group specially equipped to operate at night. At the same time, Carrier Division 11, composed of Saratoga and Ranger (CV-4), was commissioned at Pearl Harbor to train night pilots and develop night flying doctrine. Saratoga continued this important training duty for almost four months, but as early as October, her division commander was warned that "while employed primarily for training, Saratoga is of great value for combat and is to be kept potentially available for combat duty." The call came in January 1945. Light carriers like Independence had proved too small for safe night operations, and Saratoga was rushed out of Pearl Harbor on 29 January 1945 to form a night fighter task group with Enterprise for the Iwo Jima operation.

Saratoga arrived at Ulithi on 7 February and sailed three days later, with Enterprise and four other carrier task groups. After landing rehearsals with marines at Tinian on 12 February, the carrier force carried out diversionary strikes on the Japanese home islands on the night of 16 and 17 February before the landings on Iwo Jima. Saratoga was assigned to provide fighter cover while the remaining carriers launched the strikes on Japan, but, in the process, her fighters raided two Japanese airfields. The force fueled on 18 and 19 February; and, on 21 February, Saratoga was detached with an escort of three destroyers to join the amphibious forces and carry out night patrols over Iwo Jima and night heckler missions over nearby Chi-chi Jima. However, as she approached her operating area at 1700 on the 21st, an air attack developed, and taking advantage of low cloud cover and Saratoga's insufficient escort, six Japanese planes scored five hits on the carrier in three minutes. Saratoga's flight deck forward was wrecked, her starboard side was holed twice and large fires were started in her hangar deck, while she lost 123 of her crew dead or missing. Another attack at 1900 scored an additional bomb hit. By 2015, the fires were under control and the carrier was able to recover aircraft, but she was ordered to Eniwetok and then to the west coast for repairs, and arrived at Bremerton on 16 March.

On 22 May, Saratoga departed Puget Sound fully repaired, and she resumed training pilots at Pearl Harbor on 3 June. She ceased training duty on 6 September, after the Japanese surrender, and sailed from Hawaii on 9 September transporting

3,712 returning naval veterans home to the United States under Operation "Magic Carpet." By the end of her "Magic Carpet" service, Saratoga had brought home 29,204 Pacific war veterans, more than any other individual ship. At the time, she also held the record for the greatest number of aircraft landed on a carrier, with a lifetime total of 98,549 landings in 17 years.

With the arrival of large numbers of Essex-class carriers, Saratoga was surplus to postwar requirements, and she was assigned to Operation "Crossroads" at Bikini Atoll to test the effect of the atomic bomb on naval vessels. She survived the first blast, an air burst on 1 July, with only minor damage, but was mortally wounded by the second on 25 July, an underwater blast which was detonated under a landing craft 500 yards from the carrier. Salvage efforts were prevented by radioactivity, and seven and one-half hours after the blast, with her funnel collapsed across her deck, Saratoga slipped beneath the surface of the lagoon. She was struck from the Navy list on 15 August 1946.

Saratoga received seven battle stars for her World War II service.

CV-4 • USS RANGER

(Displacement. 14,500; length. 769'; beam. 81'8"; cw. 86'; draft. 19'8"; speed. 29.25 k.; complement. 1,788; armament. 8 5"; class. Ranger).

The sixth Ranger (CV-4), the first ship of the Navy to be designed and built from the keel up as an aircraft carrier, was laid down on 26 September 1931 by Newport News Shipbuilding & Dry Dock Co., Newport News, Va.; launched on 25 February 1933; sponsored by Mrs. Herbert Hoover; and commissioned at the Norfolk Navy Yard on 4 June 1934, Capt. Arthur L. Bristol in command.

Ranger conducted her first air operations off Cape Henry on 6 August 1934 and departed Norfolk on the 17th for a shakedown training cruise that took her to Rio de Janeiro, Buenos Aires, and Montevideo. She returned to Norfolk on 4 October for operations off the Virginia Capes until 28 March 1935, when she sailed for the Pacific. Transiting the Panama Canal on 7 April, she arrived in San Diego on the 15th. For nearly four years, she participated in fleet problems reaching Hawaii and in western seaboard operations that took her as far south as Callao, Peru, and north as Seattle, Wash. On 4 January 1939, she departed San Diego for winter fleet operations in the Caribbean out of Guantanamo Bay, Cuba. She then steamed north to Norfolk, Va., arriving on 18 April.

Ranger cruised along the eastern seaboard out of Norfolk and into the Caribbean Sea. In the fall of 1939, she commenced Neutrality Patrol operations, operating out of Bermuda along the trade routes of the middle Atlantic and up the eastern seaboard to Argentia, Newfoundland. She was returning to Norfolk from an ocean patrol extending to Port-of-Spain, Trinidad when the Japanese attacked Pearl Harbor. Arriving in Norfolk on 8 December, she sailed on the 21st for patrol in the South Atlantic. She then entered the Norfolk Navy Yard for repairs on 22 March 1942.

Ranger served as the flagship of Rear Adm. A. B. Cook, Commander, Carriers, Atlantic Fleet, until 6 April 1942, when he was relieved by Rear Adm. Ernest D. McWhorter, who also broke his flag in Ranger.

Steaming to Quonset Point, R.I., Ranger loaded 68 Army P-40 planes and men of the Army's 33rd Pursuit Squadron, put to sea on 22 April, and launched the Army squadron on 10 May to land at Accra, on the Gold Coast of Africa. She returned to Quonset Point on 28 May 1942, made a patrol to Argentia, then stood out of Newport on 1 July with 72 Army P-40 pursuit planes, which she launched off the coast of Africa for Accra on the 19th. After calling Trinidad, she returned to Norfolk for local battle practice until 1 October. She then based her training in Bermuda in company with four escort aircraft carriers newly converted from tankers to meet the need for naval air power in the Atlantic.

The only large carrier in the Atlantic Fleet, Ranger led the task force, comprising herself and four Sangamon-class escort carriers, that provided air superiority during the amphibious invasion of German-dominated French Morocco, which commenced on the morning of 8 November 1942.

It was still dark at 0615 that day when Ranger, stationed 30 miles northwest of Casablanca, began launching her aircraft to support the landings made at three points on the Atlantic coast of North Africa. Nine of her Wildcats attacked the Rabat and Rabat-Salé airdromes, headquarters of the French air forces in Morocco. Without loss to themselves, they destroyed seven planes on one field and 14 bombers on the other. Another flight destroyed seven planes on the Port Lyautey field. Some of Ranger's planes strafed four French destroyers in Casablanca Harbor while others strafed and bombed nearby batteries.

The carrier launched 496 combat sorties in the three-day operation. Her attack aircraft scored two direct bomb hits on the French destroyer leader Albatros, completely wrecking her forward half and causing 300 casualties. They also attacked French cruiser Primauguet as she sortied from Casablanca Harbor, dropped depth charges within lethal distance of two submarines, and knocked out coastal defense and anti-aircraft batteries. They destroyed more than 70 enemy planes on the ground and shot down 15 in aerial combat. But 16 planes from Ranger were lost or damaged beyond repair. It was estimated that 21 light enemy tanks were immobilized and some 86 military vehicles destroyed—most of them troop-carrying trucks.

Casablanca capitulated to the American invaders on 11 November 1942, and Ranger departed the Moroccan coast on 12 November, returning to Norfolk, Va., on the 23rd.

Following training in the Chesapeake Bay, the carrier underwent an overhaul in the Norfolk Navy Yard from 16 December 1942 to 7 February 1943. She next transported 75 P-40-L Army pursuit planes to Africa, arriving in Casablanca on 23 February. Then, she patrolled and trained pilots along the New England coast, steaming as far north as Halifax, Nova Scotia. Departing Halifax on 11 August, she joined the British Home Fleet at Scapa Flow, Scotland, on 19 August and patrolled the approaches to the British Isles.

The U.S.S. Ranger's flight deck bristles with more than 50 planes as the ship cruises the Atlantic in 1941. Although smaller than most major U.S. carriers of the era, the ship could handle up to 72 aircraft.

Ranger departed Scapa Flow with the Home Fleet on 2 October to attack German shipping in Norwegian waters. The objective of the force was the Norwegian port of Bodø. The task force reached the launch position off Vestfjord before dawn on 4 October, completely undetected. At 0618, Ranger launched 20 Dauntless dive bombers and an escort of eight Wildcat fighters. One division of dive bombers attacked the 8,000-ton freighter La Plata while the rest continued north to attack a small German convoy. They severely damaged a 10,000-ton tanker and a smaller troop transport. They also sank two of four small German merchantmen in the Bodø roadstead.

A second Ranger attack group of 10 Avengers and six Wildcats destroyed a German freighter and a small coaster and bombed yet another troop-laden transport. Three Ranger planes were lost to anti-aircraft fire. On the afternoon of 4

October, Ranger was finally located by three German aircraft, but her combat air patrol shot down two of the enemy planes and chased off the third.

Ranger returned to Scapa Flow on 6 October 1943. She patrolled with the British Second Battle Squadron in waters reaching Iceland and then departed Hvalfjord on 26 November, arriving in Boston on 4 December. On 3 January 1944, she became a training carrier out of Quonset Point, R.I. This duty was interrupted on 20 April when she arrived at Staten Island, N.Y., to load 76 P-38 fighter planes together with Army, Navy, and French Naval personnel for transport to Casablanca. Sailing on 24 April, she arrived in Casablanca on 4 May. There, she unloaded Army aircraft destined for stateside repairs and embarked military passengers for the return to New York.

Touching at New York on 16 May, Ranger then entered the Norfolk Navy Yard to strengthen her flight deck and install a new type of catapult, radar, and associated gear that provided her with the capacity for night fighter interceptor training. On 11 July 1944, she departed Norfolk, transited the Panama Canal five days later, and embarked several hundred Army passengers at Balboa for transportation to San Diego, arriving there on 25 July.

After embarking the men and aircraft of Night Fighting Squadron 102 and nearly a thousand marines, she sailed for Hawaiian waters on 28 July, reaching Pearl Harbor on 3 August. She conducted night carrier training operations out of Pearl Harbor for the next three months.

Ranger departed Pearl Harbor on 18 October to train pilots for combat duty. Operating out of San Diego under Commander Fleet Air, Alameda, she continued training air groups and squadrons along the California coast throughout the remainder of the war.

Departing San Diego on 30 September 1945, she embarked civilian and military passengers at Balboa and then steamed for New Orleans, arriving on 18 October. Following Navy Day celebrations there, she sailed on 30 October for brief operations at Pensacola. After calling at Norfolk, she entered the Philadelphia Naval Shipyard on 18 November for overhaul. She remained on the eastern seaboard until decommissioned at the Norfolk Naval Shipyard on 18 October 1946. Struck from the Navy list on 29 October 1946, she was sold for scrap to Sun Shipbuilding & Dry Dock Co., Chester, Pa., on 28 January 1947.

Ranger received two battle stars for World War II service.

CV-5 • USS YORKTOWN

(Displacement 19,800; length 809'6"; beam 83'1"; draft 28'0"; speed 32.5 k.; complement 2,919; armament 8 5", 22 .60-cal. mg., dct 81-85; class Yorktown).

The third Yorktown (CV-5) was laid down on 21 May 1934 at Newport News, Va., by the Newport News Shipbuilding and Drydock Co.; launched on 4 April 1936; sponsored by Mrs. Franklin D. Roosevelt; and commissioned at the Naval Operating Base (NOB), Norfolk, Va., on 30 September 1937, Capt. Ernest D. McWhorter in command.

After fitting out, the aircraft carrier trained in Hampton Roads and in the southern drill grounds off the Virginia Capes into January of 1938, conducting carrier qualifications for her newly embarked air group.

Yorktown sailed for the Caribbean on 8 January 1938 and arrived at Culebra, Puerto Rico, on 13 January. Over the ensuing month, the carrier conducted her shakedown, touching at Charlotte Amalie, St. Thomas, Virgin Islands; Gonaives, Haiti; Guantanamo Bay, Cuba, and Cristobal, Panama Canal Zone. Departing Colon Bay, Cristobal, on 1 March, Yorktown sailed for Hampton Roads and arrived there on the 6th and shifted to the Norfolk Navy Yard the next day for post-shakedown availability.

After undergoing repairs through the early autumn of 1938, Yorktown shifted from the navy yard to NOB Norfolk on 17 October and soon headed for the Southern Drill Grounds for training.

Yorktown operated off the eastern seaboard, ranging from Chesapeake Bay to Guantanamo Bay, into 1939. As flagship for Carrier Division (CarDiv) 2, she participated in her first war game, Fleet Problem XX, along with her sister ship Enterprise (CV-6) in February 1939. The scenario for the exercise called for one fleet to control the sea lanes in the Caribbean against the incursion of a foreign European power while maintaining sufficient naval strength to protect vital American interests in the Pacific. The maneuvers were witnessed, in part, by President Roosevelt, embarked in the heavy cruiser Houston (CA-30).

The critique of the operation revealed that carrier operations, a part of the scenarios for the annual exercises since the entry of Langley (CV-1) into the war games in 1925, had achieved a new peak of efficiency. Despite the inexperience of

Yorktown and Enterprise, comparative newcomers to the Fleet, both carriers made significant contributions to the success of the problem. The planners had studied the employment of carriers and their embarked air groups in connection with convoy escort, antisubmarine defense, and various attack measures against surface ships and shore installations. In short, they worked to develop the tactics that would be used when war actually came.

Following Fleet Problem XX, Yorktown returned briefly to Hampton Roads before sailing for the Pacific on 20 April. Transiting the Panama Canal a week later, Yorktown soon commenced a regular routine of operations with the Pacific Fleet. Operating out of San Diego into 1940, the carrier participated in Fleet Problem XXI that April.

Fleet Problem XXI, a two-part exercise, included some of the operations that would characterize future warfare in the Pacific. The first part of the exercise was devoted to training in making plans and estimates, in screening and scouting, in coordination of combatant units, and in employing fleet and standard dispositions. The second phase included training in convoy protection, the seizure of advanced bases, and, ultimately, the decisive engagement between the opposing fleets. The last pre-war exercise of its type, Fleet Problem XXI, contained two exercises (comparatively minor at the time) where air operations played a major role. Fleet Joint Air Exercise 114A prophetically pointed out the need to coordinate Army and Navy defense plans for the Hawaiian Islands, and Fleet Exercise 114 proved that aircraft could be used for high-altitude tracking of surface forces, a significant role for planes that would be fully realized in the war to come.

With the retention of the Fleet in Hawaiian waters after the conclusion of Fleet Problem XXI, Yorktown operated in the Pacific off the west coast of the United States and in Hawaiian waters until the following spring when the success of German U-boats preying upon British shipping in the Atlantic required a shift of American naval strength. Thus, to reinforce the Atlantic Fleet, the Navy transferred a substantial force from the Pacific including Yorktown, a battleship division, and accompanying cruisers and destroyers.

Yorktown departed Pearl Harbor on 20 April 1941 in company with Warrington (DD-383), Somers (DD-381), and Jouett (DD-396); headed southeast, transited the Panama Canal on the night of 6 and 7 May, and arrived at Bermuda on the 12th. From that time to the entry of the United States into the war, Yorktown conducted four patrols in the Atlantic, ranging from Newfoundland to Bermuda and logging 17,642 miles steamed while enforcing American neutrality.

Although Adolf Hitler had forbidden his submarines to attack American ships, the men who manned the American naval vessels were not aware of this policy and operated on a wartime footing in the Atlantic.

On 28 October, while Yorktown, battleship New Mexico (BB-41), and other American warships were screening a convoy, a destroyer picked up a submarine contact and dropped depth charges while the convoy itself made an emergency starboard turn, the first of the convoy's three emergency changes of course. Late that afternoon, engine repairs to one of the ships in the convoy, Empire Pintail, reduced the convoy's speed to 11 knots.

During the night, the American ships intercepted strong German radio signals, indicating submarines probably in the vicinity reporting the group. Rear Admiral H. Kent Hewitt, commanding the escort force, sent a destroyer to sweep astern of the convoy to destroy the U-boat or at least to drive him under.

The next day, while cruiser scout planes patrolled overhead, Yorktown and Savannah (CL-42) fueled their escorting destroyers, finishing the task just at dusk. On the 30th, Yorktown was preparing to fuel three destroyers when other escorts made sound contacts. The convoy subsequently made 10 emergency turns while Morris (DD-417) and Anderson (DD-411) dropped depth charges, and Hughes (DD-410) assisted in developing the contact. Anderson later made two more depth charge attacks, noticing "considerable oil with slick spreading but no wreckage."

The short-of-war period was becoming more like the real thing as each day went on. Elsewhere on 30 October and more than a month before Japanese planes attacked Pearl Harbor, U-562 torpedoed the destroyer Reuben James (DD-245), sinking her with a heavy loss of life, the first loss of an American warship in World War II.

After another Neutrality Patrol stint in November, Yorktown put into Norfolk on 2 December and was there five days later

when American fighting men in Hawaii were rudely awakened to find their country at war.

The early news from the Pacific was bleak: the Pacific Fleet had taken a beating. With the battle line crippled, the unhurt American carriers assumed great importance. There were, on 7 December, only three in the Pacific: Enterprise, Lexington (CV-2), and Saratoga (CV-3). While Ranger (CV-4), Wasp (CV-7), and the recently commissioned Hornet (CV-8) remained in the Atlantic, Yorktown departed Norfolk on 16 December 1941 and sailed for the Pacific, her secondary gun galleries studded with new 20-millimeter Oerlikon machine guns. She reached San Diego, Calif., on 30 December 1941 and soon became flagship for Rear Admiral Frank Jack Fletcher's newly formed Task Force (TF) 17.

The carrier's first mission in her new theater was to escort a convoy carrying Marine reinforcements to American Samoa. Departing San Diego on 6 January 1942, Yorktown and her consorts covered the movement of marines to Tutuila and Pago Pago to augment the garrison already there.

Having safely covered that troop movement, Yorktown, in company with her sister ship Enterprise, departed Samoan waters on 25 January. Six days later, TF 8, built around Enterprise, and TF 17, built around Yorktown, parted company. The former headed for the Marshall Islands, the latter for the Gilberts, each bound to take part in the first American offensive of the war, the Marshalls-Gilberts raids.

At 0517, Yorktown, screened by Louisville (CA-28) and St. Louis (CL-49) and four destroyers, launched 11 torpedo planes (Douglas TBD-1 Devastators) and 17 scout bombers (Douglas SBD-3 Dauntlesses) under the command of Comdr. Curtis W. Smiley. Those planes hit what Japanese shore installations and shipping they could find at Jaluit, but adverse weather conditions hampered the mission in which six planes were lost. Other Yorktown planes attacked Japanese installations and ships at Makin and Mili Atolls.

The attack by TF 17 on the Gilberts had apparently been a complete surprise since the American force encountered no enemy surface ships. A single, four-engined, Kawanishi E7K "Mavis" patrol-bomber seaplane attempted to attack American destroyers that had been sent astern in the hope of recovering planes overdue from the Jaluit mission. Antiaircraft fire from the destroyers drove off the intruder before he could cause any damage.

Later, another "Mavis"-or possibly the same one that had attacked the destroyers-came out of low clouds 15,000 yards from Yorktown. The carrier with- held her antiaircraft fire in order not to interfere with the combat air patrol (CAP) fighters. Presently, the "Mavis," pursued by two Wildcats, disappeared behind a cloud. Within five minutes, the enemy patrol plane fell out of the clouds and crashed in the water.

Although TF 17 was slated to make a second attack on Jaluit, it was canceled because of heavy rainstorms and the approach of darkness. Therefore, the Yorktown force retired from the area.

Admiral Chester W. Nimitz later called the Marshalls-Gilberts raids "well conceived, well planned, and brilliantly executed." The results obtained by TF's 8 and 17 were noteworthy Nimitz continued in his subsequent report, because the task forces had been obliged to make their attacks somewhat blindly, due to lack of hard intelligence data on the Japanese-mandated islands.

Yorktown subsequently returned to Pearl Harbor and replenished there before she put to sea on 14 February, bound for the Coral Sea. On 6 March, she rendezvoused with TF 11-formed around Lexington and under the command of Rear Admiral Wilson Brown-and headed towards Rabaul and Gasmata to attack Japanese shipping there in an effort to check the Japanese advance and to cover the landing of Allied troops at Noumea, New Caledonia. However, as the two flattops-screened by a powerful force of eight heavy cruisers (including the Australian HMAS Australia) and 14 destroyers-steamed toward New Guinea, the Japanese continued their advance toward Australia with a landing on 7 March at the Huon Gulf, in the Salamana-Lae area on the eastern end of New Guinea.

Word of the Japanese operation prompted Admiral Brown to change the objective of TF 11's strike from Rabaul to the Salamana-Lae sector. On the morning of 10 March 1942, American carriers launched aircraft from the Gulf of Papua. Lexington flew off her air group commencing at 0749 and, 21 minutes later, Yorktown followed suit. While the choice of the gulf as the launch point for the strike meant that the planes would have to fly some 125 miles across the Owen Stanley mountains-a range not known for the best flying conditions-that approach provided security for the task force and ensured surprise.

In the attacks that followed, Lexington's SBD's from Scouting Squadron (VS) 2 commenced dive-bombing Japanese ships at Lae at 0922. The carrier's Torpedo Squadron (VT) 2 and Bombing Squadron (VB) 2 attacked shipping at Salamaua at 0938. Her fighters from Fighter Squadron (VF) 2 split up into four-plane attack groups: one strafed Lae and the other, Salamaua. Yorktown's planes followed on the heels of those from "Lady Lex." VB-5 and VT-5 attacked Japanese ships in the Salamaua area at 0950, while VS-5 went after auxiliaries moored close in shore at Lae. The fighters of VF-42 flew over Salamana on CAP until they determined that there was no air opposition and then strafed surface objectives and small boats in the harbor.

After carrying out their missions, the American planes returned to their carriers, and 103 planes of the 104 launched were back safely on board by noon. One SB3-2 of VS-2 had been downed by Japanese antiaircraft fire. The raid on Salamana and Lae was the first attack by many pilots of both carriers; and, while the resultant torpedo and bombing accuracy was inferior to that achieved in later actions, the operation gave the fliers invaluable experience which enabled them to do so well in the Battle of the Coral Sea and the Battle of Midway.

Task Force 11 retired at 20 knots on a southeasterly course until dark, when the ships steered eastward at 15 knots and made rendezvous with Task Group (TG) 11.7 (four heavy cruisers and four destroyers) under Rear Admiral John G. Crace, Royal Navy-the group that had provided cover for the carriers on their approach to New Guinea.

Yorktown resumed her patrols in the Coral Sea area, remaining at sea into April, out of reach of Japanese land-based aircraft and ready to carry out offensive operations whenever the opportunity presented itself. After the Lae-Salamaua raid, the situation in the South Pacific seemed temporarily stabilized, and Yorktown and her consorts in TF 17 put in to the undeveloped harbor at Tongatabu, in the Tonga Islands, for needed upkeep, having been at sea continuously since departing from Pearl Harbor on 14 February.

However, the enemy was soon on the move. To Admiral Nimitz, there seemed to be "excellent indications that the Japanese intended to make a seaborne attack on Port Moresby the first week in May." Yorktown accordingly departed Tongatabu on 27 April, bound once more for the Coral Sea. TF 11-commanded by Rear Admiral Aubrey W. Fitch, who had relieved Brown in Lexington-departed Pearl Harbor to join Fletcher's TF 17 and arrived in the vicinity of Yorktown's group, southwest of the New Hebrides Islands, on 1 May.

At 1517 the next afternoon, two Dauntlesses from VS-5 sighted a Japanese submarine, running on the surface. Three Devastators took off from Yorktown sped to the scene, and carried out an attack that only succeeded in driving the submarine under.

On the morning of the 3d, TF 11 and TF 17 were some 100 miles apart, engaged in fueling operations. Shortly before midnight, Fletcher received word from Australian-based aircraft that Japanese transports were disembarking troops and equipment at Tulagi in the Solomon Islands. Arriving soon after the Australians had evacuated the place, the Japanese landed to commence construction of a seaplane base there to support their southward thrust.

Yorktown accordingly set course northward at 27 knots. By daybreak on 4 May, she was within striking distance of the newly established Japanese beachhead and launched her first strike at 0701-18 F4F-3's of VF-42, 12 TBD's of VT-5, and 28 SBD's from VS and BY-5. Yorktown's air group made three consecutive attacks on enemy ships and shore installations at Tulagi and Gavutu on the south coast of Florida Island in the Solomons. Expending 22 torpedoes and 76 1,000-pound bombs in the three attacks, Yorktown's planes sank a destroyer (Kikuzuki), three minecraft, and four barges. In addition, Air Group 5 destroyed five enemy seaplanes, all at the cost of two F4F's lost (the pilots were recovered) and one TBD (whose crew was lost).

Meanwhile, that same day, TF 44, a cruiser-destroyer force under Rear Admiral Crace (RN), joined Lexington's TF 11, thus completing the composition of the Allied force on the eve of the crucial Battle of the Coral Sea.

Elsewhere, to the northward, the enemy was on his [536] way. Eleven troop-laden transports-escorted by destroyers and

covered by the light carrier Shoho, four heavy cruisers, and a destroyer-steamed toward Port Moresby. In addition, another Japanese task force-formed around the two Pearl Harbor veterans, carriers Shokaku and Zuikaku, and screened by two heavy cruisers and six destroyers-provided additional air cover.

On the morning of the 6th, Fletcher gathered all Allied forces under his tactical command as TF 17. At daybreak on the 7th, he dispatched Crace, with the cruisers and destroyers under his command, toward the Louisiade archipelago to intercept any enemy attempt to move toward Port Moresby.\

Meanwhile, while Fletcher moved northward with his two flattops and their screens in search of the enemy, Japanese search planes located the oiler Neosho (AO-23) and her escort, Sims (DD-409) and identified the former as a "carrier." Two waves of Japanese planes-first high level bombers and then dive bombers-attacked the two ships. Sims-her antiaircraft battery crippled by gun failures-took three direct hits and sank quickly with a heavy loss of life. Neosho was more fortunate in that, even after seven direct hits and eight near-misses, she remained afloat until, on the 11th, her survivors were picked up by Henley (DD-391) and her hulk sunk by the rescuing destroyer.

In their tribulation, Neosho and Sims had performed a valuable service, drawing off the planes that might otherwise have hit Fletcher's carriers. Meanwhile, Yorktown and Lexington's planes found Shoho and punished that Japanese light carrier unmercifully, sending her to the bottom. One of Lexington's pilots reported this victory with the radio message, "Scratch one flattop."

That afternoon, Shokaku and Zuikaku-still unlocated by Fletcher's forces-launched 27 bombers and torpedo planes to search for the American ships. Their flight proved uneventful until they ran into fighters from Yorktown and Lexington, who proceeded to down nine enemy planes in the ensuing dogfight.

Near twilight, three Japanese planes incredibly mistook Yorktown for their own carrier and attempted to land. The ship's gunfire, though, drove them off; and the enemy planes crossed Yorktown's bow and turned away out of range. Twenty minutes later, when three more enemy pilots made the mistake of trying to get into Yorktown's landing circle, the carrier's gunners splashed one of the trio.

However, the Battle of the Coral Sea was far from over. The next morning, 8 May, a Lexington search plane spotted Admiral Takagi's carrier striking force-including Zuikaku and Shokaku, the flattops that had proved so elusive the day before. Yorktown planes scored two bomb hits on Shokaku, damaging

her flight deck and thus preventing her from launching aircraft; in addition, the bombs set off explosions in gasoline storage tanks and destroyed an engine repair workshop. Lexington's Dauntlesses added another hit. Between the two American air groups, the hits scored killed 108 Japanese sailors and wounded 40 more.

While the American planes were bedeviling the Japanese flattops, however, Yorktown and Lexington-alerted by an intercepted message which indicated that the Japanese knew of their whereabouts-were preparing to fight off a retaliatory strike. Sure enough, shortly after 1100, that attack came.

American CAP Wildcats slashed into the Japanese formations, downing 17 planes. Some, though, managed to slip through the fighters and the "Kates" that did so managed to launch torpedoes from both sides of Lexington's bows. Two "fish", tore into "Lady Lex" on the port side; dive bombers-"Vals"-added to the destruction with three bomb hits. Lexington developed a list with three partially-flooded engineering spaces. Several fires raged belowdecks, and the carrier's elevators were out of commission.

Meanwhile Yorktown was having problems of her own. Skillfully maneuvered by Capt. Elliott Buckmaster, her commanding officer, the carrier dodged eight torpedoes. Attacked then by "Vals," the ship managed to evade all but one bomb. That one, however, penetrated the flight deck and exploded belowdecks, killing or seriously injuring 66 men.

Yorktown's damage control parties brought the fires under control, and, despite her wounds, the ship was still able to continue her flight operations. The air battle itself ended shortly before noon on the 8th; and within an hour, "Lady Lex" was on an even keel, although slightly down by the bow. Her damage control parties had already extinguished three out of the four fires below. In addition, she was making 25 knots and was recovering her air group.

At 1247, however, disaster struck Lexington, when a heavy explosion, caused by the ignition of gasoline vapors, rocked the ship. The flames raced through the ship, and further internal explosions tore the ship apart inside. Lexington battled for survival; but, despite the valiant efforts of her crew, she had to be abandoned. Capt. Frederick C. Sherman sadly ordered "abandon ship" at 1707. Her men went over the side in an orderly fashion and were picked up by the cruisers and destroyers of the carrier's screen. Torpedoes fired by Phelps (DD-361) hastened the end of "Lady Lex."

As Yorktown and her consorts retired from Coral Sea to lick their wounds, the situation in the Pacific stood altered. The Japanese had won a tactical victory, inflicting comparatively heavy losses on the Allied force, but the Allies, in stemming the tide of Japan's conquests in the South and Southwest Pacific, had achieved a strategic victory. They had blunted the drive toward strategic Port Moresby and had saved the tenuous life-line between America and Australia.

Yorktown had not achieved her part in the victory without cost, but had suffered enough damage to cause experts to estimate that at least three months in a yard would be required to put her back in fighting trim. Unfortunately, there was little time for repairs, because Allied intelligence-most notably the cryptographic unit at Pearl Harbor-had gained enough information from decoded Japanese naval messages to estimate that the Japanese were on the threshold of a major operation aimed at the northwestern tip of the Hawaiian chain-two islets in a low coral atoll known as Midway.

Thus armed with this intelligence, Admiral Nimitz began methodically planning Midway's defense, rushing all possible reinforcement in the way of men, planes and guns to Midway. In addition, he began gathering his naval forces-comparatively meager as they were-to meet the enemy at sea. As part of those preparations, he recalled TF 16, Enterprise and Hornet (CV-8), to Pearl Harbor for a quick replenishment.

Yorktown, too, received orders to return to Hawaii; and she arrived at Pearl Harbor on 27 May. Miraculously, yard workers there-laboring around the clock-made enough repairs to enable the ship to put to sea. Her air group-for the most part experienced but weary-was augmented by planes and flyers from Saratoga (CV-3) which was then headed for Hawaiian waters after her modernization on the west coast. Ready for battle, Yorktown sailed as the central ship of TF 17 on 30 May.

Northeast of Midway, Yorktown, flying Rear Admiral Fletcher's flag, rendezvoused with TF 16 under Rear Admiral Raymond A. Spruance and maintained a position 10 miles to the northward of the latter. Over the days that ensued, as the ships proceeded toward a date with destiny, few men realized that within the next few days the pivotal battle of the war in the Pacific would be fought.

Patrols, both from Midway itself and from the carriers, proceeded apace during those days in early June. On the morning of the 4th as dawn began to streak the eastern sky, Yorktown launched a 10-plane group of Dauntlesses from VB-5 which searched a northern semicircle for a distance of 100 miles out but found nothing.

Meanwhile, PBY's flying from Midway had sighted [537] the approaching Japanese and broadcast what turned out to be the

alarm for the American forces defending the key atoll. Admiral Fletcher, in tactical command, ordered Admiral Spruance, with TF 16, to locate the enemy carrier force and strike them as soon as they were found.

Yorktown's search group returned at 0830, landing soon after the last of the six-plane CAP had left the deck. When the last of the Dauntlesses had landed, a flight deck ballet took place in which the deck was spotted for the launch of the ship's attack group-17 Dauntlesses from VB-3; 12 Devastators from VT-3, and six Wildcats from "Fighting Three." Enterprise and Hornet, meanwhile, launched their attack groups.

The torpedo planes from the three American flattops located the Japanese carrier striking force but met disaster. Of the 41 planes from VT-8, VT-6, and VT-3, only six returned to Enterprise and Yorktown, collectively. None made it back to Hornet.

The destruction of the torpedo planes, however, had served a purpose. The Japanese CAP had broken off their high-altitude cover for their carriers and had concentrated on the Devastators, flying low "on the deck." The skies above were thus left open for Dauntlesses arriving from Yorktown and Enterprise. Virtually unopposed, the SBD's dove to the attack. The results were spectacular.

Yorktown's dive-bombers pummeled Soryu, making three lethal hits with 1,000-pound bombs that turned the ship into a flaming inferno. Enterprise's planes, meanwhile, hit Akagi and Kaga-turning them, too into wrecks within a very short time. The bombs from the Dauntlesses caught all of the Japanese carriers in the midst of refueling and rearming operations, and the combination of bombs and gasoline proved explosive and disastrous to the Japanese.

Three Japanese carriers had been lost. A fourth however, still roamed at large-Hiryu. Separated from her sisters, that ship had launched a striking force of 18 "Vals" that soon located Yorktown.

As soon as the attackers had been picked up on Yorktown's radar at about 1329, she discontinued the fueling of her CAP fighters on deck and swiftly cleared for action. Her returning dive bombers were moved from the landing circle to open the area for antiaircraft fire. The Dauntlesses were ordered aloft to form a CAP. An auxiliary gasoline tank-of 800 gallons capacity-was pushed over the carrier's fantail, eliminating one fire hazard. The crew drained fuel lines and closed and secured all compartments

All of Yorktown's fighters were vectored out to intercept the oncoming: Japanese aircraft, and did so some 15 to 20 miles out. The Wildcats attacked vigorously, breaking up what appeared to be an organized attack by some 18 "Vale" and 18 "Zeroes." "Planes were flying in every direction," wrote Capt. Buckmaster after the action, "and many were falling in flames."

Yorktown and her escorts went to full speed and, as the Japanese raiders attacked, began maneuvering radically. Intense antiaircraft fire greeted the "Vals" and "Kates" as they approached their release points.

Despite the barrage, though, three "Vals" scored hits. Two of them were shot down soon after releasing their bomb loads; the third went out of control just as his bomb left the rack. It tumbled in flight and hit just abaft number two elevator on the starboard side, exploding on contact and blasting a hole about 10 feet square in the flight deck. Splinters from the exploding bomb decimated the crews of the two 1.1-inch gun mounts aft of the island and on the flight deck below. Fragments piercing the flight deck hit three planes on the hangar deck, starting fires. One of the aircraft, a Yorktown Dauntless, was fully fueled and carrying a 1,000-pound bomb. Prompt action by Lt. A. C. Emerson, the hangar deck officer, prevented a serious conflagration by releasing the sprinkler system and quickly extinguishing the fire.

The second bomb to hit the ship came from the port side,

pierced the flight deck, and exploded in the lower part of the funnel. It ruptured the uptakes for three boilers, disabled two boilers themselves, and extinguished the fires in five boilers. Smoke and gases began filling the firerooms of six boilers. The men at number one boiler, however, remained at their post despite their danger and discomfort and kept its fire going, maintaining enough steam pressure to allow the auxiliary steam systems to function.

A third bomb hit the carrier from the starboard side pierced the side of number one elevator and explode on the fourth deck, starting a persistent fire in the rag storage space, adjacent to the forward gasoline stowage and the magazines. The prior precaution of smothering the gasoline system with CO_2 undoubtedly prevented the gasoline's igniting.

While the ship recovered from the damage inflicted by the dive-bombing attack, her speed dropped to six knots; and then-at 1440, about 20 minutes after the bomb hit that had shut down most of the boilers-York- town slowed to a stop, dead in the water.

At about 1540, Yorktown prepared to get underway again; and, at 1550, the engine room force reported that they were ready to make 20 knots or better. The ship was not yet out of the fight.

Simultaneously, with the fires controlled sufficiently to warrant the resumption of fueling operations, Yorktown began fueling the gasoline tanks of the fighters then on deck. Fueling had just commenced when the ship's radar picked up an incoming air group at a distance of 33 miles away. While the ship prepared for battle-again smothering gasoline systems and stopping the fueling of the planes on her flight deck-she vectored four of the six fighters of the CAP in the air to intercept the incoming raiders. Of the 10 fighters on board, eight had as much as 23 gallons of fuel in their tanks. They accordingly were launched as the remaining pair of fighters of the CAP headed out to intercept the Japanese planes.

At 1600, Yorktown churned forward, making 20 knots. The fighters she had launched and vectored out to intercept had meanwhile made contact; Yorktown received reports that the planes were "Kates." The Wildcats downed at least three of the attacking torpedo planes, but the rest began approaching in the teeth of a heavy antiaircraft barrage from the carrier and her escorts.

Yorktown maneuvered radically, avoiding at least two torpedoes before two "fish" tore into her port side within minutes of each other—the first hit at 1620. The carrier had been mortally wounded; she lost power and went dead in the water with a jammed rudder and an increasing list to port.

As the list progressed, Comdr. C. E. Aldrich, the damage control officer, reported from central station that, without power, controlling the flooding looked impossible. The engineering officer, Lt. Comdr. J. F. Delaney, soon reported that all fires were out, all power was lost, and worse yet, it was impossible to correct the list. Faced with that situation, Capt. Buckmaster ordered Aldrich, Delaney, and their men to secure and lay up on deck to put on life jackets.

The list, meanwhile, continued to increase. When it reached 26 degrees, Buckmaster and Aldrich agreed that the ship's capsizing was only a matter of minutes. "In order to save as many of the ship's company as possible," the captain wrote later, he "ordered the ship to be abandoned."

Over the minutes that ensued, the crew left ship, lowering the wounded to life rafts and striking out for the nearby destroyers and cruisers to be picked up by boats from those ships. After the evacuation of all wounded, the executive officer, Comdr. I. D. Wiltsie left the ship down a line on the starboard side. Capt. Buckmaster, meanwhile, toured the ship for one last time, inspecting her to see if any men remained. After finding no "live personnel," Buckmaster lowered himself into the water by means of a line over the stern. By that point, water was lapping the port side of the hangar deck.

Picked up by the destroyer Hammann (DD-412), Buckmaster was transferred to Astoria (CA-34) soon thereafter and reported to Rear Admiral Fletcher, who had shifted his flag to the heavy cruiser after the first dive-bombing attack. The two men agreed that a salvage party should attempt to save the ship since she had stubbornly remained afloat despite the heavy list and imminent danger of capsizing.

Interestingly enough, while the efforts to save Yorktown had been proceeding apace, her planes were still in action, joining those from Enterprise in striking the last Japanese carrier-Hiryu-

late that afternoon. Taking four direct hits, the Japanese flattop was soon helpless. She was abandoned by her crew and left to drift out of control and manned only by her dead. Yorktown had been avenged.

Yorktown, as it turned out, floated through the night; two men were still alive on board her-one attracted attention by firing a machine gun that was heard by the sole attending destroyer, Hughes. The escort picked up the men, one of whom later died.

Meanwhile, Buckmaster had selected 29 officers and 141 men to return to the ship in an attempt to save her. Five destroyers formed an antisubmarine screen while the salvage party boarded the listing carrier, the fire in the rag storage still smoldering on the morning of the 6th. Vireo (AT-144), summoned from Pearl and Hermes Reef, soon commenced towing the ship. Progress, though, could have been faster.

Yorktown's repair party went on board with a carefully predetermined plan of action to be carried out by men from each department-damage control, gunnery air engineering, navigation, communication, supply and medical. To assist in the work, Lt. Comdr. Arnold E. True brought his ship, Hammann, alongside to starboard, aft, furnishing pumps, and electric power.

By mid-afternoon, it looked as if the gamble to save the ship was paying off. The process of reducing topside weight was proceeding well-one 5-inch gun had been dropped over the side, and a second was ready to be cast loose; planes had been pushed over the side; the submersible pumps (powered by electricity provided by Hammann) had pumped out considerable quantities of water from the engineering spaces. The efforts of the salvage crew had reduced the list by about two degrees.

Unbeknownst to Yorktown and the six nearby destroyers, the Japanese submarine I-158 had achieved a favorable firing position. Remarkably-but perhaps understandable in light of the debris and wreckage in the water in the vicinity-none of the destroyers picked up the approaching I-boat. Suddenly, at 1536, lookouts spotted a salvo of four torpedoes churning toward the ship from the starboard beam.

Hammann went to general quarters, a 20-millimeter gun going into action in an attempt to explode the "fish" in the water. One torpedo hit Hammann-her screws churning the water beneath her fantail as she tried to get underway-directly amidships and broke her back. The destroyer jackknifed and went down rapidly.

Two torpedoes struck Yorktown just below the turn of the bilge at the after-end of the island structure. The fourth torpedo passed just astern of the carrier.

Approximately a minute after Hammann's stern disappeared beneath the waves, an explosion rumbled up from the depths-possibly caused by the destroyer's depth charges going off. The blast killed many of Hammann's and a few of Yorktown's men who had been thrown into the water. The concussion battered the already-damaged carrier's hull and caused tremendous shocks that carried away Yorktown's auxiliary generator, sent numerous fixtures from the hangar deck overhead crashing to the deck below, sheared rivets in the starboard leg of the foremast, and threw men in every direction, causing broken bones and several minor injuries.

Prospects for immediate resumption of salvage work looked grim since all destroyers immediately commenced searches for the enemy submarine (which escaped) and commenced rescuing men from Hammann and Yorktown. Capt. Buckmaster decided to postpone further attempts at salvage until the following day.

Vireo cut the towline and doubled back to Yorktown to pick up survivors, taking on board many men of the salvage crew while picking up men from the water. The little ship endured a terrific pounding from the larger ship but nevertheless stayed alongside to carry out her rescue mission. Later, while on board the tug, Capt. Buckmaster conducted a burial service; two officers and an enlisted man from Hammann were committed to the deep.

The second attempt at salvage, however, would never be made. Yorktown remained stubbornly afloat throughout the night of the 6th and into the morning of the 7th. However, by 0530 on the 7th, the men in the ships nearby noted that the carrier's list was rapidly increasing to port. As if tired, the valiant flattop turned over at 0701 on her port side and sank in 3,000 fathoms of water, her battle flags flying.

Yorktown (CV-5) earned three battle stars for her World War II service, two of which were for the significant part she had played in stopping Japanese expansion and turning the tide of the war at Coral Sea and Midway.

CV-6 • USS ENTERPRISE

(Displacement. 19,800; length. 809'6"; beam. 83'1"; cw. 114'; draft. 28'; speed. 33 k.; complement. 2919; armor. 8 5", 38 cal.; class. Yorktown).

The seventh Enterprise (CV-6) was launched on 3 October 1936 by Newport News Shipbuilding and Drydock Co., Newport News, Va.; sponsored by Mrs. Claude A. Swanson, wife of the Secretary of the Navy; and commissioned on 12 May 1938, with Captain N. H. White in command.

Enterprise sailed south on a shakedown cruise which took her to Rio de Janeiro, Brazil. After her return, she operated along the east coast and in the Caribbean until April of 1939 when she was ordered to duty in the Pacific. Based first in San Diego and then at Pearl Harbor, the carrier trained herself and her aircraft squadrons for any eventuality and carried aircraft among the island bases of the Pacific. Enterprise had just completed one such mission, delivering Marine Corps Fighter Squadron 211 to Wake Island on 2 December 1941, and was en route to Hawaii when the Japanese attacked Pearl Harbor.

Enterprise's scout planes arrived over Pearl Harbor during the attack and, though surprised, immediately went into action in defense of the naval base. The carrier, meanwhile, launched her remaining aircraft in a fruitless search for the Japanese striking force. Enterprise put into Pearl Harbor for fuel and supplies on 8 December and sailed early the next morning to patrol against possible additional attacks on the Hawaiian Islands. While the group did not encounter any surface ships, Enterprise aircraft scored a kill by sinking submarine I-170 in 23° 45' N., 155° 35' W., on 10 December 1941.

During the last two weeks of December 1941, Enterprise and her group steamed to the westward of Hawaii to cover those islands while two other carrier groups made a belated attempt to relieve Wake Island. After a brief rest at Pearl Harbor, the Enterprise group sailed on 11 January to protect convoys reinforcing Samoa. On 1 February, the task force dealt a hard blow to Kwajalein, Wotje, and Maloelap in the Marshall Islands, sinking three ships, damaging eight, and destroying numerous airplanes and ground facilities. Enterprise received only minor damage in the Japanese counterattack, as her force retired to Pearl Harbor.

During the next month, Enterprise's force swept the central Pacific, blasting enemy installations on Wake and Marcus Islands, then received minor alterations and repairs at Pearl Harbor. On 8 April 1942, she departed to rendezvous with Hornet (CV-8) and sail westward to launch 16 Army B-25 bombers in a raid on Tokyo. While Enterprise fighters flew combat air patrol, the B-25s roared into the air on 18 April and raced undetected the 600 miles to their target. The task force, its presence known to the enemy, reversed course and returned to Pearl Harbor on 25 April.

Five days later, the "Big E" was speeding toward the South Pacific to reinforce the U.S. carriers operating in the Coral Sea. Distance proved too great to conquer in time, and the Battle of the Coral Sea was history before Enterprise could reach her destination. Ordered back to Hawaii, the carrier entered Pearl Harbor on 26 May and began intensive preparations to meet the expected Japanese thrust at Midway Island. Two days later, she sortied as the flagship of Rear Admiral Raymond A. Spruance, CTF 16, with orders "to hold Midway and inflict maximum damage on the enemy by strong attrition tactics." With Enterprise in TF 16 were Hornet, six cruisers, and ten destroyers. On 30 May, TF 17, Rear Admiral Frank J. Fletcher in Yorktown (CV-5), with two cruisers and six destroyers, sailed to support TF 16; as senior officer, Rear Admiral Fletcher became "Officer in Tactical Command."

Battle was joined on the morning of 4 June 1942 when four Japanese carriers, unaware of the presence of U.S. forces, launched attacks on Midway Island. Just three hours after the first bomb fell on Midway, planes from Hornet struck the enemy force, and 30 minutes later Enterprise and Yorktown aircraft streaked in to join in smashing the Japanese carriers. Each side hurled attacks at the other during the day in one of history's most decisive battles. Though the forces were in contact until 7 June, by the end of the 4th the outcome had been decided, and the tide of the war in the Pacific had been turned in the United States' favor. Yorktown and Hammann (DD-412) were the only United States ships sunk, but TFs 16 and 17 lost a total of 113 planes, 61 of them in combat, during the battle. Japanese losses, far more severe, consisted of four carriers, one cruiser, and 272 carrier aircraft. Enterprise and all other ships of TFs 16 and 17 came through undamaged, returning to Pearl Harbor on 13 June 1942.

After a month of rest and overhaul, Enterprise sailed on 15 July for the South Pacific where she joined TF 61 to support the amphibious landings in the Solomon Islands on 8 August. For the next two weeks, the carrier and her planes guarded seaborne communication lines southwest of the Solomons. On 24 August, a strong Japanese force was sighted some 200 miles north of Guadalcanal and TF 61 sent planes to the attack. An enemy light carrier was sent to the bottom, and the Japanese troops intended for Guadalcanal were forced back. Enterprise suffered most heavily of the United States ships, three direct hits and four near misses killed 74, wounded 95, and inflicted serious damage on the carrier. But well-trained damage control parties and quick hard work patched her up so that she was able to return to Hawaii under her own power.

Repaired at Pearl Harbor from 10 September to 16 October, Enterprise departed once more for the South Pacific where with Hornet, she formed TF 61. On 26 October, Enterprise scout planes located a Japanese carrier force, and the Battle of the Santa Cruz Island was underway. Enterprise aircraft struck carriers, battleships, and cruisers during the struggle, while the "Big E" herself underwent intensive attack. Hit twice by bombs, Enterprise lost 44 killed and had 75 wounded. Despite serious damage, she continued in action and took on board a large number of planes from Hornet when that carrier had to be abandoned. Though the American losses of a carrier and a destroyer were more severe than the Japanese loss of one light cruiser, the battle gained priceless time to reinforce Guadalcanal against the next enemy onslaught.

Enterprise entered Noumea, New Caledonia, on 30 October for repairs, but a new Japanese thrust at the Solomons demanded her presence and she sailed on 11 November, with repair crews from Vestal (AR-4) still on board, working vigorously. Two days later, "Big E" planes swarmed down on an enemy force and disabled a battleship which was sunk later by other American aircraft, and on 14 November, aviators from Enterprise helped to dispatch a heavy cruiser. When the Naval Battle of Guadalcanal ended on 15 November 1942, Enterprise had shared in sinking 16 ships and damaging 8 more. The carrier returned to Noumea on 16 November to complete her repairs.

Sailing again on 4 December, Enterprise trained out of Espiritu Santo, New Hebrides, until 28 January 1943 when she departed for the Solomons area. On 30 January, her fighters flew combat air patrol for a cruiser-destroyer group during the Battle of Rennell Island. Despite the destruction of a large majority of the attacking Japanese bombers by Enterprise planes, Chicago (CA-29) was sunk by aerial torpedoes. Detached after the battle, the carrier arrived at Espiritu Santo on 1 February, and for the next three months operated out of that base, covering U.S. surface forces up to the Solomons. Enterprise then steamed to Pearl Harbor where, on 27 May 1943, Admiral Chester W. Nimitz presented the ship with the first Presidential Unit Citation won by an aircraft carrier. On 20 July 1943, she entered Puget Sound Navy Yard, Bremerton, Wash., for a much-needed overhaul.

Back in action waters by mid-November, Enterprise joined in providing close air support to the Marines landing on Makin Island, from 19 to 21 November. On the night of 26 November 1943, the "Big E" introduced carrier-based night fighter operations in the Pacific when a three-plane team from the ship broke up a large group of land-based bombers attacking TG 50.2. After a heavy strike by aircraft of TF 50 against Kwajalein on 4 December, Enterprise returned to Pearl Harbor six days later.

The carrier's next operation was with TF 58 in softening up the Marshall Islands and supporting the landings on Kwajalein, from 29 January to 3 February. Then Enterprise sailed, still with TF 58, to strike the Japanese naval base at Truk in the Caroline Islands, on 17 February. Again, the "Big E" made aviation history when she launched the first night radar bombing attack from any U.S. carrier. The 12 torpedo bombers in this strike achieved

excellent results, accounting for nearly one-third of the 200,000 tons of shipping destroyed by the aircraft of the task force.

Detached from TF 58, Enterprise launched raids on Jaluit Atoll on 20 February, then steamed to Majuro and Espiritu Santo. Sailing 15 March in TG 36.1, she provided air cover and close support for the landings on Emirau Island (19-25 March). The carrier rejoined CF 58 on 26 March and for the next 12 days joined in the series of hard-hitting strikes against the Yap, Ulithi, Woleai, and the Palau Islands. After a week's rest and replenishment at Majuro, Enterprise sailed 14 April to support landings in the Hollandia area of few Guinea, and then hit Truk again (29-30 April).

On 6 June 1944, the "Big E" and her companions of TG 58.3 sortied from Majuro to strike with the rest of TF 58, the Mariana Islands. Blasting Saipan, Rota, and Guam between 11 and 14 June, Enterprise pilots gave direct support to the landings on Saipan on 15 June, and covered the troops ashore for the next 2 days. Aware of a major Japanese attempt to break up the invasion of Saipan, Admiral Raymond A. Spruance, Commander 6th Fleet, positioned TF 58 to meet the thrust. On 19 June 1944 took place the greatest carrier aircraft battle in history. For over 8 hours airmen of the United States and Imperial Japanese navies fought in the skies over TF 58 and the Marianas. By the end of the day, a United States victory was apparent, and at the conclusion of the strikes against the Japanese fleet on 20 Tune, the triumph became complete. Six American ships had been damaged, and 130 planes and a total of 76 pilots and aircrewmen had been lost. But with a major assist from U.S. submarines, 3 Japanese carriers were sunk, and 426 ship-based aircraft were destroyed. Japanese naval aviation never recovered from this blow.

The Battle of the Philippine Sea over, Enterprise and her companions continued to support the Saipan campaign through 5 July. Enterprise then sailed for Pearl Harbor and a month of rest and overhaul. Back in action waters on 24 August, the carrier sailed with TF 38 in that force's aerial assault on the Volcano and Bonin Islands from 31 August to 2 September, and Yap, Ulithi, and the Palaus from 6 to 8 September. After operating west of the Palau Islands, the "Big E" joined other units of TF 38 on 7 October, and shaped course to the northward. From 10 to 20 October her aviators roared over Okinawa, Formosa, and the Philippines, blasting enemy airfields, shore installations, and shipping in preparation for the assault on Leyte. After supporting the Leyte landings on 20 October, Enterprise headed for Ulithi to replenish but the approach of the Japanese fleet on 23 October, brought her racing back into action. In the Battle for Leyte Gulf (23-26 October), Enterprise planes struck all three groups of enemy forces, battering battleships and destroyers before the action ended. The carrier remained on patrol east of Samar and Leyte until the end of October, then retired to Ulithi for supplies. During November, her aircraft struck targets in the Manila area, and the island of Yap. The "Big E" returned to Pearl Harbor on 6 December 1944.

Sailing 24 December for the Philippine area, Enterprise carried on board an air group specially trained in night carrier operations. She joined TG 38.5 and swept the waters north of Luzon and of the China Sea during January of 1945, striking shore targets and shipping from Formosa to Indo-China. After a brief visit to Ulithi, the "Big E" joined TG 58.5 on 10 February 1945 and provided day and night combat air patrol for TF 58 as it struck Tokyo on 16 and 17 February. She then supported the marines on Iwo Jima from the day of the landings, 19 February, until 9 March when she sailed for Ulithi. During one part of that period, Enterprise kept aircraft aloft continuously over Iwo Jima for 174 hours. Departing Ulithi 15 March, the carrier continued her night work in raids against Kyushu, Honshu, and shipping in the Inland Sea of Japan. Damaged slightly by an enemy bomb on 18 March, Enterprise entered Ulithi 6 days later for repairs. Back in action on 5 April, she supported the Okinawa operation until again damaged (11 April), this time by a suicide plane, and forced back to Ulithi. Off Okinawa once more on 6 May, Enterprise flew patrols around the clock as the menace of the kamikaze increased. On 14 May 1945, the "Big E" suffered her last wound of World War II when a suicide plane destroyed her forward elevator, killing 14 and wounding 34 men. The carrier sailed for repairs at the Puget Sound Navy Yard, arriving 7 June 1945.

Restored to peak condition, Enterprise voyaged to Pearl Harbor, returning to the States with some 1,100 servicemen due for discharge, then sailed on to New York, arriving 17 October 1945. Two weeks later she proceeded to Boston for installation of additional berthing facilities, then began a series of "Magic Carpet" voyages to Europe, bringing more than 10,000 veterans home in her final service to her country. Enterprise entered the New York Naval Shipyard on 18 January 1946 for inactivation, and was decommissioned on 17 February 1947. The "Big E" was sold on 1 July 1958.

In addition to her Presidential Unit Citation, Enterprise received the Navy Unit Commendation and 20 battle stars for World War II service.

CV-7 • USS WASP

(Displacement. 14,700; length. 741'4"; beam. 80'8"; cw. 109'0"; draft. 19'11" (mean); speed. 29.5 k.; complement 2,367; armor. 8 5", 16 1", 16 .50-cal. mg.; act 80; class. Wasp).

The eighth Wasp (CV-7) was laid down on 1 April 1936 at Quincy, Mass., by the Bethlehem Shipbuilding Co., launched on 4 April 1939, sponsored by Mrs. Charles Edison, the wife of the Assistant Secretary of the Navy Charles Edison, and commissioned on 25 April 1940 at the Army Quartermaster Base, South Boston, Mass., Capt. John W. Reeves, Jr., in command.

Wasp remained at Boston through May, fitting out, before she got underway on 5 June 1940 for calibration tests on her radio direction finder gear. After further fitting out while anchored in Boston Harbor, the new aircraft carrier steamed independently to Hampton Roads, Va., anchoring there on 24 June. Four days later, she sailed for the Caribbean in company with Morris (DD-417).

En route, she conducted the first of many carrier qualification tests. Among the earliest of the qualifiers was Lt. (jg.) David T. McCampbell, who later became the Navy's top-scoring "ace" in World War II. Wasp arrived at Guantanamo Bay in time to "dress ship" in honor of Independence Day.

Tragedy marred the carrier's shakedown. On 9 July, one of her Vought SB2U-2 Vindicators crashed two miles from the ship. Wasp bent on flank speed to close, as did the plane-guarding destroyer Morris. The latter's boats recovered items from the plane's baggage compartment, but the plane itself had gone down with its crew of two.

Wasp departed Guantanamo Bay on 11 July and arrived at Hampton Roads four days later. There, she embarked planes from the 1st Marine Air Group and took them to sea for qualification trials. Operating off the southern drill grounds, the ship and her planes honed their skills for a week before the marines and their planes were disembarked at Norfolk, and the carrier moved north to Boston for post-shakedown repairs.

While civilian workmen from the Bethlehem Steel Co. [143] came on board the ship to check their workmanship and to learn how it had stood up under the rigors of shakedown, Wasp lay alongside the same pier at which she had been commissioned. While at Boston, she fired a 21-gun salute and rendered honors to President Franklin D. Roosevelt, whose yacht, Potomac (AG-25), stopped briefly at the Boston Navy Yard on 10 August.

Wasp departed the Army Quartermaster Base on the 21st to conduct steering drills and full-power trials. Late the following morning, she got underway for Norfolk. For the next few days, while Ellis (DD-164) operated as plane guard, Wasp launched and recovered her aircraft: fighters from Fighter Squadron (VF) 7 and scout-bombers from Scouting Squadron (VS) 72. The carrier put into the Norfolk Navy Yard on 28 August for repair work on her turbines—alterations which kept the ship in dockyard hands into the following month. Drydocked during the period from 12 to 18 September, Wasp ran her final sea trials in Hampton Roads on 26 September 1940.

Ready now to join the fleet and assigned to Carrier Division (CarDiv) 3, Patrol Force, Wasp shifted to Naval Operating Base (NOB), Norfolk from the Norfolk Navy Yard on 11 October. There she loaded 24 P-40's from the 8th Army Pursuit Group and nine O-47A's from the 2nd Observation Squadron, as well as her own spares and utility unit Grumman J2F's on the 12th. Proceeding to sea for maneuvering room, Wasp flew off the Army planes in a test designed to compare the take-off runs of standard Navy and Army aircraft. That experiment, the first time that Army planes had flown from a Navy carrier, foreshadowed the use of the ship in the ferry role that she performed so well in World War II.

Wasp then proceeded on toward Cuba in company with Plunkett (DD-431) and Niblack (DD-424). The carrier's planes flew routine training flights, including dive-bombing and machine gun practices, over the ensuing four days. Upon arrival at Guantanamo, Wasp's saluting batteries barked out a 13-gun salute to Rear Admiral Hayne Ellis, Commander, Atlantic Squadron, embarked in Texas (BB-35), on 19 October.

For the remainder of October and into November, Wasp trained in the Guantanamo Bay area. Her planes flew carrier qualification and refresher training flights while her gunners sharpened their skills in short-range battle practices at targets towed by the new fleet tug Seminole (AT-65). While operating in the Culebra, Virgin Islands, area, Wasp again teamed with the aviators of the 1st Marine Air Wing, giving the flying leathernecks practice in carrier take-offs and landings.

Her work in the Caribbean finished, Wasp sailed for Norfolk and arrived shortly after noon on 26 November. She remained at the Norfolk Navy Yard through Christmas of 1940. Then, after first conducting degaussing experiments with Hannibal (AG-1), she steamed independently to Cuba.

Arriving at Guantanamo Bay on 27 January 1941, Wasp conducted a regular routine of flight operations into February. With Walke (DD-416) as her plane guard, Wasp operated out of Guantanamo and Culebra, conducting her maneuvers with an impressive array of warships—Texas, Ranger (CV-4), Tuscaloosa (CA-37), Wichita (CA-45) and a host of destroyers. Wasp ran gunnery drills and exercises, as well as routine flight training evolutions, into March. Underway for Hampton Roads on 4 March, the aircraft carrier conducted a night battle practice into the early morning hours of the 5th.

During the passage to Norfolk, heavy weather sprang up on the evening of 7 March. Wasp was steaming at standard speed, 17 knots, a pace that she had been maintaining all day. Off Cape Hatteras, a lookout in the carrier spotted a red flare arcing into the stormy black night skies at 2245. The big ship swung around to head in the direction of the distress signal while a messenger notified the captain, who reached the bridge in an instant. Capt. Reeves himself took the conn, as a second set of flares was seen at 2259.

Finally, at 2329, with the aid of her searchlights probing the wet night, Wasp located the stranger in trouble. She proved to be the lumber schooner George E. Klinck, bound from Jacksonville, Fla., to Southwest Harbor, Maine.

The sea, in the meantime, worsened from a state 5 to a state 7. Wasp lay to, maneuvering alongside at 0007 on 8 March. At that time, four men from the schooner clambered up a swaying jacob's ladder buffeted by gusts of wind. Then, despite the raging tempest, Wasp lowered a boat, at 0016, and brought the remaining four men aboard from the foundering 152-foot schooner.

Later that day, Wasp disembarked her rescued mariners and immediately went into drydock at the Norfolk Navy Yard. The ship received vital repairs to her turbines. Portholes on the third deck were welded over to provide better watertight integrity, and steel splinter shielding around her 5-inch and 1.1-inch batteries was added. After those repairs and alterations were finished, Wasp got underway for the Virgin Islands on 22 March, the carrier arriving at St. Thomas three days later. She soon shifted to Guantanamo Bay and loaded marine stores for transportation to Norfolk.

Returning to Norfolk on 30 March, Wasp conducted routine flight operations out of Hampton Roads over the ensuing days, into April. In company with Sampson (DD-394), the carrier conducted an abortive search for a downed patrol plane in her vicinity on 8 April. For the remainder of the month, Wasp operated off the eastern seaboard between Newport, R.I., and Norfolk conducting extensive flight and patrol operations with her embarked air group. She shifted to Bermuda in mid-May, anchoring at Grassy Bay on the 12th. Eight days later, the ship got underway in company with Quincy (CA-39), Livermore (DD-429), and Kearny (DD-432) for exercises at sea before returning to Grassy Bay on 3 June. Wasp sailed for Norfolk three days later with Edison (DD-439) as her antisubmarine screen.

After a brief stay in the Tidewater area, Wasp headed back toward Bermuda on 20 June. Wasp and her escorts patrolled the stretch of the Atlantic between Bermuda and Hampton Roads until 5 July, as the Atlantic Fleet's neutrality patrol zones were extended eastward. Reaching Grassy Bay on that day, she remained in port a week before returning to Norfolk sailing on 12 July in company with Tuscaloosa (CA-37), Grayson (DD-435), Anderson (DD-411), and Rowan (DD-405).

Following her return to Norfolk on the 13th, Wasp and her embarked air group conducted refresher training off the Virginia capes. Meanwhile, the situation in the Atlantic had

taken on a new complexion, with American participation in the Battle of the Atlantic only a matter of time, when the United States took another step toward involvement on the side of the British. To protect American security and to free British forces needed elsewhere, the United States made plans to occupy Iceland. Wasp played an important role in the move.

Late on the afternoon of 23 July, while the carrier lay alongside Pier 7, NOB Norfolk, 32 Army Air Force (AAF) pilots reported on board "for temporary duty." At 0630 the following day, Wasp's crew watched an interesting cargo come on board, hoisted on deck by the ship's cranes: 30 Curtiss P-40C's and three PT-17 trainers from the AAF 33d Pursuit Squadron, 8th Air Group, Air Force Combat Command, home-based at Mitchell Field, N.Y. Three days later, four newspaper correspondents—including the noted journalist Fletcher Pratt—came on board.

The carrier had drawn the assignment of ferrying those vital army planes to Iceland because of a lack of British aircraft to cover the American landings. The American P-40's would provide the defensive fighter [144] cover necessary to watch over the initial increment of American occupying forces. Wasp consequently cast off from Pier 7 and slipped out to sea through the swept channel at 0932 on 28 July, with O'Brien (DD-415) and Walke as plane guards. Vincennes (CA-44) later joined the formation at sea.

Within a few days, Wasp's group joined the larger Task Force (TF) 16—consisting of Mississippi (BB-41), Quincy, Wichita, five destroyers, Semmes (AG-24), American Legion (AP-35), Mizar (AF-12), and Almaack (AK-27). Those ships, too, were bound for Iceland with the first occupation troops embarked. On the morning of 6 August, Wasp, Vincennes, Walke, and O'Brien parted company from TF 16. Soon thereafter, the carrier turned into the wind and commenced launching the planes from the

33d Pursuit Squadron. As the P-40's and the trio of trainers droned on to Iceland, Wasp headed home for Norfolk, her three escorts in company. After another week at sea, the group arrived back at Norfolk on 14 August.

Underway again on 22 August, Wasp put to sea for carrier qualifications and refresher landings off the Virginia Capes. Two days later, Rear Admiral H. Kent Hewitt, Commander of Cruisers, Atlantic Fleet, shifted his flag from Savannah (CL-42) to Wasp, while the ships lay anchored in Hampton Roads. Underway on the 25th, in company with Savannah, Monssen (DD-436), and Kearny, the aircraft carrier conducted flight operations over the ensuing days. Scuttlebutt on board the carrier had her steaming out in search of the German heavy cruiser, Admiral Hipper, which was reportedly roaming the western Atlantic in search of prey. Suspicions were confirmed for many on the 30th when the British battleship HMS Rodney was sighted some 20 miles away, on the same course as the Americans.

In any event, if they had been in search of a German raider, they did not make contact with her. Wasp and her escorts anchored in the Gulf of Paria, Trinidad, on 2 September, where Admiral Hewitt shifted his flag back to Savannah. The carrier remained in port until 6 September when she again put to sea on patrol "to enforce the neutrality of the United States in the Atlantic."

While at sea, the ship received news of a German U-boat unsuccessfully attempting to attack the destroyer Greer (DD-146). The United States had been getting more and more involved in the war; American warships were now convoying British merchantmen halfway across the Atlantic to the "mid-ocean meeting point" (MOMP).

Wasp's crew looked forward to returning to Bermuda on 18 September, but the new situation in the Atlantic meant a change in plans. Shifted to the colder climes of Newfoundland, the carrier arrived at Placentia Bay on 22 September and fueled from

Salinas (AO-19) the following day. The respite in port was a brief one, however, as the ship got underway again, late on the 23rd, for Iceland. In the company of Wichita, four destroyers, and the repair ship Vulcan (AR-6), Wasp arrived at Hvalfjordur, Iceland, on the 28th. Two days earlier, Admiral Harold R. Stark, the Chief of Naval Operations, had ordered American warships to do their utmost to destroy whatever German or Italian warships they found. The "short-of-war" operations were drawing frightfully close to the real thing!

With the accelerated activity entailed in the United States Navy's conducting convoy escort missions, Wasp put to sea on 6 October in the company of Vincennes and four destroyers. Those ships patrolled the foggy, cold North Atlantic until returning to Little Placentia Bay, Newfoundland, on the 11th, anchoring during a fierce gale that lashed the bay with high winds and stinging spray. On 17 October, Wasp set out for home—Norfolk—patrolling en route and arrived at her destination on the 20th. The carrier soon sailed for Bermuda and conducted qualifications and refresher training flights en route. Anchoring in Grassy Bay on 1 November, Wasp operated on patrols out of Bermuda for the remainder of the month.

October had seen the incidents involving American and German warships multiplying on the high seas. Kearny was torpedoed on 17 October, Salinas took a "fish" on the 28th, and in the most tragic incident that autumn, Reuben James (DD-246) was torpedoed and sunk with heavy loss of life on 30 October. Meanwhile, in the Pacific, tension between the United States and Japan increased almost with each passing day.

Wasp slipped out to sea from Grassy Bay on 3 December and rendezvoused with Wilson (DD-408). While the destroyer operated as a plane guard, Wasp's air group flew day and night refresher training missions. In addition, the two ships conducted gunnery drills before returning to Grassy Bay two days later.

Wasp lay at anchor on 7 December 1941, observing "holiday routine"—it being a Sunday. In the Pacific, the Japanese broke the Sunday morning peace in a devastating surprise attack on the Pacific Fleet at Pearl Harbor. Their daring attack plunged the United States into World War II in both oceans. On 11 December, Germany and Italy followed Japan into war against the United States.

Meanwhile, naval authorities felt considerable anxiety that French warships in the Caribbean and West Indies were prepared to make a breakout and attempt to get back to France. Accordingly, Wasp, Brooklyn (CL-40), and two destroyers, Sterett (DD-407) and Wilson, departed Grassy Bay and headed

for Martinique. Faulty intelligence gave American authorities in Washington the impression that the Vichy French armed merchant cruiser Barfleur had gotten underway for sea. The French were accordingly warned that the auxiliary cruiser would be sunk or captured unless she returned to port and resumed her internment. As it turned out, Barfleur had not departed after all but had remained in the harbor. The tense situation at Martinique eventually dissipated, and the crisis abated.

With tensions in the West Indies lessened considerably, Wasp departed Grassy Bay and headed for Hampton Roads three days before Christmas, in the company of Long Island (AVG-1) and escorted by Stack (DD-406) and Sterett. Two days later, the carrier moored at the Norfolk Navy Yard to commence an overhaul that would last into 1942.

After departing Norfolk on 14 January 1942, Wasp headed north and touched at Argentia, Newfoundland, and Casco Bay, Maine, while operating in those northern climes. On 16 March, as part of Task Group (TG) 22.6, she headed back toward Norfolk. During the morning watch the next day, visibility lessened considerably; and, at 0650, Wasp's bow plunged into Stack's starboard side, punching a hole and completely flooding the destroyer's number one fireroom. Stack was detached and proceeded to the Philadelphia Navy Yard, where her damage was repaired.

Wasp, meanwhile, made port at Norfolk on the 21st without further incident. Shifting back to Casco Bay three days later, she sailed for the British Isles on 26 March, with Task Force (TF) 39 under the command of Rear Admiral John W. Wilcox, Jr., in Washington (BB-56). That force was to reinforce the Home Fleet of the Royal Navy. While en route, Rear Admiral Wilcox was swept overboard from the battleship and drowned. Although hampered by poor visibility conditions, Wasp planes took part in the search. Wilcox's body was spotted an hour later, face down in the raging seas, but it was not recovered.

Rear Admiral Robert C. Giffen, who flew his flag in Wichita, assumed command of TF-39. The American ships were met by a force based around the light cruiser HMS Edinburgh on 3 April. Those ships escorted them to Scapa Flow in the Orkney Islands.

While the majority of TF 39 joined the British Home Fleet—being renumbered to TF 99 in the process—to cover convoys routed to North Russia, Wasp departed Scapa Flow on 9 April, bound for the Clyde estuary and Greenock, Scotland. On the following day, the carrier sailed up the Clyde River, past the John Brown Clydebank shipbuilding facilities. There, shipyard workers paused long enough from their labors to accord Wasp a tumultuous reception as she passed. Wasp's impending mission was an important one—one upon which the fate of the island bastion of Malta hung. That key isle was then being pounded daily by German and Italian planes. The British, faced with the loss of air superiority over the island, requested the use of a carrier to transport planes that could wrest air superiority from the Axis aircraft. Wasp drew ferry duty once again.

Having landed her torpedo planes and dive bombers, Wasp loaded 47 Supermarine "Spitfire" Mk. V fighter planes at the King George Dock, Glasgow, on 13 April, before she departed the Clyde estuary on the 14th. Her screen consisted of Force "W" of the Home Fleet—a group that included the battlecruiser HMS Renown and anti-aircraft cruisers HMS Cairo and HMS Charybdis. Madison (DD-425) and Lang (DD-399) also served in Wasp's screen.

Wasp and her consorts passed through the Straits of Gibraltar under cover of the pre-dawn darkness on 19 April, avoiding the possibility of being discovered by Spanish or Axis agents. At 0400 on 20 April, Wasp spotted 11 Grumman F4F-4 Wildcat fighters on her deck and quickly launched them to form a combat air patrol (CAP) over Force "W". Meanwhile, the "Spitfires" were warming up their engines in the hangar deck spaces below. With the Wildcats patrolling overhead, the Spitfires were brought up singly on the after elevator, spotted for launch, and then given the go-ahead to take off. One by one, they roared down the deck and over the forward rounddown, until each Spitfire was aloft and winging toward Malta.

When the launch was complete, Wasp retired toward England, having safely delivered her charges. Unfortunately, those "Spitfires," which flew in to augment the dwindling numbers of "Gladiator" and "Hurricane" fighters, were tracked by efficient Axis intelligence and their arrival pinpointed. The unfortunate "Spitfires" were decimated by heavy German air raids which caught many planes on the ground.

As a result, it looked as if the acute situation required a second ferry run to Malta. Accordingly, Prime Minister Winston Churchill, fearing that Malta would be "pounded to bits," asked President Roosevelt to allow Wasp to have "another good sting." Roosevelt responded in the affirmative. Rising to the occasion, Wasp loaded another contingent of "Spitfire" V's and sailed for the Mediterranean on 3 May. Again, especially vigilant for submarines, Wasp proceeded unmolested. This time, the British aircraft carrier HMS Eagle accompanied Wasp, and she, too, carried a contingent of Spitfires bound for the "unsinkable aircraft carrier," Malta.

The two Allied flattops reached their launching points early on Saturday, 9 May, with Wasp steaming in a column ahead of Eagle at a distance of 1,000 yards. At 0630, Wasp commenced launching planes—11 F4F4's of VF-71 to serve as CAP over the task force. The first "Spitfire" roared down the deck at 0643, piloted by Sergeant-Pilot Herrington, but lost power soon after takeoff and plunged into the sea. Both pilot and plane were lost.

Undaunted by the loss of Herrington, the other planes flew off safely and formed up to fly to Malta. Misfortune, however, again seemed to dog the flight when one pilot accidentally released his auxiliary fuel tank as he climbed to 2,000 feet. He obviously could not make Malta, as the slippery tank fitted beneath the belly of the plane had increased the range of the plane markedly. With that gone, he had no chance of making the island. His only alternatives were to land back on board Wasp or to ditch and take his chances in the water.

Sergeant-Pilot Smith chose the former. Wasp bent on full speed and recovered the plane at 0743. The "Spitfire" came to a stop just 15 feet from the forward edge of the flight deck, making what one Wasp sailor observed to be a "one wire" landing. With her vital errand completed, the carrier set sail for the British Isles while a German radio station broadcast the startling news that the American carrier had been sunk! Most in the Allied camp knew better, however; and, on 11 May, Prime Minister Churchill sent a witty message to the captain and ship's company of Wasp: "Many thanks to you all for the timely help. Who said a Wasp couldn't sting twice?"

While Wasp was conducting those two important missions to Malta, a train of events far to the westward beckoned the carrier to the Pacific theater. Early in May, almost simultaneously with Wasp's second Malta run—Operation "Bowery"—the Battle of the Coral Sea had been fought. That action turned back the Japanese thrust at Port Moresby. One month later, from 4 to 6 June, an American carrier force smashed its Japanese counterpart in the pivotal Battle of Midway. These two victories cost the United States two precious carriers: Lexington (CV-2) at Coral Sea and Yorktown (CV-5) at Midway. While the Japanese had suffered the damaging of two at Coral Sea and the loss of four carriers at Midway, the United States could scarcely afford to be left with only two operational carriers in the western and central Pacific—Enterprise (CV-6) and Hornet (CV-8). Saratoga (CV-3) was still undergoing repairs and modernization after being torpedoed off Oahu in early January 1942.

To prepare to strengthen the American Navy in the Pacific, Wasp was hurried back to the United States for alterations and repairs at the Norfolk Navy Yard. During the carrier's stay in the Tidewater region, Capt. Reeves—who had been promoted to flag rank—was relieved by Capt. Forrest P. Sherman on 31 May. Departing Norfolk on 6 June, the last day of the critical Battle of Midway, Wasp sailed with TF 37 which was built around the carrier and the new battleship North Carolina (BB-55) and escorted by Quincy (CA-39) and San Juan (CL-54) and a half-dozen destroyers. The group transited the Panama Canal on 10 June, at which time Wasp and her consorts became TF 18, the carrier flying the two-starred flag of Rear Admiral Leigh Noyes.

Arriving at San Diego on 19 June, Wasp embarked the remainder of her complement of aircraft, Grumman TBF-1's and Douglas SBD-3's—10 of the former and 12 of the latter conducting their carrier qualification on 22 and 23 June, respectively—the latter replacing the old Vindicators. On 1 July, she sailed for the Tonga Islands as part of the convoy for the five transports that had embarked the 2d Marine Regiment.

While TF 18 and the transports were en route to Tongatabu, Wasp received another congratulatory message—this time from Admiral Noyes, embarked in the ship. "During the two weeks my flag has been in Wasp I have been very favorably impressed by the fine spirit of her ship's company and the way that all hands have handled their many problems. Since we have been at sea, every day has shown marked improvement in operations. I am sure that when our opportunity comes to strike the enemy in this ocean, Wasp and her squadrons will add more glory to the name she bears." Noyes' hopes were to be realized, but for all too brief a time.

Four days out of Nukualofa harbor, Wasp developed serious engine trouble. The ship's "black gang," however, worked diligently to do the preliminary work in lifting, repairing, and replacing the ship's starboard high-pressure turbine. The work done en route substantially helped enough to allow speedy completion of the repairs after the ship dropped her hook at Tongatabu on 18 July.

Meanwhile, preparations to invade the Solomon Islands were proceeding apace. Up to that point, the Japanese had been on the offensive, establishing their defensive perimeter around the edge of their "Greater East Asia Co-Prosperity Sphere."

On 4 July, while Wasp was en route to the South Pacific, the Japanese landed on Guadalcanal. Allied planners realized that if the enemy operated land-based aircraft from that key island, then it immediately imperiled Allied control of the New Hebrides and New Caledonia area. Rather than wait until the Japanese were firmly entrenched, they proposed to evict the Japanese before they got too deeply settled. Vice Admiral Robert L. Ghormley—who had attained a sterling record in London as Special Naval Observer—was detailed to take command of the operation; and he established his headquarters at Auckland, New Zealand. Since the Japanese had gotten a foothold on Guadalcanal, time was of the essence. Preparations for the invasion proceeded apace with the utmost secrecy and speed.

Wasp—together with the carriers Saratoga and Enterprise—was assigned to the Support Force under Vice Admiral Frank Jack Fletcher. Under the tactical command of Rear Admiral Noyes, embarked in Wasp, the carriers were to provide air support for the invasion.

Wasp and her airmen worked intensively practicing day and night operations to hone their skills to a high degree. Pilot qualification and training, necessitated by the ship's recent operations in the Atlantic and by the re-equipment of her air group and newer types of planes, proceeded at an intensive

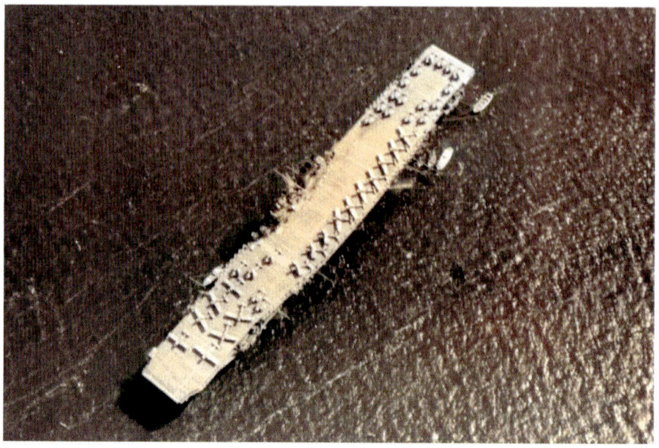

pace and, by the time the operations against Guadalcanal were pushed into high gear, Capt. Sherman was confident that his airmen could perform their mission. "D-day" had originally been set for 1 August, but the late arrival of some of the transports carrying marines pushed the date to 7 August.

Wasp, screened by San Francisco (CA-38), Salt Lake City (CA-25), and four destroyers, steamed westward toward Guadalcanal on the evening of 6 August until midnight. Then, she changed course to the eastward to reach her launch position 84 miles from Tulagi one hour before the first rays of sunlight crept over the horizon. A fresh breeze whipped across the carrier's darkened flight deck as the first planes were brought up to prepare for launch. The night offshore was bright, but clouds hung heavily over the assigned objective. So far, so good. No Japanese patrols had been spotted.

At 0530, the first planes from Wasp's air group barreled down the deck: 16 F4F-4s under Lt. Comdr. Courtney Shands. Then, 15 SBD-3s under Lt. Comdr. John Eldridge, Jr., and the TBF-1 flown by the air group commander, Lt. Comdr. Wallace M. Beakley, fitted with a larger gasoline tank in its bomb bay to lengthen its time in the air, followed seven minutes later. At 0557, the first combat air patrol fighter took off.

The early flights of F4Fs and SBDs were assigned specific targets: Tulagi, Gavutu, Tanambogo, Halavo, Port Purvis, Haleta, Bungana, and the radio station dubbed "Asses' Ears." After taking off, the 16 Wildcats split up into sections and raced off to their respective hunting areas. At about 0600, the planes passed over the transport area off Lunga Point, as the ships were preparing to disembark their troops. In the pre-dawn darkness, the ships were almost invisible until the fighters passed directly over them. Soon, the fourth division of Shands' flight climbed to

5,000 feet above Tulagi to serve as CAP for the strafers. The third division broke off and headed for their target—Haleta—before Shands took three planes around the northwest tip of Tulagi.

Shands and his wingman, Ens. S. W. Forrer, then swung down the north coast toward Gavatu. The other two headed for Tanambogo, to work over the seaplane facilities there. The Japanese appeared to be caught flat-footed, and the Grummans, arriving simultaneously at daybreak, shot up all of the patrol planes and fighter seaplanes that were in the area. Fifteen Kawanishi flying boats and seven Nakajima floatplane fighters—the seaplane derivative of the Mitsubishi "Zero"—were destroyed by Shands' fighters that flew almost "on the deck." Shands himself bagged at least four Nakajima single-float fighter seaplanes and one four-engined flying boat. His wingman, Forrer, bagged three floatplane fighters and one patrol plane. Lt. Wright and Ens. Kenton bagged three patrol planes apiece and destroyed a motorboat apparently attempting to tend the flying boats; Ensigns Reeves and Conklin each bagged two and shared a fifth patrol plane between them. In addition, the strafing F4Fs destroyed an aviation fuel truck and a truck loaded with spare parts.

The SBDs, too, laid their bombs "on the money." Post-attack assessment estimated that the antiaircraft and shore battery sites pinpointed by intelligence had been destroyed by the dive bombers in their first attack. So complete was the enemy's unpreparedness that none of Wasp's planes was shot down. Only one plane from the 16 Grummans failed to return, and, in that case, its pilot, Ensign Reeves, put her down on board Enterprise after having run low on fuel.

That was not all, however. At 0704, 12 Grumman TBF-1s, led by Lt. H. A. Romberg, rolled ponderously down the deck, loaded with bombs for use against land targets. Having encountered resistance, the initial landing forces called for help. Romberg's dozen Avengers blasted enemy troop concentrations east of the knob of land known as Hill 281, in the Makambo-Sasapi sector, and the prison on Tulagi Island. "All enemy resistance," the official report later stated, was "apparently effectively silenced by this flight."

The first day's operations against Guadalcanal had proved successful. Some 10,000 men had been put ashore there and met only slight resistance. On Tulagi, however, the Japanese resisted stoutly, retaining about one-fifth of the island by nightfall. Wasp, Saratoga, and Enterprise, with their screens, retired to the southward at nightfall.

Wasp returned the next morning, 8 August, to maintain a continuous CAP over the transport area until noon. These fighters were led by Lt. C. S. Moffett. Meanwhile, she also launched a scouting flight of 12 SBD-3s led by Lt. Comdr. E. M. Snowden. The Dauntlesses searched a sector to a radius of 220 miles from their carrier, extending it to include all of Santa Isabel Island and the New Georgia group.

The Dauntless pilots sighted nothing that morning and made no contact with the enemy during their two hours in the air. But that was soon to change for the flight leader. At 0815, Snowden sighted a "Rufe" some 40 miles from Rekata Bay and gave chase. The Japanese airman, seeing that he had been spotted, had no stomach for a fight. He pulled up and attempted to use the clouds for cover. Each time the dogged dive bomber pilot gunned the SBD-3 after him: twice the "Rufe" headed for the clouds. Snowden finally pulled within close range, and, using his two fixed .50-caliber guns, fired a short burst that hit home, causing the "Rufe" to spin into the Solomon Sea.

Meanwhile, a large group of Japanese planes approached from Bougainville, apparently bent upon attacking the transports off Lunga Point. Upon learning of their approach, Rear Admiral Richmond K. Turner ordered all transports to get underway and to assume cruising disposition. The Americans accordingly cleared the decks for action. Wasp's planes took part in the melee that followed—some planes by accident.

Lt. Comdr. Eldridge—again leading a formation of SBD-3s from VS-71—had led his planes against Mbangi Island, off Tulagi, the site of some still fierce Japanese resistance. Eldridge's rear seat gunner, Aviation Chief Radioman L. A. Powers, suddenly spotted a formation of planes coming in from the northeast: but thinking them to be a relief flight, Eldridge continued on his present course. The Americans did a double-take, however, and discovered that the planes were, in fact, enemy. At that instant, six "Zeroes" showed up and bounced the first section, but showed remarkably little skill in the attack, for they made 12 firing passes but could not down any of the Dauntlesses.

Meanwhile, the leader of the last section of VS-71, Lt. (jg.) Robert L. Howard, spotted a cluster of twin-engined G4M1 "Betty" bombers heading for the American transports. Howard dove to the attack; but, in his excitement, failed to flip his armament switch to "on." After two runs during which his guns had failed to fire—thinking that the guns needed to be recharged—he discovered his error—but too late to do anything about the Mitsubishi bombers. At that moment, four "Zeroes," escorts for the bombers, attacked the single SBD.

Howard's rear gunner, Seaman 2d Class Lawrence P. Lupo, handled his twin 30-caliber mount magnificently and kept the enemy fighters at arm's length, his bullets scoring several hits on them as well. After about eight passes, one "Zero" veered up sharply and made a head-on run that Howard met with simultaneous fire from his fixed .50s. The "Zero" caught fire like a flying tinderbox, passed close aboard the Dauntless' left wing, and crashed in flames amidst the American landing craft far below. At the same time Howard was downing the "Zero" ahead, Seaman Lupo was firing on another "Zero" making an attack from the stern. Lupo kept the enemy away, but he had to shoot through his own plane's vertical stabilizer to do it. Eventually, the enemy tired of sporting with the SBD and retired to leave Howard and his squadron mates in VS-71 to return safely to their carrier.

At 1807 on 8 August, Vice Admiral Frank Jack Fletcher recommended to Ghormley, at Noumea, that the air support force be withdrawn. Fletcher, concerned by the large numbers of enemy planes that had attacked on the 8th, reported that he had only 78 fighters left (he had started with 99) and that fuel for the carriers was running low. Ghormley approved the recommendation, and Wasp joined Enterprise and Saratoga in retiring from Guadalcanal. By midnight on 8 August, the landing had been a success, having attained the immediate objectives of the landing. All Japanese resistance—but a few snipers—on Gavutu and Tanombogo had been overcome. Early on 9 August, a Japanese surface force engaged an American one off Savo Island and retired at very little cost to themselves. The Allied force suffered the loss of four heavy cruisers off Savo Island, including two that had served with Wasp in the Atlantic: Vincennes and Quincy. The early and unexpected withdrawal of the support force, including Wasp, when coupled with Allied

losses in the Battle of Savo Island, jeopardized the success of the operation in the Solomons.

After the initial day's action in the Solomons campaign, the carrier spent the next month engaged in patrol and covering operations for convoys and resupply units headed for Guadalcanal. The Japanese, while reacting sluggishly to the initial thrust at Guadalcanal, soon began pouring reinforcements down to contest the Allied forces.

Wasp was ordered south by Vice Admiral Fletcher to refuel and did not participate in the Battle of Eastern Solomons on 24 August 1942. That engagement cost the American force the use of the valuable Enterprise. Saratoga was torpedoed a week later and departed the South Pacific war zone for repairs as well. That left only two carriers in the southwest Pacific: Hornet—commissioned for only a year—and Wasp.

On Tuesday, 16 September, those two carriers and North Carolina—with 10 other warships—were escorting the transports carrying the 7th Marine Regiment to Guadalcanal as reinforcements. Wasp had drawn the job of ready-duty carrier and was operating some 150 miles southeast of San Cristobal Island. Her gasoline system was in use, as planes were being refueled and rearmed for antisubmarine patrol missions; and Wasp had been at general quarters from an hour before sunrise until the time when the morning search returned to the ship at 1000. There was no contact with the enemy during the day, except for a Japanese four-engined flying boat downed by a Wasp Wildcat at 1215.

About 1420, the carrier turned into the wind to launch eight fighters and 18 SBD-3s and to recover eight F4F-3s and three SBDs that had been airborne since before noon. The ship rapidly completed the recovery of the 11 planes, then turned easily to starboard, the ship heeling slightly as the course change was made.

The air department at flight quarters worked coolly at refueling and respotting the ship's planes for the afternoon mission. Suddenly, at 1444, a lookout called out, "Three torpedoes...three points forward of the starboard beam!"

A spread of four torpedoes, fired from the tubes of the Japanese submarine I-19, churned inexorably closer. Wasp put over her rudder hard-a-starboard, but it was too late. Two torpedoes smashed home in quick succession while a fourth passed ahead. Both hit in the vicinity of gasoline tanks and magazines.

In quick succession, fiery blasts ripped through the forward part of the ship. Aircraft on the flight and hangar decks were thrown about as if they were toys and dropped on the deck with such force that landing gears snapped. Planes triced up in the hangar overheads fell and landed upon those on the hangar deck; fires broke out almost simultaneously in the hangar and below decks. Soon, the heat of the intense gasoline fires detonated the ready ammunition at the forward antiaircraft guns on the starboard side, and fragments showered the forward part of the ship. The number two 1.1-inch mount was blown overboard and the corpse of the gun captain was thrown onto the bridge where it landed next to Capt. Sherman.

Water mains in the forward part of the ship proved useless, since they had been broken by the force of the explosions. There was no water available to fight the conflagration forward, and the fires continued to set off ammunition, bombs, and gasoline. As the ship listed to starboard between 10 and 15 degrees, oil and gasoline, released from the tanks by the torpedo hit, caught fire on the water.

Sherman slowed to 10 knots, ordering the rudder put to port to try to get the wind on the starboard bow; he then went astern with right rudder until the wind was on the starboard quarter, in an attempt to keep the fire forward. At that point, some flames made the central station untenable, and communication circuits went dead. Soon, a serious gasoline fire broke out in the forward portion of the hangar. Within 24 minutes of the initial attack, three additional major gasoline vapor explosions occurred. Ten minutes later, Capt. Sherman consulted with his executive officer, Comdr. Fred C. Dickey. The two men saw no course but to abandon, as all fire-fighting was proving ineffectual. The survivors would have to be gotten off quickly if unnecessary loss of life was not to be incurred.

Reluctantly, after consulting with Rear Admiral Noyes, Capt. Sherman ordered "abandon ship" at 1520. All badly injured men were lowered into rafts or rubber boats. Many unwounded men had to abandon from aft because the forward fires were burning with such intensity. The departure, as Capt. Sherman observed it, looked "orderly," and there was no panic. The only delays occurred when many men showed reluctance to leave until all the wounded had been taken off. The abandonment took nearly 40 minutes; and, at 1600—satisfied that no one was left on deck, in the galleries, or in the hangar aft—Capt. Sherman swung over the lifeline on the fantail and slid into the sea.

Although the submarine hazard caused the accompanying destroyers to lie well clear or to shift position, the "tin cans" carried out the rescue efforts with persistence and determination until Laffey (DD-459), Lansdowne (DD-486), Helena (CL-50), and Salt Lake City had 1,946 men embarked. The abandoned ship drifted with her crew of remaining dead. The fires greedily traveled aft; four more violent explosions boomed as night began to fall. Lansdowne drew the duty of destruction, and she fired five torpedoes into the dying ship's fire-gutted hull. Three hit, but she remained afloat. By now, the orange flames had enveloped the stern. The carrier literally floated in a burning pool of gasoline and oil. She sank at 2100 by the bow.

Wasp received two battle stars for her World War II service.

CV-8 • USS HORNET

(Displacement 19,800; length 809'9"; ew. 144'; draft 21'8"; speed 33 k.; complement 1,889; armor 8 5", 16 1.1 mg.; class Hornet)

The seventh Hornet (CV-8) was launched on 14 December 1940 by the Newport News Ship Building & Dry Dock Co., Newport News, Va.; sponsored by Mrs. Frank M. Knox, wife of the Secretary of the Navy; and commissioned at Norfolk on 20 October 1941, Captain Marc A. Mitscher in command.

During the uneasy period before Pearl Harbor, Hornet trained out of Norfolk. A hint of a future mission occurred on 2 February 1942 when Hornet departed Norfolk with two Army B-25 medium bombers on deck. Once at sea, the planes were launched to the surprise and amazement of Hornet's crew. Her men were unaware of the meaning of this experiment, as Hornet returned to Norfolk, prepared to leave for combat, and on 4 March sailed for the West Coast via the Panama Canal.

Hornet arrived in San Francisco on 20 March. With her own planes on the hangar deck, she loaded 16 Army B-25 bombers on the flight deck. Under the command of Lieutenant Colonel James H. Doolittle, 70 officers and 64 enlisted men reported aboard. In the company of escort ships, Hornet departed San Francisco on 2 April and embarked on her mission under sealed orders. That afternoon, Captain Mitscher informed his men of their mission: a bombing raid on Japan.

Eleven days later, Hornet joined Enterprise off Midway and Task Force 16 turned toward Japan. With Enterprise providing air combat cover, Hornet was to steam deep into enemy waters where Colonel Doolittle would lead the B-25s in a daring strike on Tokyo and other important Japanese cities. Originally, the task force intended to proceed to within 400 miles of the Japanese coast; however, on the morning of 18 April, a Japanese patrol boat, No. 23 Nitto Maru, sighted Hornet. The cruiser Nashville sank the craft which had already informed the Japanese of the presence and location of the American task force. Though some 600 miles from the Japanese coast, confirmation of the patrol boat's warning prompted Admiral William F. Halsey at 0800 to order the immediate launching of the "Tokyo Raiders."

As Hornet swung about and prepared to launch the bombers which had been readied for take-off the previous day, a gale of more than 40 knots churned the sea with 30-foot crests;

heavy swells caused the ship to pitch violently, shipping sea and spray over the bow, wetting the flight deck and drenching the deck crews. The lead plane, commanded by Colonel Doolittle, had but 467 feet of flight deck while the last B-25 hung far out over the fantail. The first of the heavily-laden bombers lumbered down the flight deck, circled Hornet after take-off, and set course for Japan. By 0920, all 16 of the bombers were airborne, heading for the first American air strike against the heart of Japan.

Hornet brought her own planes on deck and steamed at full speed for Pearl Harbor. Intercepted broadcasts, both in Japanese and English, confirmed at 1446 the success of the raids. Exactly one week to the hour after launching the B-25s, Hornet sailed into Pearl Harbor. Hornet's mission was kept an official secret for a year; until then, President Roosevelt referred to the origin of the Tokyo raid only as "Shangri-La."

Hornet steamed from Pearl Harbor on 30 April to aid Yorktown and Lexington at the Battle of the Coral Sea. But that battle was over before she reached the scene. She returned to Hawaii on 26 May and sailed 2 days later with her sister carriers to repulse an expected Japanese fleet assault on Midway.

Japanese carrier-based planes were reported headed for Midway on the early morning of 4 June 1942. Hornet, Yorktown, and Enterprise launched strikes as the Japanese carriers struck their planes below to prepare for a second strike on Midway. Hornet dive bombers missed contact, but 15 planes comprising her Torpedo Squadron 8 found the enemy and pressed home their attacks. They were met by overwhelming fighter opposition about 8 miles from three enemy carriers and followed all the way in to be shot down one by one. Ens. George H. Gay, USNR, the only surviving pilot, reached the surface as his plane sank. He hid under a rubber seat cushion to avoid strafing and witness the greatest carrier battle in history.

Of 41 torpedo planes launched by the American carriers, only six returned. Their sacrifices drew enemy fighters away from dive bombers of Enterprise and Yorktown, who sank three Japanese carriers with an assist from submarine Nautilus. The fourth Japanese carrier, Hiryu, was sunk the following day; gallant Yorktown was lost to a combined aerial and submarine attack.

Hornet planes attacked the fleeing Japanese fleet on 6 June 1942 to assist in sinking cruiser Mikuma, damaged a destroyer, and left cruiser Mogami aflame and heavily damaged. Hits were also made on other ships. Hornet's attack on Mogami wrote the finis to one of the decisive battles of history that had far-reaching and enduring results on the Pacific War. Midway was saved as an important base for operations into the western Pacific. Likewise saved was Hawaii. Of greatest importance was the crippling of Japan's carrier strength, a severe blow from which she never fully recovered. The four large aircraft carriers sent to the bottom of the sea carried with them some 250 planes along with a high percentage of Japan's most highly trained and battle-experienced carrier pilots. This great victory by Hornet and our other ships at Midway spelled the doom of Japan.

Following the Battle of Midway, Hornet had new radar installed and trained out of Pearl Harbor. She sailed on 17 August 1942 to guard the sea approach to bitterly contested Guadalcanal in the Solomons. Bomb damage to Enterprise (24

August), torpedo damage to Saratoga (31 August), and the loss of WASP (15 September) reduced carriers in the South Pacific to one: Hornet. She bore the brunt of air cover in the Solomons until 24 October 1942 when she joined Enterprise northwest of the New Hebrides Islands and steamed to intercept a Japanese carrier-battleship force bearing down on Guadalcanal.

The Battle of Santa Cruz Island took place on 26 October 1942 without contact between surface ships of the opposing forces. That morning, Enterprise planes bombed carrier Zuiho. Planes from Hornet severely damaged carrier Shokaku and cruiser Chikuma. Two other cruisers were also attacked by Hornet aircraft. Meanwhile, Hornet herself was fighting off a coordinated dive bombing and torpedo plane attack which left her so severely damaged that she had to be abandoned. Commented one sailor, awaiting rescue, when asked if he planned to re-enlist, "Dammit, yes-on the new Hornet!" Captain Mason, the last man on board, climbed over the side and survivors were soon picked up by destroyers.

The abandoned Hornet, ablaze from stem to stern, refused to accept her intended fate from friends. She still floated after receiving nine torpedoes and more than 400 rounds of 5-inch shellfire from destroyers Mustin and Anderson. Japanese destroyers hastened the inevitable by firing four 24-inch torpedoes at her blazing hull. At 0135, 27 October 1942, she finally sank off the Santa Cruz Islands. Her proud name was struck from the Navy List on 13 January 1943.

Hornet (CV-8) received four battle stars for World War II service. Her famed Torpedo Squadron 8 was awarded the Presidential Unit Citation "for extraordinary heroism and distinguished service beyond the call of duty" in the Battle of Midway.

CV-9 • USS ESSEX

(Displacement 27,100; length 872'; beam 93'; ew. 147'6"; draft 28'7"; speed 33 k.; complement 3,448; armor 12 5"; class Essex)

The fourth Essex (CV-9) was launched on 31 July 1942 by Newport News Shipbuilding and Dry Dock Co., sponsored by Mrs. Artemus L. Gates, wife of the Assistant Secretary of the Navy for Air, and commissioned on 31 December 1942, Captain D. B. Duncan commanding. She was reclassified (CVA-9) on 1 October 1952 and (CVS-9) on 8 March 1960.

Following her shakedown cruise, Essex sailed to the Pacific in May 1943 to begin a succession of victories which would bring her to Tokyo Bay. Departing Pearl Harbor, she participated with TF 15 in carrier operations against Marcus Island (31 August 1943); was designated flagship of TF 14 and struck Wake Island (5-6 October); launched an attack with TG 50.3 against the Gilbert Islands where she also took part in her first amphibious assault, the landing on Tarawa (18-23 November). Refueling at sea, she cruised as flagship of TG 50.3 to attack Kwajalein (4 December). Her second amphibious assault, delivered in company with TG 58.2, was against the Marshalls (29 January-2 February 1944).

Essex in TG 58.2 now joined with TG 58.1 and 58.3 to constitute the most formidable carrier striking force to date, launching an attack against Truk (17-18 February) during which eight Japanese ships were sunk. En route to the Marianas to sever Japanese supply lines, the carrier force was detected and received a prolonged aerial attack which it repelled in a businesslike manner and then continued with the scheduled attack upon Saipan, Tinian, and Guam (23 February).

After this operation, Essex proceeded to San Francisco for her single wartime overhaul. She then joined carriers Wasp (CV-18) and San Jacinto (CVL-30) in TG 12.1 to strike Marcus Island (19-20 May) and Wake (23 May). She deployed with TF 58 to support the occupation of the Marianas (12 June—10 August); sortied with TG 38.3 to lead an attack against the Palau Islands (6-8 September), and Mindanao (9-10 September) with enemy shipping as the main target, and remained in the area to support landings on Peleliu. On 2 October she weathered a typhoon and 4 days later departed with TF 38 for the Ryukyus.

For the remainder of 1944, she continued her frontline action, participating in strikes against Okinawa (10 October) and

Formosa (12-14 October), covering the Leyte landings, taking part in the battle for Leyte Gulf (24-25 October), and continuing the search for enemy fleet units until 30 October when she returned to Ulithi, Caroline Islands, for replenishment. She resumed the offensive and delivered attacks on Manila and the northern Philippine Islands during November. On 25 November, for the first time in her far-ranging operations and destruction to the enemy, Essex received injury. A kamikaze hit the port edge of her flight deck landing among planes gassed for takeoff, causing extensive damage, killing 15, and wounding 44.

This "cramped her style" very little. Following quick repairs, we find her with the 3rd Fleet off Luzon supporting the occupation of Mindoro (14-16 December). She rode out the typhoon of 18 December and made a special search for survivors afterward. With TG 38.3, she participated in the Lingayen Gulf operations, launched strikes against Formosa, Sakishima, Okinawa, and Luzon. Entering the South China Sea in search of enemy surface forces, the task force pounded shipping and conducted strikes on Formosa, the China coast, Hainan, and Hong Kong. Essex withstood the onslaught of the third typhoon in 4 months (20-21 January 1945) before striking again at Formosa, Miyako Shima, and Okinawa (26-27 January).

During the remainder of the war, she operated with TF 58, conducting attacks against the Tokyo area (16-17, and 25 February) both to neutralize the enemy's airpower before the landings on Iwo Jima and to cripple the aircraft manufacturing industry. She sent support missions against Iwo Jima and neighboring islands, but from 23 March to 28 May was employed primarily to support the conquest of Okinawa.

In the closing days of the war, Essex took part in the final telling raids against the Japanese home islands (10 July-15 August). Following the surrender, she continued defensive combat air patrols until 3 September when she was ordered to Bremerton, Wash., for inactivation. On 9 January 1947, she was placed out of commission in reserve.

Modernization endowed Essex with a new flight deck and a streamlined island superstructure, on 15 January 1951 when recommissioned, Captain A. W. Wheelock commanding.

After a brief cruise in Hawaiian waters, she began the first of three tours in Far Eastern waters during the Korean War. She served as flagship for Carrier Division 1 and TF 77. She was the first carrier to launch F2H "Banshee" twin-jet fighters on combat missions; on 16 September 1951, one of these planes, damaged in combat, crashed into aircraft parked on the forward flight deck, causing an explosion and fire which killed seven. After repairs

at Yokosuka, she returned to front-line action on 3 October to launch strikes up to the Yalu River and provide close air support for U.N. troops.

On 1 December 1953, she started her final tour of the war, sailing the China Sea with the Peace Patrol. From November 1954 to June 1955, she engaged in training exercises, operated for 3 months with the 7th Fleet, assisted in the Tachen Islands evacuation, and engaged in air operations and fleet maneuvers off Okinawa.

In July 1955, Essex entered Puget Sound Naval Shipyard for repairs and extensive alterations, including the installation of an angled flight deck. Modernization completed, she rejoined the Pacific Fleet in March 1956. For the next 14 months, the carrier operated off the west coast, except for a 6-month cruise with the 7th Fleet in the Far East. Ordered to join the Atlantic Fleet for the first time in her long career, she sailed from San Diego on 21 June 1957, rounded Cape Horn, and arrived in Mayport, Fla., on 1 August.

In the fall of 1957, Essex participated as an antisubmarine carrier in the NATO exercises, "Strike Back," and in February 1958 deployed with the 6th Fleet until May when she shifted to the eastern Mediterranean. Alerted to the Middle East crisis on 14 July 1958, she sped to support the U.S. Peace Force landing in Beirut, Lebanon, launching reconnaissance and patrol missions until 20 August. Once again she was ordered to proceed to Asian waters and transited the Suez Canal to arrive in the Taiwan operational area where she joined TF 77 in conducting flight operations before rounding the Horn and proceeding back to Mayport.

Essex joined with the 2nd Fleet and British ships in Atlantic exercises and with NATO forces in the eastern Mediterranean during the fall of 1959. In December, she aided victims of a disastrous flood at Frejus, France.

In the spring of 1960, she was converted into an ASW Support Carrier and was thereafter homeported at Quonset Point, R.I. Since that time, she has operated as flagship of Carrier Division 18 and Antisubmarine Carrier Group Three. She conducted rescue and salvage operations off the New Jersey coast for a downed blimp, cruised with midshipmen, and was deployed on NATO and CENTO exercises. In November, she joined the French navy in Operation "Jet Stream" and has since continued her widespread activities in the protection of freedom and peace.

Essex received the Presidential Unit Citation, and 13 battle stars for World War II service; 4 battle stars and the Navy Unit Commendation for Korean War service.

CV-10 • USS YORKTOWN

(Displacement. 27,100; length. 872'0"; beam. 93'0"; e.w. 147'6"; draft. 28'7"; speed. 32.1 k.; complement. 3,448; armor. 12 5"; 32 40mm., 46 20mm.; class. Essex)

The fourth Yorktown (CV-10) was laid down on 1 December 1941 at Newport News, Va., by the Newport News Shipbuilding & Drydock Co. as Bon Homme Richard; renamed Yorktown on 26 September 1942; launched on 21 January 1943; sponsored by Mrs. Eleanor Roosevelt; and commissioned on 15 April 1943 at the Norfolk Navy Yard, Capt. Joseph J. ("Jocko") Clark in command.

Yorktown remained in the Norfolk area until 21 May, at which time she got underway for shakedown training in the vicinity of Trinidad. She returned to Norfolk on 17 June and began post-shakedown availability. The aircraft carrier completed repairs on 1 July and began air operations out of Norfolk until the 6th. On that day, she exited Chesapeake Bay on her way to the Pacific Ocean. She transited the Panama Canal on 11 July and departed Balboa on the 12th. The warship arrived in Pearl Harbor on 24 July and began a month of exercises in the Hawaiian Islands. On 22 August, she stood out of Pearl Harbor, bound for her first combat of the war. Her task force, TF 15, arrived at the launching point about 128 miles from Marcus Island early on the morning of 31 August. She spent most of that day launching fighter and bomber strikes on Marcus Island before beginning the retirement to Hawaii that evening. The aircraft carrier reentered Pearl Harbor on 7 September and remained there for two days.

On the 9th, she stood out to sea, bound for the west coast of the United States. She arrived in San Francisco on 13 September, loaded aircraft and supplies, and returned to sea on the 15th. Four days later, the aircraft carrier reentered Pearl Harbor. After 10 days in the Hawaiian Islands, Yorktown returned to sea to conduct combat operations on the 29th. Early on the morning of 5 October, she began two days of air strikes on Japanese installations on Wake Island. After retiring to the east for the night, she resumed those air raids early on the morning of the 6th and continued them through most of the day. That evening, the task group began its retirement to Hawaii. Yorktown arrived at Oahu on 11 October and, for the next month, conducted air training operations out of Pearl Harbor.

On 10 November, Yorktown departed Pearl Harbor in company with Task Force (TF) 50—the Fast Carrier Forces, Pacific Fleet—to participate in her first major assault operation, the occupation of certain of the Gilbert Islands. On the 19th, she arrived at the launch point near Jaluit and Mili and, early that morning, launched the first of a series of raids to suppress enemy airpower during the amphibious assaults on Tarawa, Abemama, and Makin. On the 20th, she not only sent raids back to the airfield at Jaluit, but some of her planes also supported the troops wresting Makin from the Japanese. On 22 November, her air group concentrated upon installations and planes at Mili once again. Before returning to Pearl Harbor, the aircraft carrier made passing raids on the installations at Wotje and Kwajalein Atolls on 4 December. The warship reentered Pearl Harbor on 9 December and began a month of air training operations in the Hawaiian Islands.

On 16 January 1944, the warship exited Pearl Harbor once again to support an amphibious assault—Operation "Flintlock," the Marshall Islands operation. Her task group, Task Group (TG) 58.1, arrived at its launching point early on the morning of 29 January, and its carriers—Yorktown, Lexington (CV-16), and Cowpens (CVL-25)—began sending air strikes aloft at about 0520 for attacks on Taroa airfield located on Maloelap Atoll. Throughout the day, her aircraft hit Maloelap in preparation for the assaults on Majuro and Kwajalein scheduled for the 31st. On the 30th, Yorktown and her sister carriers shifted targets to Kwajalein to begin softening up one of the targets itself. When the troops stormed ashore on the 31st, Yorktown aviators continued their strikes on Kwajalein in support of the troops attacking that atoll. The same employment occupied the Yorktown air group during the first three days in February. On the 4th, however, the task group retired to the Fleet anchorage at recently secured Majuro Atoll.

Over the next four months, Yorktown participated in a series of raids in which she ranged from the Marianas in the north to New Guinea in the south. After eight days at Majuro, she sortied with her task group on 12 February to conduct air strikes on the main Japanese anchorage at Truk Atoll. Those highly sucessful raids occurred on 16 and 17 February. On the 18th, the carrier set a course for the Marianas and, on the 22d, conducted a single day of raids on enemy airfields and installations on Saipan. That same day, she cleared the area on her way back to Majuro. The warship arrived in Majuro lagoon on 26 February and remained there, resting and replenishing until 8 March. On the latter day, the carrier stood out of Majuro, rendezvoused with the rest of TF 58, and shaped a course for Espiritu Santo in

the New Hebrides. She reached her destination on 13 March and remained there for 10 days before getting underway for another series of raids on the Japanese middle defense line. On 30 and 31 March, she launched air strikes on enemy installations located in the Palau Islands; and, on 1 April, her aviators went after the island of Woleai. Five days later, she returned to her base at Majuro for a week of replenishment and recreation.

On 13 April, Yorktown returned to sea once more. On this occasion, however, she laid in a course for the northern coast of New Guinea. On 21 April, she began launching raids in support of General Douglas Mac-Arthur's assault on the Hollandia area. That day, her aviators attacked installations in the Wakde-Sarmi area of northern New Guinea. On the 22d and 23d, they shifted to the landing areas at Hollandia themselves and began providing direct support for the assault troops. After those attacks, she retired from the New Guinea coast for another raid on Truk lagoon, which her aircraft carried out on 29 and 30 April. The aircraft carrier retuned to Majuro on 4 May; however, two days later she got underway again, bound for Oahu. The warship entered Pearl Harbor on 11 May and, for the next 18 days, conducted training operations in the Hawaiian Islands. On 29 May, she headed back to the Central Pacific. Yorktown entered Majuro lagoon again on 3 June and began preparations for her next major amphibious support operation—the assault on the Marianas.

On 6 June, the aircraft carrier stood out of Majuro with TF 58 and set a course for the Mariana Islands. After five days steaming, she reached the launch point and began sending planes aloft for the preliminary softening up of targets in preparation for the invasion of Saipan. Yorktown aircrews concentrated primarily upon airfields located on Guam. Those raids continued until the 13th when Yorktown, with two of the task groups of TF 58, steamed north to hit targets in the Bonin Islands. That movement resulted in a one-day raid on the 16th before the two task groups headed back to the Marianas to join in the Battle of the Philippine Sea. Task Force 58 reunited on 18 June and began a short wait for the approaching Japanese Fleet and its aircraft.

On the morning of 19 June, Yorktown aircraft began strikes on Japanese air bases on Guam in order to deny them to their approaching carrier-based air and to keep the land-based planes out of the fray. Duels with Guam-based aircraft continued until mid-morning. At about 1017, however, she got her first indication of the carrier plane attacks when a large bogey appeared on her radar screen. At that point she divided her attention, sending part of her air group back to Guam and another portion of it out to meet the raid closing from the west. Throughout the battle, Yorktown's planes continued both to strike the Guam airfields and intercept the carrier raids. During the first day of the Battle of the Philippine Sea, Yorktown aircraft claimed 37 enemy planes destroyed and dropped 21 tons of bombs on the Guam air bases.

On the morning of the 20th, Yorktown steamed generally west with TF 58 while search planes scouted for the fleeing enemy task force. Contact was not made until about 1540 that afternoon when a Hornet (CV-12) pilot spotted the retiring Combined Fleet units. Yorktown launched a 40-plane strike between 1623 and 1643, sending it winging after the Japanese.

Her planes found Admiral Ozawa's force at about 1840 and began a 20-minute attack, during which they targeted Zuikaku and scored some hits, but failed to sink the carrier. They also attacked several other ships in the Japanese force, though no records show a confirmed sinking credited to the Yorktown air group. On 21 June, the carrier joined in the futile pursuit of the enemy by TF 58 but gave up that evening when air searches failed to locate the Japanese. Yorktown returned to the Marianas area and resumed air strikes on Pagan on the 22nd and 23rd. On the 24th, she launched another series of raids on Iwo Jima. On 25 June, she set a course for Eniwetok and arrived there two days later. On the 30th, the aircraft carrier headed back to the Marianas and the Bonins. She renewed combat operations on 3 and 4 July with a series of attacks on Iwo Jima and Chichi Jima. On the 6th, the warship resumed strikes in the Marianas and continued them for the next 17 days. On 23 July, she headed off to the west for a series of raids on Yap, Ulithi, and the Palaus, carrying out those attacks on 25 July and arriving back in the Marianas on the 29th.

On the 31st, she cleared the Mariana Islands and headed—via Eniwetok and Pearl Harbor—back to the United States. Yorktown arrived at the Puget Sound Navy Yard on 17 August and began a two-month overhaul. She completed repairs on 6 October and departed Puget Sound on the 9th. She stopped at the Alameda Naval Air Station from 11 to 13 October to load planes and supplies and then set a course back to the western Pacific. After a stop at Pearl Harbor from 18 to 24 October, Yorktown arrived back in Eniwetok on 31 October. She departed the lagoon on 1 November and arrived at Ulithi on the 3rd. There, she reported for duty with TG 38.4. That task group left Ulithi on 5 November, and Yorktown departed with it.

On 7 November, the aircraft carrier changed operational control to TG 38.1 and, for the next two weeks, launched air strikes on targets in the Philippines in support of the Leyte invasion. Detached from the task force on 23 November, Yorktown arrived back in Ulithi on the 24th. She remained there until 10 December, at which time she put to sea to rejoin TF 38. She rendezvoused with the other carriers on 13 December and began launching air strikes on targets on the island of Luzon in preparation for the invasion of that island scheduled for the second week in January. On the 17th, the task force began its retirement from the Luzon strikes. During that retirement, TF 38 steamed through the center of the famous typhoon of December 1944. That storm sank three destroyers—Spence (DD-512), Hull (DD-350), and Monaghan (DD-354)—and Yorktown participated

in some of the rescue operations for the survivors of those three destroyers. She did not finally clear the vicinity of Luzon until the 23rd. The warship arrived back in Ulithi on 24 December.

The aircraft carrier fueled and provisioned at Ulithi until 30 December, at which time she returned to sea to join TF 38 on strikes at targets in the Philippines in support of the landings at Lingayen. The carriers opened the show on 3 January 1945 with raids on airfields on the island of Formosa. Those raids continued on the 4th, but a fueling rendezvous occupied Yorktown's time on the 5th. She sent her planes against Luzon targets and on antishipping strikes on the 6th and 7th. The 8th brought another fueling rendezvous, and, on the 9th, she conducted her last attack—on Formosa—in direct support of the Lingayen operation. On 10 January, Yorktown and the rest of TF 38 entered the South China Sea via Bashi Channel to begin a series of raids on Japan's inner defenses. On 12 January, her planes visited the vicinity of Saigon and Tourane Bay, Indochina, in hopes of catching major units of the Japanese fleet. Though foiled in their primary desire, TF 38 aviators still managed to rack up a stupendous score—44 enemy ships, of which 15 were combatants. She fueled on the 13th and, on the 15th, launched raids on Formosa and Canton in China. The following day, her aviators struck at Canton again and paid a visit to Hong Kong. Fueling took up her time on 17, 18, and 19 January; and, on the 20th, she exited the South China Sea with TF 38 via Balintang Channel. She participated in a raid on Formosa on the 21st and another on Okinawa on the 22nd before clearing the area for Ulithi. On the morning of 26 January, she reentered Ulithi lagoon with TF 38.

Yorktown remained at Ulithi arming, provisioning, and conducting upkeep until 10 February. At that time, she sortied with TF 58, the 3rd Fleet becoming the 5th Fleet when Spruance relieved Halsey, on a series of raids on the Japanese and thence to support the assault on and occupation of Iwo Jima. On the morning of 16 February, the aircraft carrier began launching strikes on the Tokyo area of Honshu. On the 17th, she repeated those strikes before heading toward the Bonins. Her aviators bombed and strafed installations on Chichi Jima on the 18th. The landings on Iwo Jima went forward on 19 February, and Yorktown aircraft began support missions over the island on the 20th. Those missions continued until the 23rd, at which time Yorktown cleared the Bonins to resume strikes on Japan proper. She arrived at the launch point on the 25th and sent two raids aloft to bomb and strafe airfields in the vicinity of Tokyo. On the 26th, Yorktown aircrewmen conducted a single sweep of installations on Kyushu before TG 58.4 began its retirement to Ulithi. Yorktown reentered the anchorage at Ulithi on 1 March.

She remained in the anchorage for about two weeks. On 14 March, the aircraft carrier departed the lagoon on her way to resume raids on Japan and to begin preliminary support work for the Okinawa operations scheduled for 1 April. On 18 March, she arrived in the operating area off Japan and began launching strikes on airfields on Kyushu, Honshu, and Shikoku. The task group came under air attack almost as soon as operations began. At about 0800, a twin-engine bomber, probably a "Frances," attacked from her port side. The ship opened fire almost immediately and began scoring hits quickly. The plane began to burn but continued its run, passing over Yorktown's bow and splashing in the water on her starboard side. Just seven minutes later, another "Frances" tried his luck, but he too went down, a victim of the combined fire of the formation. No further attacks developed until that afternoon, and in the meantime, Yorktown continued air operations. That afternoon, three "Judy's" launched attacks on the carrier. The first two failed in their attacks and were shot down for their trouble. The third succeeded in planting his bomb on the signal bridge. It passed through the first deck and exploded near the ship's hull. It punched two large holes through her side, killed five men, and wounded another 26. Yorktown, however, remained fully operational, and her antiaircraft gunners brought the offender down. She continued air operations against the three southernmost islands of Japan on the 19th but retired for fueling operations on the 20th.

On the 21st, she headed for Okinawa, where she began softening-up strikes on the 23rd. Those attacks continued until the 28th when she started back to Japanese waters for an additional strike on the home islands. On the 29th, the carrier launched two raids and one photographic reconnaissance mission over Kyushu. That afternoon, at about 1410, a single "Judy" made an apparent suicide dive on Yorktown. Her antiaircraft gunners opened up on him and scored numerous hits. He passed over the ship, very near to her "island," and splashed about 60 feet from her portside.

On 30 March, Yorktown and the other carriers of her task group began to concentrate solely on the island of Okinawa and its surrounding islets. For two days, the 30th and 31st, they pounded the island in softening-up strikes. On 1 April, the assault troops stormed ashore, and for almost six weeks, she sent her planes to the island to provide direct support for the troops operating ashore. About every three days, she retired to the east to conduct fueling rendezvous or to rearm and reprovision. The only exception to that routine came on 7 April when it was discovered that a Japanese task force built around the elusive battleship Yamato was steaming south for one last desperate offensive. Yorktown and the other carriers quickly launched strikes to attack that valued target. Air Group 9 aviators claimed several torpedo hits on Yamato just before the battleship exploded and sank, as well as at least three 500-pound bomb hits on light cruiser Yahagi before that warship followed her big sister to the bottom. The pilots also made strafing runs on the escorting destroyers and claimed to have left one afire in a sinking condition. After that action, Yorktown and her planes resumed their support for the troops on Okinawa. On 11 April, she came under air attack again when a single-engine plane sped in on her. Yorktown's antiaircraft gunners proved equal to the test, splashing him just inside 2,000 yards' range. Sporadic air attacks continued until her 11 May departure from the Ryukyus, but Yorktown sustained no additional damage and claimed only one further kill with her antiaircraft battery. On 11 May, TG 58.4 was detached to proceed to Ulithi for upkeep, rest, and relaxation.

Yorktown entered the lagoon at Ulithi on 14 May and remained there until 24 May, at which time she sortied with TG 58.4 to rejoin the forces off Okinawa. On 28 May, TG 58.4 became TG 38.4 when Halsey relieved Spruance and the 5th Fleet again became the 3rd Fleet. That same day, the carrier resumed air support missions over Okinawa. That routine lasted until the beginning of June when she moved off with TF 38 to resume strikes on the Japanese homeland. On 3 June, her aircraft made four different sweeps of airfields. The following day, she returned to Okinawa for a day of additional support missions before steaming off to evade a typhoon. On the 6th and 7th, she resumed Okinawa strikes. She sent her aviators back to the Kyushu airfields and, on the 9th, launched them on the first of two days of raids on Minami Daito Shima. After the second day's strikes on the 10th, Yorktown began retirement with TG 38.4 toward Leyte. She arrived in San Pedro Bay at Leyte on 13 June and began replenishment, upkeep, rest, and relaxation.

The warship remained at Leyte until 1 July when she and TG 38.4 got underway to join the rest of the fast carriers in the final series of raids on the Japanese home islands. By 10 July, she was off the coast of Japan launching air strikes on the Tokyo area of Honshu. After a fueling rendezvous on the 11th and 12th, she resumed strikes on Japan, focusing on the southern portion of the northernmost island—Hokkaido. Those strikes lasted from the 13th to the 15th. A fueling retirement and heavy weather precluded air operations until the 18th, at which time her aviators returned to the Tokyo area. From the 19th to the 22nd, she made a fueling and underway replenishment retirement and then,

on the 24th, resumed air attacks on Japan. For two days, planes from her air group pounded installations around the Kure naval base. Another fueling retirement came on the 26th, but the 27th and 28th found her planes in the air above Kure again. On the 29th and 30th, she shifted targets back to the Tokyo area before another fueling retirement and another typhoon took her out of action until the beginning of the first week in August. On 8 and 9 August, the carrier launched her planes at northern Honshu and southern Hokkaido. On the 10th, she sent them back to Tokyo. The 11th and 12th brought another fueling retirement and a typhoon evasion; but, on the 13th, her aircraft hit Tokyo for the last time. On the 14th, she retired to fuel destroyers again, and on the 15th, Japan agreed to capitulate, so all strikes planned for that day were canceled.

From 16 to 23 August, Yorktown and the other carriers of TF 58 steamed around more or less aimlessly in waters to the east of Japan, awaiting instructions while peace negotiations continued. Then, on the 23rd, she received orders to head for waters east of Honshu where her aircraft were to provide cover for the forces occupying Japan. She began providing that air cover on the 25th and continued to do so until mid-September. After the formal surrender on board Missouri (BB-63) on 2 September, the aircraft carrier also began air-dropping supplies to Allied prisoners of war still in their prison camps. On 16 September, Yorktown entered Tokyo Bay with TG 38.1. She remained there, engaged in upkeep and crew recreation, through the end of the month. On 1 October, the carrier stood out of Tokyo Bay on her way to Okinawa. She arrived in Buckner Bay on 4 October, loaded passengers on the 5th, and got underway for the United States on the 6th.

After a non-stop voyage, Yorktown entered San Francisco Bay on 20 October, moored at the Alameda Naval Air Station, and began discharging passengers. She remained at the air station until 31 October, at which time she shifted to Hunters Point Navy Yard to complete minor repairs. On 2 November, while still at the navy yard, she reported to the Service Force, Pacific Fleet, for duty in conjunction with the return of American servicemen to the United States. That same day, she stood out of San Francisco Bay, bound for Guam on just such a mission. She arrived in Apra Harbor on 15 November and, two days later, got underway with a load of passengers. She arrived back in San Francisco on 30 November and remained there until 8 December. On the latter day, the warship headed back to the Far East. Initially routed to Samar in the Philippines, she was diverted to Manila en route. She arrived in Manila on 26 December and departed there on the 29th. She reached San Francisco again on 13 January 1946. Later that month, she moved north to Bremerton, Wash., where she was placed in commission, in reserve, on 21 June. She remained there in that status through the end of the year. On 9 January 1947, Yorktown was placed out of commission and was berthed with the Bremerton Group, Pacific Reserve Fleet.

Yorktown remained in reserve for almost five years. In June of 1952, she was ordered reactivated, and work began on her at Puget Sound. On 15 December 1952, she was placed in commission, in reserve, at Bremerton. Her conversion continued into 1953, and she conducted post-conversion trials late in January. On 20 February 1953, Yorktown was placed in full commission, Capt. William M. Nation in command. The aircraft carrier conducted normal operations along the west coast through most of the summer of 1953. On 3 August, she departed San Francisco on her way to the Far East. She arrived in Pearl Harbor and remained there until the 27th, at which time she continued her voyage west. On 5 September, the carrier arrived in Yokosuka, Japan. She put to sea again on the 11th to join TF 77 in the Sea of Japan. The Korean War armistice had been signed two months earlier; therefore, the carrier conducted training operations rather than combat missions. She served with TF 77 until 18 February 1954, at which time she stood out of Yokosuka on her way home. She made a stop at Pearl Harbor along the way and then moored at Alameda once more on 3 March. After a brief repair period at Hunters Point Naval Shipyard, Yorktown put to sea to serve as a platform for the filming of the movie "Jet Carrier." She conducted further, more routine, operations along the west coast until 1 July, at which time she headed back to the Orient. She stopped at Pearl Harbor from 8 to 28 July before continuing on to Manila, where she arrived on 4 August.

Yorktown operated out of the Manila-Subic Bay area, conducting 7th Fleet maneuvers, for the duration of the deployment. She did, however, take periodic breaks from that schedule to make frequent port visits to Yokosuka, and during the Christmas holidays, she made a liberty call at Hong Kong on the Chinese coast. In January of 1955, she was called upon to help cover the evacuation of Nationalist Chinese from the Tachen Islands, located near the communist-controlled mainland. Yorktown entered Yokosuka for the last time on 16 February 1955 but departed again on the 18th to return home. After an overnight stop at Pearl Harbor on 23 and 24 February, she resumed her voyage east and arrived in Alameda on 28 February. On 21 March 1955, she was placed in commission, in reserve, at the Puget Sound Naval Shipyard, where she was to receive extensive modifications—most significantly, an angled flight deck to increase her jet aircraft launching capability. She completed her conversion that fall and, on 14 October, was placed back in full commission.

The aircraft carrier resumed normal operations along the west coast soon after recommissioning. That assignment lasted until mid-March 1956. On the 19th, she stood out of San Francisco Bay on her way to her third tour of duty with the 7th Fleet since her reactivation in 1953. Yorktown stopped at Pearl Harbor from 24 March to 9 April and then continued her voyage west. She arrived in Yokosuka, Japan, on 18 April and departed again on the 29th. The warship operated with the 7th Fleet for the next five months. During that time, she conducted operations in the Sea of Japan, the East China Sea, and the South China Sea. She also visited such places as Sasebo, Manila, Subic Bay, and Buckner Bay at Okinawa. On 7 September, the aircraft carrier stood out of Yokosuka and pointed her bow to the east. After a non-stop voyage, she arrived back at Alameda on 13 September. She resumed west coast operations for about two months. On 13 November, she embarked upon a round-trip to Pearl Harbor, from which she returned to Alameda on 11 December. Yorktown resumed normal operations out of Alameda upon her return and remained so employed until March of 1957. On 9 March, she departed Alameda for yet another tour of duty in the Far East. She made stops at Oahu and Guam along the way and arrived at Yokosuka on 19 April. She put to sea to join TF 77 on 25 April and served with that task force for the next three months. On 13 August, the warship departed Yokosuka for the last time, made a brief pause at Pearl Harbor, and arrived in Alameda on the 25th.

On 1 September, her home port was changed from Alameda to Long Beach, and she was reclassified an antisubmarine warfare (ASW) aircraft carrier with the new designation CVS-10. On the 23d, she departed Alameda and, four days later, entered the Puget Sound Naval Shipyard for overhaul and for modification to an ASW carrier. That yard period lasted until the beginning of February 1958. She departed the naval ammunition depot at Bangor, Wash., on 7 February and entered Long Beach five days later. For the next eight months, Yorktown conducted normal operations along the west coast. On 1 November, she departed San Diego to return to the western Pacific. After a stop at Pearl Harbor from the 8th to the 17th, Yorktown continued her voyage west and arrived in Yokosuka on the 25th. During that deployment, the aircraft carrier qualified for the Armed Forces Expeditionary Medal on three occasions. The first time came on 31 December and 1 January 1959 when she participated

in an American show of strength in response to the communist Chinese shelling of the offshore islands, Quemoy and Matsu, held by Nationalist Chinese forces. During January, she also joined contingency forces off Vietnam during internal disorders caused by communist guerrillas in the southern portion of that country. That month also saw her earn the expeditionary medal for service in the Taiwan Strait. The remainder of the deployment—save for another visit to Vietnamese waters late in March—consisted of a normal round of training evolutions and port visits. She concluded that tour of duty at San Diego on 21 May. The warship resumed normal operations along the west coast, and that duty consumed the remainder of 1959.

In January of 1960, Yorktown headed back to the Far East via Pearl Harbor. During that deployment, she earned additional stars for her Armed Forces Expeditionary Medal for duty in Vietnamese waters at various times in March, April, May, and June. She returned to the west coast late in the summer and, late in September, bee-an a four-month overhaul at the Puget Sound Naval Shipyard.

Yorktown emerged from the shipyard in January 1961 and returned to Long Beach on the 27th. She conducted refresher training and then resumed normal west coast operations until late July. On 29 July, the aircraft carrier stood out of Long Beach, bound once again for the Orient. She made an extended stopover in the Hawaiian Islands in August and, consequently, did not arrive in Yokosuka until 4 September. That tour of duty in the Far East consisted of a normal schedule of antiair and antisubmarine warfare exercises as well as the usual round of port visits. She concluded the deployment at Long Beach on 2 March 1962. Normal west coast operations occupied her time through the summer and into the fall. On 26 October, the warship left Long Beach in her wake and set a course for the Far East. During that deployment, she served as flagship for Carrier Division (CarDiv) 19. She participated in a number of ASW and

AAW exercises, including the SEATO ASW exercise, Operation "Sea Serpent." The deployment lasted until 6 June 1963 at which time the carrier set a course back to Long Beach.

Yorktown arrived back in her home port on 18 June and resumed normal operations for the remainder of the year. Those operations continued throughout most of 1964 as well. However, on 22 October, she pointed her bow westward again and set out for a tour of duty with the 7th Fleet. Another period of operations in the Hawaiian Islands delayed her arrival in Japan until 3 December. The 1964 and 1965 deployment brought Yorktown her first real involvement in the Vietnamese civil war. In February, March, and April, she conducted a series of special operations in the South China Sea in waters near Vietnam—presumably ASW services for the fast carriers conducting air strikes against targets in Vietnam in support of the increased American involvement in the civil war in that country. She concluded her tour of duty in the Far East on 7 May 1965 when she departed Yokosuka to return to the United States. The carrier arrived in Long Beach on 17 May.

For the remainder of her active career, Yorktown's involvement in combat operations in Vietnam proved a dominant feature of her activities. After seven months of normal operations out of Long Beach, she got underway for the western Pacific again on 5 January 1966. She arrived in Yokosuka on 17 February and joined TF 77 on Yankee Station later that month. Over the next five months, the aircraft carrier spent three extended tours of duty on Yankee Station providing ASW and sea-air rescue services for the carriers of TF 77. She also participated in several ASW exercises, including the major

SEATO exercise, Operation "Sea Imp." The warship concluded her last tour of duty on Yankee Station early in July and, after a stop at Yokosuka, headed home on the 15th. She disembarked her air group at San Diego on 27 July and reentered Long Beach that same day. She resumed normal operations—carrier qualifications and ASW exercises—for the remainder of the year and during the first two months of 1967.

On 24 February 1967, Yorktown entered the Long Beach Naval Shipyard for a seven-month overhaul. She completed repairs early in October and, after refresher training, resumed normal west coast operations for most of what remained of 1967. On 28 December, she stood out of Long Beach, bound for her last tour of duty in the western Pacific. After a stop at Pearl Harbor, she arrived in the Far East late in January. Instead of putting in at a Japanese port for turnover, Yorktown headed directly to the Sea of Japan to provide ASW and search and rescue (SAR) support for the contingency force assembled in the wake of the North Korean capture of Pueblo (AGER-2). She remained on that assignment for 30 days. On 1 March, she was released from that duty, and the warship headed for Subic Bay in the Philippines. During the remainder of the deployment, the aircraft carrier did another three tours of duty with TF 77 on Yankee Station. In each instance, she provided ASW and SAR support for the fast carriers launching air strikes on targets in Vietnam. She concluded her last tour of duty in Vietnamese waters on 16 June and set a course for Yokosuka where she stopped from 19 to 21 June before heading back to the United States.

Yorktown arrived back in Long Beach on 5 July and entered the Long Beach Naval Shipyard that same day for almost three months of repairs. She completed repairs on 30 September and resumed normal operations. Late in November and early in December, she served as a platform for the filming of another movie, "Tora! Tora! Tora!," which recreated the Japanese attack on Pearl Harbor. In December, she served as one of the recovery ships for the Apollo 8 space shot. The two unique missions mentioned above were conducted out of Pearl Harbor. She departed Pearl Harbor of 2 January 1969 and, after a two-week stop in Long Beach, continued her voyage to join the Atlantic Fleet. Steaming all the way around South America, the aircraft carrier arrived in her new home port—Norfolk, Va.—on 28 February. She conducted operations along the east coast and in the West Indies until late summer. On 2 September, Yorktown departed Norfolk for a northern European cruise and participation in the major fleet exercise Operation "Peacekeeper." During the exercise, she provided ASW and SAR support for the task force.

The exercise ended on 23 September, and Yorktown began a series of visits to northern European ports. After a visit each to Brest, France, and Rotterdam in the Netherlands, Yorktown put to sea for a series of hunter/killer ASW exercises between 18 October and 11 November. She resumed her itinerary of port visits on 11 November at Kiel, Germany. After that, she stopped at Copenhagen, Denmark, and at Portsmouth, England, before getting underway for home on 1 December. She reentered Norfolk on 11 December and began her holiday leave period.

During the first half of 1970, Yorktown operated out of Norfolk and began preparations for inactivation. On 27 June 1970, Yorktown was decommissioned at Philadelphia, Pa., and was berthed with the Philadelphia Group, Atlantic Reserve Fleet. She remained there almost three years before her name was struck from the Navy list on 1 June 1973. During 1974, the Navy Department approved the donation of Yorktown to the Patriot's Point Development Authority, Charleston, S.C. She was towed from Bayonne, N.J., to Charleston, S.C., in June of 1975. She was formally dedicated as a memorial on the 200th anniversary of the Navy, 13 October 1975. As of April 1980, she was still on display at Patriot's Point, S.C.

Yorktown (CV-10) earned 11 battle stars and the Presidential Unit Citation during World War II and five battle stars for Vietnam service.

CV-11 • USS INTREPID

The fourth Intrepid (CV-11) was launched on 26 April 1943 by Newport News Shipbuilding & Dry Dock Co., Newport News, Va., sponsored by Mrs. John Howard Hoover, and commissioned on 16 August with Captain Thomas L. Sprague in command.

After training in the Caribbean, Intrepid departed Norfolk on 3 December 1943 for San Francisco and then proceeded to Hawaii, arriving at Pearl Harbor on 10 January. She prepared for the invasion of the Marshall Islands and sortied from Pearl Harbor with carriers Cabot and Essex on 16 January. They raided islands at the northeastern corner of Kwajalein Atoll on 29 January 1944, continuing the attack until 2 February. Intrepid's planes destroyed 83 Japanese planes on Roi and Namur before the first landings on adjacent islets on 31 January. After securing Ennuebing Island, marines set up artillery to support their assault on Roi.

Intrepid then headed for Truk, where three fast carrier groups arrived undetected at daybreak on 17 February, sinking two destroyers and 200,000 tons of merchant shipping in two days of nearly continuous attacks. The night of 17 February, an aerial torpedo struck Intrepid's starboard quarter, jamming her rudder. Captain Sprague kept her on course using improvised methods until strong winds forced her back. A makeshift sail fashioned from hatch covers and scrap canvas allowed her to maintain course to Pearl Harbor, where she arrived on 24 February 1944 for temporary repairs. After further repairs on the West Coast, she returned to operations out of Pearl Harbor and then to the Marshalls.

Intrepid's planes struck Japanese positions in the Palaus on 6 and 7 September and then moved to airfields in the southern Philippines. After supporting marines in the Palaus, she struck throughout the Philippines, Okinawa, and Formosa, neutralizing Japanese air threats to Leyte.

During the Battle for Leyte Gulf, Intrepid's planes helped sink the battleship Musashi and damaged other Japanese ships. On 30 October, a kamikaze crashed into Intrepid's port gun tub, killing 10 men. She resumed operations after quick repairs. On 25 November, two kamikazes hit Intrepid, killing 11. She sailed to San Francisco for repairs, returning to action in mid-February 1945.

Intrepid supported strikes against airfields on Kyushu, Japan, in March 1945. On 16 April, a kamikaze crashed into her flight deck, killing eight. She returned to San Francisco for repairs, arriving on 19 May. Intrepid resumed operations in June, supporting the occupation of Japan after receiving the cease-offensive order on 15 August.

After the war, Intrepid decommissioned in March 1947 and joined the Pacific Reserve Fleet. She recommissioned in 1952, underwent modernization, and served in the Atlantic and Mediterranean, earning the Battle Efficiency "E" for antisubmarine warfare.

Intrepid participated in several significant missions, including the recovery of astronauts from Project Mercury and Gemini missions. She underwent a major overhaul in 1965 and returned to action in Vietnam in 1966, earning commendations for her performance.

Intrepid was struck from the Naval Register in 1980 and now serves as the Intrepid Sea, Air & Space Museum in New York City.

CV-12 • USS HORNET

The eighth Hornet (CV-12) was launched on 30 August 1943 by the Newport News Shipbuilding & Dry Dock Co., Newport News, Va.; sponsored by Mrs. Frank M. Knox, wife of the Secretary of the Navy, and commissioned on 29 November 1943, with Captain Miles M. Browning in command.

Hornet conducted shakedown training off Bermuda before departing Norfolk on 14 February 1944 to join the Fast Carrier Task Force on 20 March at Majuro Atoll in the Marshalls. After lending air support to protect the invasion beaches in New Guinea, she conducted massive aerial raids against Japanese bases in the Caroline Islands and prepared to support the amphibious assault for the occupation of the Marianas Islands.

On 11 June 1944, Hornet launched raids on Tinian and Saipan. The following day, she conducted heavy bombing attacks on Guam and Rota. During 15 to 16 June, she blasted enemy airfields at Iwo and Chichi Jima to prevent air attacks on troops invading Saipan in the Marianas. On the afternoon of 18 June 1944, Hornet formed with the Fast Carrier Task Force to intercept the Japanese First Mobile Fleet, headed through the Philippine Sea for Saipan. The Battle of the Philippine Sea opened on 19 June 1944 when Hornet launched strikes to destroy as many land-based Japanese planes as possible before the carrier-based Japanese aircraft came in.

The enemy approached the American carriers in four massive waves. But fighter aircraft from Hornet and other carriers did a magnificent job and broke up all the attacks before the Japanese aerial raiders reached the task force. Nearly every Japanese aircraft was shot down in the great air battles of 19 June 1944, which became commonly known as "The Marianas Turkey Shoot." As the Japanese Mobile Fleet fled in defeat on 20 June, the carriers launched long-range air strikes that sank the Japanese carrier Hiji and so damaged two tankers that they were abandoned and scuttled. Admiral Ozawa's own flag log for 20 June 1944 showed his surviving carrier air power as only 35 operational aircraft out of the 430 planes with which he had commenced the Battle of the Philippine Sea.

Hornet, basing from Eniwetok in the Marshalls, raided enemy installations ranging from Guam to the Bonins, then turned her attention to the Palaus, throughout the Philippine Sea, and to enemy bases on Okinawa and Formosa. Her aircraft gave direct support to the troops invading Leyte on 20 October 1944. During the Battle for Leyte Gulf, she launched raids for damaging hits to the Japanese center force in the Battle off Samar and hastened the retreat of the enemy fleet through the Sibuyan Sea towards Borneo.

In the following months, Hornet attacked enemy shipping and airfields throughout the Philippines. This included participation in a raid that destroyed an entire Japanese convoy in Ormac Bay. On 30 December 1944, she departed Ulithi in the Carolines for raids against Formosa, Indo-China, and the Pescadores Islands. En route back to Ulithi, Hornet planes made photo reconnaissance of Okinawa on 22 January 1945 to aid the planned invasion of that "last stepping-stone to Japan."

Hornet again departed Ulithi on 10 February for full-scale aerial assaults on Tokyo, then supported the amphibious landing assault on Iwo Jima from 19 to 20 February 1945.

Repeated raids were made against the Tokyo plains industrial complex, and Okinawa was hard hit. On 1 April 1946, Hornet planes gave direct support to the amphibious assault landings on Okinawa. On 6 April, her aircraft joined in attacks that sank the mighty Japanese battleship Yamato and her entire task force as it closed Okinawa. The following two months found Hornet alternating between close support to ground troops on Okinawa and hard-hitting raids to destroy the industrial capacity of Japan. She was caught in a howling typhoon on 4 to 5 June 1945, which collapsed some 25 feet of her forward flight deck.

Hornet was routed back to the Philippines and from there to San Francisco, arriving on 7 July 1945. Her overhaul was complete by 13 September 1945 when she departed as part of the "Magic Carpet" operation that saw her return home troops

from the Marianas and Hawaiian Islands. She returned to San Francisco on 9 February 1946. She decommissioned there on 15 January 1947 and joined the Pacific Reserve Fleet.

Hornet recommissioned on 20 March 1951, then sailed from San Francisco for the New York Naval Shipyard, where she decommissioned on 12 May 1951 for conversion to an attack aircraft carrier (CVA-12). She recommissioned on 11 September 1953 and trained in the Caribbean Sea before departing from Norfolk on 11 May 1954 on a month-long global cruise.

After operations in the Mediterranean Sea and the Indian Ocean, Hornet joined the mobile 7th Fleet in the South China Sea, where on 25 July, search planes from her task group shot down two attacking Chinese Communist fighter planes. She returned to San Francisco on 12 March 1954, trained out of San Diego, then sailed on 4 May 1955 to join the 7th Fleet in the Far East.

Hornet helped cover the evacuation of Vietnamese from the Communist-controlled north to freedom in South Vietnam. Then, in readiness training with the 7th Fleet, she ranged from Japan to Formosa, Okinawa, and the Philippines. She returned to San Diego on 10 March 1955 and entered the Puget Sound Naval Shipyard the following month for conversion, including a hurricane bow and the installation of an angled flight deck that permits the simultaneous launching and recovery of aircraft.

Following her modernization overhaul, Hornet operated along the California Coast. She departed San Diego on 21 January 1957 to bolster the strength of the 7th Fleet until her return from the troubled Far East on 25 July. Following a similar cruise, from 6 January to 2 July 1958, she was converted to an Antisubmarine Warfare Support Carrier (CVS-12) in the Puget Sound Naval Shipyard. On 3 April 1959, she sailed from Long Beach to join the 7th Fleet in antisubmarine warfare tactics ranging from Japan to Okinawa and the Philippines. She returned home in October, for training along the western seaboard.

In the following years, Hornet was regularly deployed to the 7th Fleet for operations ranging from the coast of South Vietnam to the shores of Japan, the Philippines, and Okinawa. On 25 August 1966, she was on recovery station for the unmanned Apollo moonship that rocketed three-quarters of the way around the globe in 93 minutes before splashdown near Wake Island. Scorched from the heat of its re-entry into the Earth's atmosphere, the Apollo space capsule, designed to carry American astronauts to the moon, was brought aboard Hornet after its test.

Hornet returned to Long Beach on 8 September but headed back to the Far East on 27 March 1967. She reached Japan exactly a month later and departed Sasebo on 19 May for the war zone. She operated in Vietnamese waters throughout the remainder of spring and much of the summer of 1967, aiding in the struggle to keep freedom alive in Southeast Asia.

Hornet was the recovery carrier for the Apollo 11 moon mission, during which astronauts Neil Armstrong and Edwin Aldrin Jr. landed and walked on the moon in July 1969. Fellow astronaut Michael Collins remained in orbit around the moon. On 24 November, the Apollo 12 astronauts—all Naval Aviators—Richard F. Gordon, Charles Conrad Jr., and Alan L. Bean were recovered by Helicopter Antisubmarine Squadron Four (HS 4) and returned to Hornet.

Hornet was decommissioned on 26 June 1970. Following nearly two decades in mothballs, she was stricken from the Naval Vessel Register on 25 July 1989 and sold for breaking up in April 1993. However, the old carrier was saved from the scrap heap by the efforts of historically-minded citizens and was donated to The Aircraft Carrier Hornet Foundation for use as a museum at Alameda, Calif., on 26 May 1998.

Hornet received the Presidential Unit Citation and seven battle stars for service in World War II.

CV-13 • USS FRANKLIN

(Displacement. 27,100; length. 872'; beam. 93'; width of flight deck. 147'6"; draft. 28'7"; speed. 33 k.; complement. 3,448; a. 12 5"; cl. Essex)

The fifth Franklin (CV-13) was laid down on 7 December 1942 at Newport News, Virginia, by Newport News Shipbuilding and Dry Dock Company; launched on 14 October 1943; sponsored by Lieutenant Commander Mildred A. McAfee, USNR, Director of the WAVES; and commissioned at the Norfolk Navy Yard, Portsmouth, Virginia, on 31 January 1944, with Captain James M. Shoemaker in command.

Franklin cruised to Trinidad, British West Indies, for shakedown and soon thereafter departed in Task Group (TG) 27.7 for San Diego, California, to engage in intensive training. In June, she sailed via Pearl Harbor, Territory of Hawaii, for Eniwetok, in the Marshall Islands, where she joined TG 58.2, part of the fast carrier striking force.

On the last day of June 1944, she sortied for carrier strikes on the Bonins in support of the subsequent assault on the Marianas. Her planes destroyed aircraft on the ground and in the air, as well as pounded gun installations, airfields, and enemy shipping. On 4 July, she hurled strikes against Iwo Jima, Chichi Jima, and Ha Ha Jima, with her planes not only battering shore installations but sinking a large cargo vessel in the harbor and setting three smaller ships afire.

On 6 July, she began strikes on Guam and Rota to soften them up for the invasion forces, and continued them until the 21st when she lent direct support to enable the safe landing of the first assault waves. Two days of replenishment at Saipan permitted her to steam in Task Force 58 for photographic reconnaissance and air strikes against the islands of the Palau group. Her planes completed their mission on the 25th and 26th, exacting a heavy toll in enemy planes, ground installations, and shipping. She departed on 28 July en route to Saipan and the following day shifted to TG 58.1.

Although high seas prevented taking on needed bombs and rockets, Franklin proceeded to carry out another raid against the Bonins. The 4th of August bode well, for her fighters launched against Chichi Jima and her dive bombers and torpedo planes against a convoy north of Ototo Jima; they rained destruction against the radio stations, seaplane base, airstrips, and ships.

A period of upkeep and recreation from 9 to 28 August ensued at Eniwetok before she departed in the company of carriers Enterprise (CV-6), Belleau Wood (CVL-24), and San Jacinto (CVL-30) for neutralization and diversionary attacks against the Bonins. From 31 August to 2 September, spirited and productive strikes from Franklin inflicted much ground damage, sank two cargo ships, bagged numerous enemy planes in flight, and accomplished photographic surveys.

On 4 September, she replenished at Saipan and steamed in TG 38.4 for an attack against Yap (3-6 September), which included direct air coverage of the Peleliu invasion on the 15th. The group resupplied at Manus Island, in the Admiralty Islands, from 21-25 September.

Franklin, as flagship of TG 38.4, returned to the Palaus where she launched daily patrols and night fighters. On 9 October, she rendezvoused with carrier groups cooperating in air strikes in support of the coming occupation of Leyte. At twilight on the 13th, the Task Group came under attack by four bombers; torpedoes narrowly missed Franklin twice. An enemy plane attempted to crash Franklin, but only succeeded in glancing off the flight deck abaft the island structure; the unsuccessful suicider slid across the deck and into the water on the carrier's starboard beam, the pilot failing in his attempt to destroy his larger adversary.

Early on the 14th, the fast carriers sent a fighter sweep against Aparri, Luzon, following which Franklin steamed to the east of Luzon to neutralize installations to the east prior to invasion landings on Leyte. On the 15th, she came under attack by three enemy planes, one of which scored a bomb hit on the after outboard corner of the deck edge elevator, killing three and wounding 22. The tenacious carrier continued her daily operations, however, striking hard at Manila Bay on 19 October when her planes sank a number of ships, damaged many, destroyed a floating drydock, and bagged 11 planes.

During the initial landings on Leyte (20 October), her aircraft hit surrounding airstrips and launched search patrols in anticipation of the approach of a reported enemy attack force. On the morning of 24 October, in the Battle of the Sibuyan Sea, her planes formed part of the waves that attacked the Japanese "First Raiding Force" (Vice Admiral Takeo Kurita), in so doing helping to sink Japanese superbattleship Musashi south of Luzon, and damage battleships Fuso and Yamashiro, and sink destroyer Wakaba. As further enemy threats materialized in another quarter, Franklin, with TGs 38.4, 38.3, and 38.2, sped to intercept the advancing Japanese carrier force and attack at dawn. Franklin's strike groups combined with those from the other carriers on 25 October in the Battle off Cape Engano to damage the carrier Chiyoda (she would be sunk by American cruiser gunfire subsequently) and sink the small carrier Zuiho.

Retiring in her task group to refuel, she returned to the Leyte action on 27 October, her planes concentrating on a heavy cruiser and two destroyers south of Mindoro. She was underway about 1,000 miles off Samar on 30 October when enemy bombers appeared, bent on a suicide mission. Three doggedly pursued Franklin, the first plummeting off her starboard side; the second hitting the flight deck and crashing through to the gallery deck, showering destruction, killing 56 and wounding 60; the third discharging another near miss at Franklin before diving into the flight deck of the small carrier Belleau Wood. Both carriers retired to Ulithi for temporary repairs and Franklin proceeded to Puget Sound Navy Yard, Bremerton, Washington, arriving 28 November 1944 for repairs to her battle damage.

She departed Bremerton on 2 February 1945 and after training exercises and pilot qualification, joined TG 58.2 for strikes on the Japanese homeland in support of the Okinawa landings. On 15 March, she rendezvoused with Task Force 58 units and three days later launched sweeps and strikes against Kagoshima and Izumi on southern Kyushu.

Before dawn on 19 March 1945, Franklin, with Captain Leslie E. Gehres commanding, launched a fighter sweep against Honshu and later a strike against shipping in Kobe Harbor. Suddenly, a single enemy plane pierced the cloud cover and made a low-level run on the gallant ship to drop two semi-armor piercing bombs. One struck the flight deck centerline, penetrating to the hangar deck, effecting destruction and igniting fires through the second and third decks, and knocking out the combat information center and air plot. The second hit aft, tearing through two decks and fanning fires that triggered ammunition, bombs, and rockets. Franklin, within 50 miles of the Japanese mainland, lay dead in the water, took a 13° starboard list, lost all radio communications, and broiled under the heat from enveloping fires.

Many of the crew had been blown overboard, driven off by fire, or had been killed or wounded, but the 106 officers and 604 enlisted who voluntarily remained on board saved their ship through sheer valor and tenacity. The casualties totaled 724 killed and 265 wounded and would have far exceeded this number except for the heroic work of many survivors. Among these were Lieutenant Commander Joseph T. O'Callahan, ChC (SJ) USNR, the ship's Roman Catholic chaplain, who emerged "as a soul-stirring sight. He seemed to be everywhere," an eyewitness recounted later, "giving Extreme Unction to the dead and dying, urging the men on and himself handling hoses, jettisoning ammunition and doing everything he could to help save our ship. He was so conspicuous not only because of the cross daubed with paint across his helmet but because of his seemingly detached air as he went from place to place with head slightly bowed as if in meditation or prayer." Lieutenant (junior grade) Donald A. Gary also emerged a hero, calming anxious men seemingly trapped in a smoke-filled compartment. After finding an exit after repeated attempts, he led some 300 of his shipmates to safety. He later organized and led firefighting parties to battle the blazing inferno on the hangar deck, and entered number three fireroom to raise steam in one boiler, braving extreme hazards in so doing. Both men subsequently received Medals of Honor for their bravery; ships were also named for them. Light cruiser Santa Fe (CL-60) similarly rendered vital assistance in rescuing crewmen from the sea and closing Franklin to take off the numerous wounded.

Franklin was taken in tow by heavy cruiser Pittsburgh (CA-72), but she managed to work up to 14 knots and ultimately reach Pearl Harbor, where a cleanup job permitted her to proceed under her own power to the United States, ultimately reaching Brooklyn, New York, on 28 April. Following the end of the war, Franklin was opened to the public for Navy

Day celebrations in October 1945, and on 17 February 1947 was placed out of commission at Bayonne, New Jersey.

While Franklin lay "mothballed" at Bayonne, never returning to active service, she was redesignated to an attack aircraft carrier (CVA-13) on 1 October 1952, to an antisubmarine warfare support carrier (CVS-13) on 8 August 1953 and, ultimately, to an aircraft transport (AVT-8) on 15 May 1959. She was struck from the Naval Vessel Register on 1 October 1964.

Although the Navy initially sold the ship to Peck Iron and Metal Company, Portsmouth, Virginia, it re-possessed her due to an urgent Bureau of Ships requirement for the use of her four turbo-generators. Ultimately, however, she was sold, for scrapping, to Portsmouth Salvage Company, Chesapeake, Virginia, on 27 July 1966. She departed naval custody under tow (Red Star Towing Company) on the evening of 1 August 1966.

Franklin received four battle stars for her World War II service.

CV-14 • USS TICONDEROGA

(Displacement: 27,100; length: 888'; beam: 93'0"; width of flight deck: 147'6"; draft: 28'7"; speed: 33 k.; complement: 3,448; armor: 12 5"; 72 40mm.; ac. 80+; class: Essex)

The fourth Ticonderoga (CV-14) was laid down as Hancock on 1 February 1943 at Newport News, Va., by the Newport News Shipbuilding & Dry Dock Co.; renamed Ticonderoga on 1 May 1943, launched on 7 February 1944, sponsored by Miss Stephanie Sarah Pell, and commissioned at the Norfolk Navy Yard on 8 May 1944, Capt. Dixie Kiefer in command.

Ticonderoga remained at Norfolk for almost two months outfitting and embarking Air Group 89. On 26 June, the carrier shaped a course for the British West Indies. She conducted air operations and drills en route and reached Port of Spain, Trinidad, on the 30th. For the next 15 days, Ticonderoga trained intensively to weld her air group and crew into an efficient wartime team. She departed the West Indies on 16 July and headed back to Norfolk where she arrived on the 22nd. Eight days later, the carrier headed for Panama. She transited the canal on 4 September and steamed up the coast to San Diego the following day. On the 13th, the carrier moored at San Diego where she loaded provisions, fuel, aviation gas, and an additional 77 planes, as well as the Marine Corps aviation and defense units that went with them. On the 19th, she sailed for Hawaii where she arrived five days later.

Ticonderoga remained at Pearl Harbor for almost a month. She and Carina (AK-74) conducted experiments in the underway transfer of aviation bombs from cargo ship to aircraft carrier. Following those tests, she conducted air operations—day and night landing and antiaircraft defense drills—until 18 October when she exited Pearl Harbor and headed for the western Pacific. After a brief stop at Eniwetok, Ticonderoga arrived at Ulithi Atoll in the Western Carolines on the 29th. There she embarked Rear Admiral A. W. Radford, Commander, Carrier Division 6, and joined Task Force (TF) 38 as a unit of Rear Admiral Frederick C. Sherman's Task Group (TG) 38.3.

The carrier sortied from Ulithi with TF 38 on 2 November. She joined the other carriers as they resumed their extended air cover for the ground forces capturing Leyte. She launched her first air strike on the morning of the 5th. The planes of her air

group spent the next two days pummeling enemy shipping near Luzon and air installations on that island. Her planes bombed and strafed the airfields at Zablan, Mandaluyong, and Pasig. They also joined those of other carriers in sending the heavy cruiser Nachi to a watery resting place. In addition, Ticonderoga pilots claimed six Japanese aircraft shot down and one destroyed on the ground, as well as 23 others damaged.

Around 1600 on the 5th, the enemy retaliated by sending up a flock of planes piloted by members of the suicide corps dubbed kamikaze, or "Divine Wind," in honor of the typhoon that had destroyed a Chinese invasion fleet four centuries previously. Two of the suicide planes succeeded in slipping through the American combat air patrol and antiaircraft fire to crash into Lexington (CV-16). Ticonderoga emerged from that airborne banzai charge unscathed and claimed a tally of two splashes. On 6 November, the warship launched two fighter sweeps and two bombing strikes against the Luzon airfields and enemy shipping in the vicinity. Her airmen returned later that day claiming the destruction of 35 Japanese aircraft and attacks on six enemy ships in Manila Bay. After recovering her planes, the carrier retired to the east for a fueling rendezvous.

She refueled and received replacement planes on the 7th and then headed back to continue pounding enemy forces in the Philippines. Early on the morning of 11 November, her planes combined with others of TF 38 to attack a Japanese reinforcement convoy, just as it was preparing to enter Ormoc Bay from the Camotes Sea. Together, the planes accounted for all the enemy transports and four of the seven escorting destroyers. On the 12th and 13th, Ticonderoga and her sisters launched strikes at Luzon airfields and docks and shipping around Manila. This raid tallied an impressive score: light cruiser Kiso, four destroyers, and seven merchant ships. At the conclusion of the raid, TF 38 retired eastward for a refueling breather. Ticonderoga and the rest of TG 38.3, however, continued east to Ulithi where they arrived on the 17th to replenish, refuel, and rearm.

On 22 November, the aircraft carrier departed Ulithi once more and steamed back toward the Philippines. Three days later, she launched air strikes on central Luzon and adjacent waters. Her pilots finished off the heavy cruiser Kumano, damaged in the Battle off Samar. Later, they attacked an enemy convoy about 15 miles southwest of Kumano's not-so-safe haven in Dasol Bay. Of this convoy, cruiser Yasoshima, a merchantman, and three landing ships went to the bottom. Ticonderoga's air group rounded out their day of destruction with an aerial rampage which cost the Japanese 15 planes shot down and 11 destroyed on the ground.

While her air group busily pounded the Japanese, Ticonderoga's ship's company also made their presence felt. Just after noon, a torpedo launched by an enemy plane broached in Langley's (CVL-27) wake to announce the approach of an air raid. Ticonderoga's gunners raced to their battle stations as the raiders made both conventional and suicide attacks on the task group. Her sister ship Essex (CV-9) erupted in flames when one of the kamikazes crashed into her. When a second suicide plane tried to finish off the stricken carrier, Ticonderoga's gunners joined those firing from other ships in cutting his approach abruptly short. That afternoon, while damage control parties dressed Essex's wounds, Ticonderoga extended her hospitality to that damaged carrier's homeless airmen as well as to Intrepid (CV-11) pilots in similar straits. The following day, TF 38 retired to the east.

TF 38 stood out of Ulithi again on 11 December and headed for the Philippines. Ticonderoga arrived at the launch point early in the afternoon of the 13th and sent her planes aloft to blanket Japanese airbases on Luzon while Army planes took care of those in the central Philippines. For three days, Ticonderoga airmen and their comrades wreaked havoc with a storm of destruction on enemy airfields. She withdrew on the 16th with the rest of TF 38 in search of a fueling rendezvous. While attempting to find calmer waters in which to refuel, TF 38 steamed directly through a violent, but unheralded, typhoon. Though the storm cost Admiral Halsey's force three destroyers and over 800 lives, Ticonderoga and the other carriers managed to ride it out with a minimum of damage. Having survived the tempest's fury, Ticonderoga returned to Ulithi on Christmas Eve.

Repairs occasioned by the typhoon kept TF 38 in the anchorage almost until the end of the month. The carriers did not return to sea until 30 December 1944 when they steamed north to hit Formosa and Luzon in preparation for the landings on the latter island at Lingayen Gulf. Severe weather limited the Formosa strikes on 3 and 4 January 1945 and, in all likelihood, obviated the need for them. The warships fueled at sea on the 5th. Despite rough weather on the 6th, the strikes on Luzon airfields were carried out. That day, Ticonderoga's airmen and their colleagues of the other air groups increased their score by another 32 enemy planes. The 7th brought more strikes on Luzon installations. After a fueling rendezvous on the 8th, Ticonderoga sped north at night to get into position to blanket Japanese airfields in the Ryukyus during the Lingayen assault the following morning. However, foul weather, the bugaboo of TF 38 during the winter of 1944 and 1945, forced TG 38.3 to abandon the strikes on the Ryukyu airfields and join TG 38.2 in pounding Formosa.

During the night of 9 and 10 January, TF 38 steamed boldly through the Luzon Strait and then headed generally southwest, diagonally across the South China Sea. Ticonderoga provided combat air patrol coverage on the 11th and helped to bring down four enemy planes which attempted to snoop the formation. Otherwise, the carriers and their consorts proceeded unmolested to a point some 150 to 200 miles off the coast of Indochina. There, on the 12th, they launched their approximately 850 planes and made a series of anti-shipping sweeps during which they sank a whopping 44 ships, totaling over 130,000 tons. After recovering planes in the late afternoon, the carriers moved off to the northeast. Heavy weather hindered fueling operations on the 13th and 14th, and air searches failed to turn up any tempting targets. On the 15th, fighters swept Japanese airfields on the Chinese Coast while the flattops headed for a position from which to strike Hong Kong. The following morning, they launched anti-shipping bombing raids and fighter sweeps of air installations. Weather prevented air operations on the 17th and again made fueling difficult. It worsened the next day and stopped replenishment operations altogether, so that they were not finally concluded until the 19th. The force then shaped a course generally northward to re-transit Luzon Strait via Balintang Channel.

The three task groups of TF 38 completed their transit during the night of 20 and 21 January. The next morning, their planes hit airfields on Formosa, in the Pescadores, and at Sakishima Gunto. The good flying weather brought mixed blessings. While it allowed American flight operations to continue through the day, it also brought new gusts of the "Divine Wind." Just after noon, a single-engined Japanese plane scored a hit on Langley with a glide-bombing attack. Seconds later, a kamikaze swooped out of the clouds and plunged toward Ticonderoga. He crashed through her flight deck abreast of the No. 2 5-inch mount, and his bomb exploded just above her hangar deck. Several planes stowed nearby erupted into flames. Death and destruction abounded, but the ship's company fought valiantly to save the threatened carrier. Capt. Kiefer conned his ship smartly. First, he changed course to keep the wind from fanning the blaze. Then, he ordered magazines and other compartments flooded to prevent further explosions and to correct a 10-degree starboard list. Finally, he instructed the damage control party to continue flooding compartments on Ticonderoga's port side. That operation induced a 10-degree port list which neatly dumped the fire overboard! Firefighters and plane handlers completed the job by dousing the flames and jettisoning burning aircraft.

Wounded denizens of the deep often attract predators. Ticonderoga was no exception. The other kamikazes pounced on her like a school of sharks in a feeding frenzy. Her antiaircraft gunners struck back with desperate but methodical ferocity and quickly swatted three of her tormentors into the sea. A fourth plane slipped through her barrage and smashed into the carrier's starboard side near the island. His bomb set more planes on fire, riddled her flight deck, and injured or killed another 100 sailors—including Capt. Kiefer. Yet, Ticonderoga's crew refused to submit. Spared further attacks, they brought her fires completely under control not long after 1400; and Ticonderoga retired painfully.

The stricken carrier arrived at Ulithi on 24 January but remained there only long enough to move her wounded to the hospital ship Samaritan (AH-10), to transfer her air group to Hancock (CV-19), and to embark passengers bound for home. Ticonderoga cleared the lagoon on 28 January and headed for the

United States. The warship stopped briefly at Pearl Harbor en route to the Puget Sound Navy Yard where she arrived on 15 February.

Her repairs were completed on 20 April, and she cleared Puget Sound the following day for the Alameda Naval Air Station. After embarking passengers and aircraft bound for Hawaii, the carrier headed for Pearl Harbor where she arrived on 1 May. The next day, Air Group 87 came on board and, for the next week, trained in preparation for the carrier's return to combat. Ticonderoga stood out of Pearl Harbor and shaped a course for the western Pacific. En route to Ulithi, she launched her planes for what amounted to training strikes on Japanese-held Taroa in the Marshalls. On 22 May, the warship arrived in Ulithi and rejoined the Fast Carrier Task Force as an element of Rear Admiral Radford's TG 58.4.

Two days after her arrival, Ticonderoga sortied from Ulithi with TF 68 and headed north to spend the last weeks of the war in Japanese home waters. Three days out, Admiral Halsey relieved Admiral Spruance, the 5th Fleet reverted back to 3d Fleet, and TF 68 became TF 38 again for the duration. On 2 and 3 June, Ticonderoga fighters struck at airfields on Kyushu in an effort to neutralize the remnants of Japanese air power—particularly the Kamikaze Corps—and to relieve the pressure on American forces at Okinawa. During the following two days, Ticonderoga rode out her second typhoon in less than six months and emerged relatively unscathed. She provided combat air patrol cover for the 6 June refueling rendezvous, and four of her fighters intercepted and destroyed three Okinawa-bound kamikazes. That evening, she steamed off at high speed with TG 38.4 to conduct a fighter sweep of airfields on southern Kyushu on the 8th. Ticonderoga's planes then joined in the aerial bombardment of Minami Daito Shima and Kita Daito Shima before the carrier headed for Leyte where she arrived on the 13th.

During the two-week rest and replenishment period she enjoyed at Leyte, Ticonderoga changed task organizations from

TG 38.4 to Rear Admiral Gerald F. Bogan's TG 38.3. On 1 July, she departed Leyte with TF 38 and headed north to resume raids on Japan. Two days later, a damaged reduction gear forced her into Apra Harbor, Guam, for repairs. She remained there until the 19th when she steamed off to rejoin TF 38 and resume her role in the war against Japan. On the 24th, her planes joined those of other fast carriers in striking ships in the Inland Sea and airfields at Nagoya, Osaka, and Miko. During those raids, TF 38 planes found the sad remnants of the once-mighty Japanese Fleet and bagged battleships Ise, Hyuga, and Haruna as well as an escort carrier, Kaiyo, and two heavy cruisers. On 28 July, her aircraft directed their efforts toward the Kure Naval Base, where they pounded an aircraft carrier, three cruisers, a destroyer, and a submarine. She shifted her attention to the industrial area of central Honshu on the 30th, then to northern Honshu and Hokkaido on 9 and 10 August. The latter attacks thoroughly destroyed the marshaling area for a planned airborne suicide raid on the B-29 bases in the Marianas. On the 13th and 14th, her planes returned to the Tokyo area and helped to subject the Japanese capital to another severe drubbing.

The two atomic bombs dropped on Hiroshima and Nagasaki on the 6th and 9th, respectively, convinced the Japanese of the futility of continued resistance. On the morning of 16 August, Ticonderoga launched another strike against Tokyo. During or just after that attack, word reached TF 38 to the effect that Japan had capitulated.

The shock of peace, though not so abrupt as that of war almost four years previously, took some getting used to. Ticonderoga and her sister ships remained on a full war footing. She continued patrols over Japanese territory and sent reconnaissance flights in search of camps containing Allied prisoners of war so that air-dropped supplies could be rushed to them. On 6 September—four days after the formal surrender ceremony on board Missouri (BB-63)—Ticonderoga entered Tokyo Bay.

Her arrival at Tokyo ended one phase of her career and began another. She embarked homeward-bound passengers and put to sea again on the 20th. After a stop in Pearl Harbor, the carrier reached Alameda, Calif., on 5 October. She disembarked her passengers and unloaded cargo before heading out on the 9th to pick up another group of veterans. Ticonderoga delivered over a thousand soldiers and sailors to Tacoma, Wash., and remained there through the 28th for the Navy Day celebration. On 29 October, the carrier departed Tacoma and headed back to Alameda. En route, all of the planes of Air

Group 87 were transferred ashore so that the carrier could be altered to accommodate additional passengers in the "Magic-Carpet" voyages to follow. Following the completion of those modifications at the Pearl Harbor Naval Shipyard in November, the warship headed for the Philippines and arrived at Samar on 20 November. She returned to Alameda on 6 December and debarked almost 4,000 returning servicemen. The carrier made one more "Magic-Carpet" run in December 1946 and January 1946 before entering the Puget Sound Naval Shipyard to prepare for inactivation. Almost a year later on 9 January 1947, Ticonderoga was placed out of commission and berthed with the Bremerton Group of the Pacific Reserve Fleet.

On 31 January 1962, Ticonderoga came out of reserve and went into reduced commission for the transit from Bremerton to New York. She departed Puget Sound on 27 February and reached New York on 1 April. Three days later, she was decommissioned at the New York Naval Shipyard to begin an extensive conversion. During the ensuing 29 months, the carrier received the numerous modifications—steam catapults to launch jets, a new nylon barricade, a new deck-edge elevator, and the latest electronic and fire control equipment—necessary for her to become an integral unit of the fleet. On 11 September 1954, Ticonderoga was recommissioned at New York, Capt. William A. Schoech in command.

In January 1955, the carrier shifted to her new home port—Norfolk, Va.—where she arrived on the 6th. Over the next month, she conducted carrier qualifications with Air Group 6 in the Virginia Capes operating area. On 3 February, she stood out of Hampton Roads for shakedown near Cuba, after which she returned via Norfolk to New York for additional alterations. During the late summer, the warship resumed carrier qualifications in the Virginia Capes area. After a visit to Philadelphia early in September, she participated in tests of three new planes—the A4D-1 "Skyhawk," the F4D-1 "Skyray," and the F3H-2N "Demon." Ticonderoga then returned to normal operations along the east coast until 4 November when she departed Mayport, Fla., and headed for Europe. She relieved Intrepid at Gibraltar 10 days later and cruised the length of the Mediterranean during the following eight months. On 2 August 1956, Ticonderoga returned to Norfolk and entered the shipyard to receive an angled flight deck and an enclosed hurricane bow.

Those modifications were completed by early 1957 and, in April, she got underway for her new home port—Alameda, Calif. She reached her destination on 30 May, underwent repairs, and finished out the summer with operations off the California coast.

On 16 September, she stood out of San Francisco Bay and shaped a course for the Far East. En route, she stopped at Pearl Harbor before continuing west to Yokosuka, Japan, where she arrived on 15 October. For six months, Ticonderoga cruised Oriental waters from Japan in the north to the Philippines in the south. Upon arriving at Alameda on 25 April 1958, she completed her first deployment to the western Pacific since recommissioning.

Between 1958 and 1963, Ticonderoga made four more peacetime deployments to the western Pacific. During each, she conducted training operations with other units of the 7th Fleet and made goodwill and liberty port calls throughout the Far East. Early in 1964, she began preparations for her sixth cruise to the western Pacific and, following exercises off the west coast and in the Hawaiian Islands, the carrier cleared Pearl Harbor on 4 May for what began as another peaceful tour of duty in the Far East. The first three months of that deployment brought normal operations—training and port calls. However, on 2 August, while operating in international waters in the Gulf of Tonkin, Maddox (DD-731) reported being attacked by units of the North Vietnamese Navy. Within minutes of her receipt of the message, Ticonderoga dispatched four rocket-armed F8E "Crusaders" to the destroyer's assistance. Upon arrival, the "Crusaders" launched Zuni rockets and strafed the North Vietnamese craft with their 20-millimeter cannons. The Ticonderoga airmen teamed up with Maddox gunners to thwart the North Vietnamese attack, leaving one boat dead in the water and damaging the other two.

Two days later, late in the evening of the 4th, Ticonderoga received urgent requests from Turner Joy (DD-951)—by then on patrol with Maddox—for air support in resisting what the destroyer alleged to be another torpedo boat foray. The carrier again launched planes to aid the American surface ships, and

Turner Joy directed them. The Navy surface and air team believed it had sunk two boats and damaged another pair. President Johnson responded with a reprisal to what he felt at the time to be two unprovoked attacks on American seapower and ordered retaliatory air strikes on selected North Vietnamese motor torpedo boat bases. On 5 August, Ticonderoga and Constellation (CV-46) launched 60 sorties against four bases and their supporting oil storage facilities. Those attacks reportedly resulted in the destruction of 25 PT-type boats, severe damage to the bases, and almost complete razing of the oil storage depot. For her quick reaction and successful combat actions on those three occasions, Ticonderoga received the Navy Unit Commendation.

After a return visit to Japan in September, the aircraft carrier resumed normal operations in the South China Sea until winding up the deployment late in the year. She returned to the Naval Air Station, North Island, Calif., on 15 December 1964. Following post-deployment and holiday stand-down, Ticonderoga moved to the Hunter's Point Naval Shipyard on 27 January 1965 to begin a five-month overhaul. She completed repairs in June and spent the summer operating along the coast of southern California. On 28 September, the aircraft carrier put to sea for another deployment to the Orient. She spent some time in the Hawaiian Islands for an operational readiness exercise then continued on to the Far East. She reached "Dixie Station" on 5 November and immediately began combat air operations.

Ticonderoga's winter deployment of 1965 and 1966 was her first total combat tour of duty during American involvement in the Vietnam War. During her six months in the Far East, the carrier spent a total of 116 days in air operations off the coast of Vietnam, dividing her time almost evenly between "Dixie" and "Yankee Stations," the carrier operating areas off South and North Vietnam, respectively. Her air group delivered over 8,000 tons of ordnance in more than 10,000 combat sorties, with a loss of 16 planes, but only five pilots. For the most part, her aircraft hit enemy installations in North Vietnam and interdicted supply routes into South Vietnam, including river-borne and coastwise junk and sampan traffic as well as roads, bridges, and trucks on land. Specifically, they claimed the destruction of 35 bridges as well as numerous warehouses, barracks, trucks, boats, and railroad cars and severe damage to a major North Vietnamese thermal power plant located at Uong Bi north of Haiphong. After a stop at Sasebo, Japan, from 25 April to 3 May 1966, the warship put to sea to return to the United States. On 13 May, she pulled into port at San Diego to end the deployment.

Following repairs, she stood out of San Diego on 9 July to begin a normal round of west coast training operations. Those and similar evolutions continued until 15 October, when

Ticonderoga departed San Diego, bound via Hawaii for the western Pacific. The carrier reached Yokosuka, Japan, on 30 October and remained there until 5 November when she headed south for an overnight stop at Subic Bay in the Philippines on the 10th and 11th. On the 13th, Ticonderoga arrived in the Gulf of Tonkin and began the first of three combat tours during her 1966-67 deployment. She launched 11,650 combat sorties, all against enemy targets located in North Vietnam. Again, her primary targets were logistics and communications lines and transportation facilities. For their overall efforts in the conduct of day and night strikes on enemy targets, Ticonderoga and her air group earned their second Navy Unit Commendation. She completed her final line period on 27 April 1967 and returned to Yokosuka, from which she departed again on 19 May to return to the United States. Ten days later, the carrier entered San Diego and began a month-long post-deployment stand-down. At the beginning of July, the warship shifted to Bremerton, Wash., where she entered the Puget Sound Naval Shipyard for two months of repairs. Upon the completion of yard work, she departed Bremerton on 6 September and steamed south to training operations off the coast of southern California.

On 28 December, Ticonderoga sailed for her fourth combat deployment to the waters off the Indochinese coast. She made Yokosuka on 17 January 1968 and after two days of upkeep continued on to the Gulf of Tonkin where she arrived on station on the 26th and began combat operations. Between January and July, Ticonderoga was on the line off the coast of Vietnam for five separate periods totaling 120 days of combat duty. During that time, her air wing flew just over 13,000 combat sorties against North Vietnamese and Viet Cong forces, most frequently in the continuing attempts to interdict the enemy lines of supply. In mid-April, following her second line period, she made a port visit to Singapore and then, after upkeep at Subic Bay, returned to duty off Vietnam. On 9 July, during her fifth and final line period, Lt. Comdr. J. B. Nichols claimed Ticonderoga's first MiG kill. The carrier completed that line period and entered Subic Bay for upkeep on 25 July.

On the 27th, she headed north to Yokosuka where she spent a week for upkeep and briefings before heading back to the United States on 7 August. Ticonderoga reached San Diego on the 17th and disembarked her air group. On the 22nd, she entered the Long Beach Naval Shipyard for post-deployment repairs. She completed those repairs on 21 October, conducted sea trials on the 28th and 29th, and began normal operations out of San Diego early in November. For the remainder of the year, she conducted refresher training and carrier qualifications along the coast of southern California.

During the first month of 1969, Ticonderoga made preparations for her fifth consecutive combat deployment to the Southeast Asia area. On 1 February, she cleared San Diego and headed west. After a brief stop at Pearl Harbor a week later, she continued her voyage to Yokosuka where she arrived on the 20th. The carrier departed Yokosuka on the 28th for the coast of Vietnam where she arrived on 4 March. Over the next four months, Ticonderoga served four periods on the line off Vietnam, interdicting communist supply lines and making strikes against their positions.

During her second line period, however, her tour of duty off Vietnam came to an abrupt end On 16 April when she was shifted north to the Sea of Japan. North Korean aircraft had shot down a Navy reconnaissance plane in the area, and Ticonderoga was called upon to beef up the forces assigned to the vicinity. However, the crisis abated, and Ticonderoga entered Subic Bay on 27 April for upkeep. On 8 May, she departed the Philippines to return to "Yankee Station" and resumed interdiction operations. Between her third and fourth line periods, the carrier visited Sasebo and Hong Kong.

The aircraft carrier took station off Vietnam for her last line period of the deployment on 26 June and there followed 37 more days of highly successful air sorties against enemy targets. Following that tour, she joined TF 71 in the Sea of Japan for the remainder of the deployment. Ticonderoga concluded the deployment—a highly successful one, for she received her third Navy Unit Commendation for her operations during that tour of duty—when she left Subic Bay on 4 September.

Ticonderoga arrived in San Diego on 18 September. After almost a month of post-deployment stand-down, she moved to the Long Beach Naval Shipyard in mid-October to begin conversion to an antisubmarine warfare (ASW) aircraft carrier. Overhaul and conversion work began on 20 October, and Ticonderoga was redesignated CVS-14 on the 21st. She completed overhaul and conversion on 28 May 1970 and conducted exercises out of Long Beach for most of June. On the 26th, the new ASW support carrier entered her new home port, San Diego. During July and August, she conducted refresher training, refresher air operations, and carrier landing qualifications. The warship operated off the California coast for the remainder of the year and participated in two exercises—HUKASWEX 4-70 late in October and COMPUTEX 23-70 between 30 November and 3 December.

During the remainder of her active career, Ticonderoga made two more deployments to the Far East. Because of her change in mission, neither tour of duty included combat operations off Vietnam. Both, however, included training exercises in the Sea of Japan with ships of the Japanese Maritime Self Defense Force. The first of these two cruises also brought operations in the Indian Ocean with units of the Thai Navy and a transit of Sunda Strait, during which a ceremony was held to commemorate the loss of Houston (CA-30) and HMAS Perth in 1942.

In between these two last deployments, she operated in the eastern Pacific and participated in the recovery of the Apollo 16 moon mission capsule and astronauts near American Samoa during April of 1972. The second deployment came in the summer of 1972, and, in addition to the training exercises in the Sea of Japan, Ticonderoga also joined ASW training operations in the South China Sea. That fall, she returned to the eastern Pacific and, in November, practiced for the recovery of Apollo 17. The next month, Ticonderoga recovered her second set of space voyagers near American Samoa. The carrier then headed back to San Diego, where she arrived on 28 December.

Ticonderoga remained active for nine more months, first operating out of San Diego and then making preparations for inactivation. On 1 September 1973, the aircraft carrier was decommissioned after a board of inspection and survey found her to be unfit for further naval service. Her name was struck from the Navy list on 16 November 1973, and arrangements were begun to sell her for scrap.

Ticonderoga received five battle stars during World War II and three Navy Unit Commendations, one Meritorious Unit Commendation, and 12 battle stars during the Vietnam War.

CV-15 • USS RANDOLPH

(Displacement: 36,380 tons; length: 888 feet; beam: 93 feet; extreme width at flight deck: 147.5 feet; draft: 28 feet 7 inches; speed: 32 knots; complement: 3,448 crew; armament: 12 five-inch guns, 68 40mm guns, 59 20mm guns; aircraft: 80 to 100; class: Essex.)

The second Randolph (CV-15) was laid down on 10 May 1943 by Newport News Shipbuilding & Dry Dock Co., Newport News, Va.; launched on 28 June 1944, sponsored by Mrs. Guy M. Gillette, and commissioned on 9 October 1944, with Capt. Felix Baker in command.

Following shakedown off Trinidad, Randolph got underway for the Panama Canal and the Pacific. On 31 December, she reached San Francisco where Air Group 87 was detached and Air Group 12 reported on board for four months of duty. On 20 January 1945, Randolph departed San Francisco for Ulithi whence she sortied on 10 February with TF 58. She launched attacks on 16 and 17 February against Tokyo airfields and the Tachikawa engine plant. The following day she made a strike on the island of Chichi Jima. On 20 February, she launched three aerial sweeps in support of ground forces invading Iwo Jima and two against Haha Jima. During the next four days, further strikes hit Iwo Jima and combat air patrols were flown almost continuously. Three sweeps against airfields in the Tokyo area and one against Hachijo Jima followed on 25 February before the carrier returned to Ulithi.

A kamikaze Frances, a twin-engine bomber, hit Randolph on the starboard side aft just below the flight deck, killing 25 men and wounding 106, as the carrier was riding at anchor at Ulithi on 11 March 1945. Repaired at Ulithi, Randolph joined the Okinawa Task Force on 7 April. Combat air patrols were flown daily until 14 April, when strikes were sent against Okinawa, Ie Shima, and Kakeroma Island. The following day, an air support mission of fighters, bombers, and torpedo planes hit Okinawa and a fighter sweep struck an airfield in southern Kyushu. Under daily air attack from 17 April onward, Randolph continued to send her aircraft on CAP and support missions throughout the month. During May, planes from the carriers hit the Ryukyus and southern Japan, Kikai-Amami Island naval base and airfields, and Kyushu airfields. Becoming flagship TF 58 on 15 May, Randolph continued her support of the occupation of Okinawa Shima until 29 May, when she retired via Guam to the Philippines.

On her next war cruise, as a part of Admiral Halsey's famed 3rd Fleet, Randolph made a series of strikes up and down the Japanese home islands. With Air Group 16 replacing Air Group 12, the ship launched eight raids on 10 July against airfields in the Tokyo area, principally those on the peninsula east of Tokyo Bay. On the 14th, her planes struck the airfields and shipping in and near Tsugaru Strait. In this attack, two of the important Honshu-Hokkaido train ferries were sunk and three were damaged. Attacks on the Japanese home islands continued for the next few days, and, on 18 July 1945, Nagato, lying camouflaged alongside a pier at the Yokosuka Naval Base, was bombed.

Moving southwest, Randolph and other carriers were off the coast of Shikoku on 24 July for an anti-shipping sweep of the Inland Sea, during which the carrier-battleship Hyuga was heavily damaged and airfields and industrial installations on Kyushu, Honshu, and Shikoku were hit hard. Randolph's pilots estimated that from 10 to 25 July they had destroyed 25 to 30 ships, ranging in size from small luggers to a 6,000-ton freighter, and had damaged 35 to 40 others. Randolph's strikes continued right up to the morning of the 15 August 1945 surrender, when her planes hit Kisarazu Airfield and surrounding installations.

Following the end of the war, Randolph headed home. Transiting the Panama Canal in late September, she arrived at Norfolk on 15 October, where she was rigged for the "Magic Carpet" service. Before the end of the year, she completed two trips to the Mediterranean area to return American servicemen. Then, in 1946, she became a training ship for reservists and midshipmen, and made a Mediterranean cruise in the latter half of the year. After another voyage to the Caribbean, she embarked

midshipmen in the early summer of 1947 for a cruise to northern European waters. Randolph was placed out of commission, in reserve, on 25 February 1948 and berthed at Philadelphia.

Reclassified CVA-15 on 1 October 1952, Randolph recommissioned on 1 July 1953. After shakedown off Guantanamo Bay with Carrier Air Group 10, she took on Carrier Air Group 14, departed Norfolk for the Mediterranean, and joined the 6th Fleet on 3 February 1954. Deployed to the Mediterranean for 6 months of Fleet and NATO exercises during 1954 and 1955, Randolph entered the Norfolk Navy Yard on 18 June 1955 for the installation of an angled deck and other modernization. Leaving the yard in January 1956, Randolph conducted air operations off the east coast for the next 6 months and was the first Atlantic Fleet carrier to launch a Regulus guided missile from her flight deck.

On 14 July 1956, Randolph again steamed east for a seven-month tour of duty with the 6th Fleet in the Mediterranean. When Israel, Britain, and France invaded the United Arab Republic in October of that year, Randolph stood ready. Operating near the Suez Canal, her aircraft provided air cover and surface and air reconnaissance for the evacuation of U.S. nationals from Alexandria. She returned to the United States on 19 February 1957.

After a few months operating off the east coast, Randolph deployed to the Mediterranean again on 1 July 1957. Between August and December, as political turmoil in Syria threatened to further disturb the already turbulent Mideast, she patrolled the eastern Mediterranean. Back in the United States on 24 February 1958, the flattop made her 5th Mediterranean deployment from 2 September 1958 to 12 March 1959. Randolph was reclassified CVS-15 on 31 March 1959, and conducted ASW operations off the east coast throughout that year and the next, receiving her fourth Battle Efficiency "E" in a row in September 1960. After overhaul at Norfolk, Randolph sailed for operations in the Caribbean and served as the recovery ship for Astronaut Virgil Grissom on America's second manned space flight, a suborbital shot. In February 1962, Randolph was the primary recovery ship for Astronaut John Glenn on his flight, the first American orbital voyage in space. After his historic three-orbit flight, he landed safely near the destroyer USS Noa (DD 841), from which he was transferred by helicopter to Randolph.

In the summer of 1962, Randolph again steamed to the Mediterranean. Returning to the western Atlantic as the Cuban missile crisis broke, she operated in the Caribbean from the end of October through November. After a Norfolk overhaul, Randolph resumed her station in the Atlantic. During the next five years, she made two Mediterranean cruises and a northern European cruise, while spending most of her time off the east coast and in the Caribbean.

On 7 August 1968, the Defense Department announced that it would inactivate Randolph and 49 other ships to reduce fiscal expenditures in 1969. Randolph was placed out of commission, in reserve, berthed at Philadelphia, on 13 February 1969, where she remained until 1 June 1973 when she was stricken from the Navy list. Randolph earned three battle stars for World War II service.

CV-16 • USS LEXINGTON

(Displacement: 27,100 tons; length: 872 feet; beam: 93 feet; extreme beam: 192 feet; draft: 28 feet 7 inches; speed: 32.7 knots; complement: 3,748 crew; armament: 12 five-inch guns, 68 40mm guns; aircraft: 103; class: Essex.)

The fifth Lexington (CV-16) was laid down as Cabot on 15 July 1941 by Bethlehem Steel Co., Quincy, Mass.; renamed Lexington on 16 June 1942; launched on 23 September 1942, sponsored by Mrs. Theodore D. Robinson; and commissioned on 17 February 1943, Capt. Felix B. Stump in command.

After Caribbean shakedown and yard work at Boston, Lexington sailed for Pacific action via the Panama Canal, arriving at Pearl Harbor on 9 August 1943. She raided Tarawa in late September and Wake in October, then returned to Pearl Harbor to prepare for the Gilbert Islands operation. From 19 to 24 November, she made searches and flew sorties in the Marshalls, covering the landings in the Gilberts. Her aviators downed 29 enemy aircraft on 23 and 24 November.

Lexington sailed to raid Kwajalein on 4 December 1943. Her morning strike destroyed a cargo ship, damaged two cruisers, and accounted for 30 enemy aircraft. Her gunners splashed two of the enemy torpedo planes that attacked at midday, and opened fire again at 1920 that night when a major air attack began. At 2322, parachute flares silhouetted the carrier, and 10 minutes later she was hit by a torpedo to starboard, knocking out her steering gear. Settling five feet by the stern, the carrier began circling to port amidst dense clouds of smoke pouring from ruptured tanks aft. An emergency hand-operated steering unit was quickly devised, and Lexington made Pearl Harbor for emergency repairs, arriving on 9 December. She reached Bremerton, Wash., on 22 December for full repairs, completed on 20 February 1944.

Lexington sailed via Alameda, Calif., and Pearl Harbor for Majuro, where Rear Adm. Marc Mitscher commanding TF 58 broke his flag in her on 8 March. After a warm-up strike against Mille, TF 58 operated against the major centers of resistance in Japan's outer empire, supporting the Army landing at Hollandia on 13 April, and hitting supposedly invulnerable Truk on 28 April. Heavy counterattack left Lexington untouched, her planes splashing 17 enemy fighters; but, for the second time, Japanese propaganda announced her sunk.

A surprise fighter strike on Saipan on 11 June virtually eliminated all air opposition over the island, then battered from the air for the next five days. On 16 June 1944, Lexington fought off a fierce attack by Japanese torpedo planes based on Guam, once again emerging unhurt but sunk a third time by propaganda pronouncements. As Japanese opposition to the Marianas operation provoked the Battle of the Philippine Sea on 19 and 20 June, Lexington played a major role in TF 58's great victory. With over 300 enemy aircraft destroyed the first day, and a carrier, a tanker, and a destroyer sunk the second day, American aviators virtually knocked Japanese naval aviation out of the war; for with the planes went the trained and experienced pilots without whom Japan could not continue air warfare at sea.

Using Eniwetok as her base, Lexington flew sorties over Guam and against the Palaus and Bonins into August. She arrived in the Carolinas on 6 September for three days of strikes against Yap and Ulithi, then began attacks on Mindanao, the Visayas, the Manila area, and shipping along the west coast of Luzon, preparing for the coming assault on Leyte. Her task force then blasted Okinawa on 10 October and Formosa two days later to destroy bases from which opposition to the Philippines campaign might be launched. She was again unscathed through the air battle fought after the Formosa assault. Now covering the Leyte landings, Lexington's planes scored importantly in the Battle for Leyte Gulf, the climactic American naval victory over Japan. While the carrier came under constant enemy attack in the engagement in which USS Princeton (CVL 23) was sunk, her planes joined in sinking Japan's super-battleship Musashi and scored hits on three cruisers on 24 October 1944. The next day, with Essex aircraft, they sank the carrier Chitose, and alone sank Zuikaku. Later in the day, they aided in sinking a third carrier, Zuiho. As the retiring Japanese were pursued, her planes sank the heavy cruiser Nachi with four torpedo hits on 5 November off Luzon.

But in the same action, she was introduced to the kamikaze as a flaming Japanese plane crashed near her island, destroying most of the island structure and spraying fire in all directions. Within 20 minutes major blazes were under control, and she was able to continue normal flight actions, her guns knocking down a would-be kamikaze heading for the carrier USS Ticonderoga (CV 14) as well. On 9 November, Lexington arrived at Ulithi to repair battle damage and learn that Tokyo once again claimed her destroyed.

Chosen flagship for TG 58.2 on 11 December, she struck at the airfields of Luzon and Formosa during the first 9 days of January 1945, encountering little enemy opposition. The task force then entered the China Sea to strike enemy shipping and

air installations. Strikes were flown against Saipan, Camranh Bay in then Indochina, Hong Kong, the Pescadores, and Formosa. Task force planes sank four merchant ships and four escorts in one convoy and destroyed at least 12 in another, at Camranh Bay on 12 January. Leaving the China Sea on 20 January, Lexington sailed north to strike Formosa again on 21 January and Okinawa again on 22 January.

After replenishing at Ulithi, TG 58.2 sailed on 10 February to hit airfields near Tokyo on 16 and 17 February to minimize opposition to the Iwo Jima landings on 19 February. Lexington flew close support for the assaulting troops from 19 to 22 February, then sailed for further strikes against the Japanese home islands and the Nansei Shoto before heading for overhaul at Puget Sound.

Lexington was combat-bound again on 22 May, sailing via Alameda and Pearl Harbor for San Pedro Bay, Leyte, where she joined Rear Adm. T. L. Sprague's task force for the final round of airstrikes which battered the Japanese home islands through July until 15 August, when the last strike was ordered to jettison its bombs and return to Lexington on receiving word of Japanese surrender. During this period, she had launched attacks on Honshu and Hokkaido airfields, and Yokosuka and Kure naval bases to destroy the remnants of the Japanese fleet. She had also flown bombing attacks on industrial targets in the Tokyo area. After hostilities ended, she continued to fly precautionary patrols over Japan, and dropped supplies to prisoner of war camps on Honshu. She supported the occupation of Japan until leaving Tokyo Bay on 3 December 1945 with homeward-bound veterans for transportation to San Francisco, where she arrived on 16 December.

After west coast operations, Lexington decommissioned at Bremerton, Wash., on 23 April 1947 and entered the Reserve Fleet there. Designated attack carrier CVA-16 on 1 October 1952, she began conversion and modernization in Puget Sound Naval Shipyard on 1 September 1953, receiving the new angled flight deck.

Lexington recommissioned on 15 August 1955, Capt. A. S. Heyward, Jr., in command. Assigned San Diego as her homeport, she operated off California until May 1956, sailing then for a six-month deployment with the 7th Fleet. She based on Yokosuka for exercises, maneuvers, and search and rescue missions off the coast of China, and called at major Far Eastern ports until returning to San Diego on 20 December. She next trained Air Group 12, which deployed with her on the next 7th Fleet deployment. Arriving in Yokosuka on 1 June 1957, Lexington embarked Rear Adm. H. D. Riley, Commander Carrier Division 1, and sailed as his flagship until returning to San Diego on 17 October.

Following overhaul at Bremerton, her refresher training was interrupted by the Lebanon crisis. On 14 July 1958, she was ordered to embark Air Group 21 at San Francisco and sail to reinforce the 7th Fleet off Taiwan, arriving on station on 7 August. With another peacekeeping mission of the U.S. Navy successfully accomplished, she returned to San Diego on 19 December. Now the first carrier whose planes were armed with air-to-surface Bullpup guided missiles, Lexington left San Francisco on 26 April 1959 for another tour of duty with the 7th Fleet. She was on standby alert during the Laotian crisis of late August and September, then exercised with British forces before sailing from Yokosuka on 16 November for San Diego, arriving on 2 December. Through early 1960, she overhauled at Puget Sound Naval Shipyard.

Lexington's next Far Eastern tour began late in 1960 and was extended well into 1961 by renewed tension in Laos. Returning to west coast operations, she was ordered in January 1962 to prepare to relieve USS Antietam (CVS-36) as an aviation training carrier in the Gulf of Mexico, and she was redesignated CVS-16 on 1 October 1962. However, during the Cuban Missile Crisis, she resumed duty as an attack carrier, and it was not until 29 December 1963 that she relieved Antietam at Pensacola. Lexington operated out of her home port, Pensacola, as well as Corpus Christi and New Orleans, qualifying student aviators and maintaining the high state of training of both active-duty and reserve naval aviators. Lexington marked her 200,000th arrested landing on 17 October 1967 and was redesignated CVT-16 on 1 January 1969. She continued as a training carrier for the next 22 years until decommissioned on 8 November 1991. On 15 June 1992, the ship was donated as a museum and now operates as such in Corpus Christi, Tex.

Lexington received the Presidential Unit Citation and 11 battle stars for World War II service.

(Displacement: 27,100 tons; length: 872 feet; beam: 93 feet; draft: 28 feet 7 inches; speed: 32.7 knots; complement: 3,448 crew; armament: 12 five-inch guns; class: Essex.)

CV-17 • USS BUNKER HILL

Bunker Hill (CV-17) was launched on 7 December 1942 by Bethlehem Steel Co., Quincy, Mass.; sponsored by Mrs. Donald Boynton; and commissioned on 24 May 1943, with Captain J. J. Ballentine in command.

Reporting to the Pacific in the fall of 1943, Bunker Hill participated in carrier operations during the Rabaul strike (11 November 1943); Gilbert Islands operation, including support of the landings on Tarawa (13 November - 8 December); the Kavieng strikes in support of the Bismarck Archipelago operation (25 December 1943, 1 and 4 January 1944); Marshall Islands operation (29 January - 8 February); strikes against Truk (17-18 February), during which eight Japanese combatant vessels were sunk; Marianas raid (23 February); Palau-Yap-Ulithi-Woleai raids (30 March - 1 April); Truk-Satawan-Ponape raids (29 April - 1 May); Hollandia operation (21-28 April); and Marianas operation (12 June - 10 August), including the Battle of the Philippine Sea.

On 19 June 1944, during the opening phases of the battle, Bunker Hill was damaged when an enemy near miss scattered shrapnel fragments across the ship. Two men were killed and over 80 were wounded. Bunker Hill continued to do battle, and her planes aided in sinking one Japanese carrier and destroying a part of the 476 Japanese aircraft that were downed. During September, she participated in the Western Caroline Islands operation and then launched strikes at Okinawa, Luzon, and Formosa until November.

On 6 November, Bunker Hill retired from the forward area and steamed to Bremerton, Wash., for a period of yard availability. Repairs completed, she departed the west coast on 24 January 1945 and returned to the war front. During the remaining months of World War II, Bunker Hill participated in the Iwo Jima operation and the 5th Fleet raids against Honshu and the Nansei Shoto (15 February - 4 March); and the 5th and 3rd Fleet raids in support of the Okinawa operation. On 7 April 1945, Bunker Hill's planes took part in a fast carrier task force attack on a Japanese naval force in the East China Sea. The enemy battleship Yamato, one cruiser, and four destroyers were sunk.

On the morning of 11 May 1945, while supporting the Okinawa invasion, Bunker Hill was hit and severely damaged by two suicide planes. Gasoline fires flamed up, and several explosions took place. The ship suffered the loss of 346 men killed, 43 missing, and 264 wounded. Although badly crippled, she managed to return to Bremerton via Pearl Harbor.

In September, Bunker Hill reported for duty with the "Magic Carpet" fleet. She remained on this duty as a unit of TG 16.12, returning veterans from the Pacific until January 1946 when she was ordered to Bremerton for inactivation. She was placed out of commission in reserve there on 9 January 1947.

While laid up, she was reclassified three times, becoming CVA-17 in October 1952, CVS-17 in August 1953, and AVT-9 in May 1959, the latter designation indicating that any future commissioned duty would be as an aircraft transport. However, she was one of the two Essex-class carriers that saw no Cold War active service. Stricken from the Naval Vessel Register in November 1966, Bunker Hill was used as a stationary electronics test platform at San Diego during the 1960s and early 1970s. She was sold for scrapping in May 1973.

Bunker Hill received the Presidential Unit Citation for the period 11 November 1943 to 11 May 1945. In addition, she received 11 battle stars for her World War II service.

CV-18 • USS WASP

(Displacement: 27,100 tons; length: 872 feet; beam: 93 feet; extreme width at flight deck: 147.5 feet; draft: 28 feet 7 inches; speed: 32.7 knots; complement: 3,448 crew; armament: 40 40mm guns, 55 20mm guns; aircraft: 80 - 100; class: Essex.)

The ninth Wasp (CV-18) was laid down as Oriskany on 18 March 1942 at Quincy, Mass., by the Bethlehem Steel Co.; renamed Wasp on 13 November 1942; launched on 17 August 1943; sponsored by Miss Julia M. Walsh, the sister of Senator David I. Walsh of Massachusetts, and commissioned on 24 November 1943, Capt. Clifton A. F. Sprague in command.

Following a shakedown cruise which lasted through the end of 1943, Wasp returned to Boston for a brief yard period to correct minor flaws discovered during her time at sea.

On 10 January 1944, the new aircraft carrier departed Boston, steamed to Hampton Roads, Va., and remained there until the last day of the month, when she sailed for Trinidad, her base of operations through 22 February. She returned to Boston five days later and prepared for service in the Pacific. Early in March, the ship sailed south, transited the Panama Canal, arrived at San Diego, Calif., on 21 March, and reached Pearl Harbor on 4 April.

Following training exercises in Hawaiian waters, Wasp steamed to the Marshall Islands and at Majuro Rear Adm. Alfred E. Montgomery's newly formed Task Group (TG) 58.6 of Vice Adm. Marc A. Mitscher's Fast Carrier Task Force (TF 58). On 14 May, she and her sister carriers of TG 58.6, USS Essex (CV-9) and USS San Jacinto (CV-30), sortied for raids on Marcus and Wake Islands to give the new task group combat experience, to test a recently devised system of assigning, before takeoff, each pilot a specific target, and to neutralize those islands for the forthcoming Marianas campaign. As the force neared Marcus, it split, sending San Jacinto north to search for Japanese picket boats while Wasp and Essex launched strikes on the 19th and 20th, aimed at installations on the island. American planes encountered heavy antiaircraft fire but still managed to do enough damage to prevent Japanese forces on the island from interfering with the impending assault on Saipan.

When weather canceled launches planned for 21 May 1944, the two carriers rejoined San Jacinto and steamed to Wake. Planes from all three carriers pounded that island on the 24th and were sufficiently effective to neutralize that base. However, the system of pre-selecting targets for each plane fell short of the Navy's expectations, and, thereafter, tactical air commanders resumed responsibility for directing the attacks of their planes.

After the strike on Wake, TG 58.6 returned to Majuro to prepare for the Marianas campaign. On 6 June, Wasp, reassigned to TG 58.2, which was also commanded by Rear Adm. Montgomery, sortied for the invasion of Saipan. During the afternoon of the 11th, she and her sister carriers launched fighters for strikes against Japanese air bases on Saipan and Tinian. They were challenged by some 30 land-based fighters, which they promptly shot down. Antiaircraft fire was heavy, but the American planes braved it as they went on to destroy many Japanese aircraft which were still on the ground.

During the next three days, the American fighters, now joined by bombers, pounded installations on Saipan to soften up Japanese defenses for American assault troops who would go ashore on the 15th. That day and thereafter until the morning of June 17, planes from TG 58.2 and TG 58.3 provided close air support for Marines fighting on the Saipan beachhead.

The fast carriers of those task groups then turned over to escort carriers the responsibility for providing air support for the American ground forces, refueled, and steamed to rendezvous with TG 58.1 and 58.4, which were returning from strikes against Chichi Jima and Iwo Jima to prevent Japanese air bases on those islands from being used to launch attacks against American forces on or near Saipan.

Meanwhile, Japan, determined to defend Saipan no matter the cost, was sending Adm. Jisaburo Ozawa's powerful First Mobile Fleet from the Sulu Islands to the Marianas to sink the warships of Adm. Spruance's 5th Fleet and to annihilate the American troops who had fought their way ashore on Saipan. Soon after the Japanese task force sortied from Tawi Tawi on the morning of 13 June, American submarine USS Redfin (SS-272) spotted and reported it. Other submarines, which from time to time made contact with Ozawa's warships, kept Spruance posted on their progress as they wended their way through the Philippine Islands, transited San Bernardino Strait, and entered the Philippine Sea.

All day on the 18th, each force sent out scout planes in an effort to locate its adversary. Because of their greater range, the Japanese aircraft managed to obtain some knowledge of Spruance's ships, but American scout planes were unable to find Ozawa's force. Early the following morning, 19 June 1944, aircraft from Mitscher's carriers headed for Guam to neutralize

that island for the coming battle and, in a series of dogfights, destroyed many Japanese land-based planes.

During the morning, carriers from Ozawa's fleet launched four massive raids against their American counterparts, but all were thwarted almost completely. Nearly all of the Japanese warplanes were shot down while failing to sink a single American ship. They did manage to score a single bomb hit on USS South Dakota (BB-57), but that solitary success did not even put the tough Yankee battleship out of action.

That day, Mitscher's planes did not find the Japanese ships, but American submarines succeeded in sending two enemy carriers to the bottom. In the evening, three of Mitscher's four carrier task groups headed west in search of Ozawa's retiring fleet, leaving only TG 58.4 and a gun line of old battleships in the immediate vicinity of the Marianas to cover ground forces on Saipan. Planes from the American carriers failed to find the Japanese force until mid-afternoon on June 20 when an Avenger pilot reported spotting Ozawa almost 300 miles from the American carriers. Mitscher daringly ordered an all-out strike even though he knew that night would descend before his planes could return.

Over two hours later, the American aviators caught up with their quarry. They damaged two oilers so severely that they had to be scuttled, sank the carrier Hiyo, and scored damaging but non-lethal hits on the carriers Ryuho, Junyo, and Zuikaku and several other Japanese ships. However, during the sunset attack, the fuel gauges in many of the American planes registered half empty or more, presaging an anxious flight back to their now distant carriers.

When the carriers spotted the first returning plane at 2030 that night, Rear Adm. J. J. Clark bravely defied the menace of

Japanese submarines by ordering all lights to be turned on to guide the weary fliers home.

After a plane from Hornet landed on Lexington, Mitscher gave pilots permission to land on any available deck. Despite these unusual efforts to help the Navy's airmen, a good many planes ran out of gasoline before they reached the carriers and dropped into the water.

When fuel calculations indicated that no aircraft which had not returned could still be aloft, Mitscher ordered the carriers to reverse course and resume the stern chase of Ozawa's surviving ships, more in the hope of finding any downed fliers who might still be alive and pulling them from the sea than in the expectation of overtaking Japan's First Mobile Fleet before it reached the protection of the Emperor's land-based planes. During the chase, Mitscher's ships picked up 36 pilots and 26 crewmen.

At mid-morning of 21 June, Adm. Spruance detached Wasp and USS Bunker Hill (CV-17) from their task group and sent them with Adm. Lee's battleships in Ozawa's wake to locate and destroy any crippled enemy ships. The ensuing two-day hunt failed to flush out any game, so this ad hoc force headed toward Eniwetok for replenishment and well-earned rest.

The respite was brief, for, on 30 June 1944, Wasp sortied in TG 58.2-with TG 58.1-for strikes at Iwo Jima and Chichi Jima. Planes from the carriers pounded those islands on 3 and 4 July and, during the raids, destroyed 75 enemy aircraft, for the most part in the air. Then, as a grand finale, cruisers from the force's screen shelled Iwo Jima for two and one-half hours. The next day, 5 July, the two task groups returned to the Marianas and attacked Guam and Rota to begin more than a fortnight's effort to soften the Japanese defenses there in preparation for landings on Guam. Planes from Wasp and her sister carriers provided close air support for the Marines and soldiers who stormed ashore on 21 July.

The next day, Wasp's task group, TG 58.2, sortied with two other groups of Mitscher's carriers headed southwest toward the Western Carolines and launched raids against the Palaus on the 25th. The force then parted, with TG 58.1 and TG 58.3 steaming back north for further raids to keep the Bonin and Volcano Islands neutralized while Wasp in TG 58.2 was retiring toward the Marshalls for replenishment at Eniwetok, which she reached on 2 August 1944.

Toward the end of Wasp's stay at that base, Adm. Halsey relieved Adm. Spruance on 26 August and the 5th Fleet became the 3rd Fleet. Two days later, the Fast Carrier Task Force, redesignated TF 38, sortied for the Palaus. On 6 September, Wasp, now assigned to Vice Adm. John S. McCain's TG 38.1, began three days of raids on the Palaus. On the 9th, she headed—with her task group, TG 38.2, and TG 38.3—for the southern Philippines to neutralize air power there during the American conquest of Morotai, Peleliu, and Ulithi, three islands needed as advanced bases during the impending campaign to liberate the Philippines. Planes from these carriers encountered little resistance as they lashed Mindanao airfields that day and on the 10th. Raids against the Visayan Islands on 12 and 13 September were carried out with impunity and were equally successful.

D-day in the Palaus, 15 September 1944, found Wasp's TG 38.1 some 50 miles off Morotai, launching air strikes. It then returned to the Philippines for revisits to Mindanao and the Visayas before retiring to the Admiralties on 29 September for replenishment at Manus in preparation for the liberation of the Philippines. Ready to resume battle, she got underway again on 4 October 1944 and steamed to the Philippine Sea where TF 38 reassembled at twilight on the evening of 7 October, some 375 miles west of the Marianas. Its mission was to neutralize airbases within operational air distance of the Philippines to keep Japanese warplanes out of the air during the American landings on Leyte scheduled to begin on 20 October. The carriers steamed north to rendezvous with a group of nine oilers and spent the next day, 8 October 1944, refueling. They then followed a generally northwesterly course toward the Ryukyus until the 10th, when their planes raided Okinawa, Amami, and Miyako. That day, TF 38 planes destroyed a Japanese submarine tender, 12 sampans, and over 100 planes. But for Lt. Col. Doolittle's Tokyo raid from USS Hornet (CV-8) on 18 April 1942 and the daring war patrols of Pacific Fleet submarines, this carrier foray was the United States Navy's closest approach to the Japanese home islands up to that point in the war.

Beginning on the 12th, Formosa—next on the agenda—received three days of unwelcome attention from TF 38 planes.

In response, the Japanese Navy made an all-out effort to protect that strategic island, even though doing so meant denuding its remaining carriers of aircraft. Yet, the attempt to thwart the ever-advancing American Pacific Fleet was futile. At the end of a three-day air battle, Japan had lost more than 500 planes and 20-odd freighters. Many other merchant ships were damaged, as were hangars, barracks, warehouses, industrial plants, and ammunition dumps. However, the victory was costly to the United States Navy, for TF 38 lost 79 planes and 64 pilots and air crewmen, while cruisers USS Canberra (CA-70) and USS Houston (CL-81) and the carrier USS Franklin (CV-13) received damaging but non-lethal bomb hits. From Formosa, TF 38 shifted its attention to the Philippines. After steaming to waters east of Luzon, Wasp's TG 58.1 began to launch strikes against that island on 18 October 1944 and continued the attack the following day, hitting Manila for the first time since it was occupied by the Japanese early in the war.

On the 20th, the day the first American troops waded ashore on Leyte, Wasp had moved south to the station off that island whence she and her sister carriers launched some planes for close air support missions to assist MacArthur's soldiers, while sending other aircraft to destroy airfields on Mindanao, Cebu, Negros, Panay, and Leyte. Task Group 38.1 refueled the following day and, on the 22nd, set a course for Ulithi to rearm and provision.

While McCain's carriers were steaming away from the Philippines, great events were taking place in the waters of that archipelago. Adm. Soemu Toyoda, the Commander in Chief of Japan's Combined Fleet, activated plan Sho-Go-1, a scheme for bringing about a decisive naval action off Leyte. The Japanese strategy called for Ozawa's carriers to act as a decoy to lure TF 38 north of Luzon and away from the Leyte beachhead. Then, with the American fast carriers out of the way, heavy Japanese surface ships were to debouch into Leyte Gulf from two directions: from the south through Surigao Strait and from the north through San Bernardino Strait. During much of 24 October, planes from Halsey's carrier task groups still in Philippine waters pounded Adm. Kurita's powerful Force "A," or Center Force, as it steamed across the Sibuyan Sea toward San Bernardino Strait. When darkness stopped their attack, the American aircraft had sunk the superbattleship Musashi and had damaged several other Japanese warships. Moreover, Halsey's pilots reported that Kurita's force had reversed course and was moving away from San Bernardino Strait.

That night, Adm. Nishimura's Force "C," or Southern Force, attempted to transit Surigao Strait but met a line of old battleships commanded by Rear Adm. Jesse B. Oldendorf. The venerable American men-of-war crossed Nishimura's "T" and all but annihilated his force. Adm. Shima, who was following in Nishimura's wake to lend support, realized that disaster had struck and wisely withdrew.

Meanwhile, late in the afternoon of 24 October 1944, after Kurita's Center Force had turned away from San Bernardino Strait in apparent retreat, Halsey's scout planes finally located Ozawa's carriers a bit under 200 miles north of TF 38. This intelligence prompted Halsey to head north toward Ozawa with his Fast Carrier Task Force. However, at this point, he did not recall McCain's TG 68.1 but allowed it to continue steaming toward Ulithi.

After dark, Kurita's Center Force again reversed course and once more headed for San Bernardino Strait. About half an hour past midnight, it transited that narrow passage, turned to starboard, and steamed south, down the east coast of Samar. Since Halsey had dashed north in pursuit of Ozawa's carriers, only three 7th Fleet escort carrier groups and their destroyer and destroyer escort screens were available to challenge Kurita's mighty battleships and heavy cruisers and to protect the American amphibious ships which were supporting the troops fighting on Leyte.

Remembered by their call names, "Taffy 1," "Taffy 2," and "Taffy 3," these three American escort-carrier groups were deployed along Samar's east coast with "Taffy 3," commanded by Wasp's first captain, Rear Adm. Clifton A. F. Sprague, in the northernmost position, about 40 miles off Paninihian Point. "Taffy 2" was covering Leyte Gulf, and "Taffy 1" was still farther south watching Surigao Strait.

At 0645 on 25 October 1944, lookouts on "Taffy 3" ships spotted bursts of antiaircraft fire blossoming in the northern sky as Center Force gunners opened fire on an American anti-submarine patrol plane. Moments later, "Taffy 3" made both radar and visual contact with the approaching Japanese warships. Shortly before 0700, Kurita's guns opened fire on the hapless "baby flattops" and their comparatively tiny but incredibly courageous escorts. For more than two hours, "Taffy 3's" ships and planes, aided by aircraft from sister escort-carrier groups to the south, fought back with torpedoes, guns, bombs, and consummate seamanship. Then, at 0311, Kurita, shaken by the loss of three heavy cruisers and thinking that he had been fighting TF 38, ordered his remaining warships to break off the action.

Meanwhile, at 0848, Adm. Halsey had radioed McCain's TG 38.1, then refueling en route to Ulithi, calling that carrier group back to Philippine waters to help "Taffy 3" in its fight for survival. Wasp and her consorts raced toward Samar at flank speed until 1030 when they began launching planes for strikes at Kurita's ships, which were still some 330 miles away. While these raids did little damage to the Japanese Center Force, they did strengthen Kurita's decision to retire from Leyte.

While his planes were in the air, McCain's carriers continued to speed westward to lessen the distance of his pilots' return flight and to be in an optimum position at dawn to launch more warplanes at the fleeing enemy force. With the first light of 26 October, TG 38.1 and Rear Adm. Bogan's TG 38.2, which finally had been sent south by Halsey, launched the first of their strikes that day against Kurita. The second left the carriers a little over two hours later. These fliers sank light cruiser Noshiro and damaged, but did not sink, heavy cruiser Kumano. The two task groups launched a third strike in the early afternoon, but it did not add to their score.

Following the Battle for Leyte Gulf, which ended the Japanese Fleet as a serious challenge to American supremacy at sea in the Far East, TG 38.1 operated in the Philippines for two more days providing close air support before again heading for Ulithi on the 28th. However, the respite, during which Rear Adm. Montgomery took command of TG 38.1 when McCain fleeted up to relieve Mitscher as CTF 38, was brief since Japanese land-based planes attacked troops on the Leyte beachhead on 1 November.

Wasp participated in raids against Luzon air bases on 5 and 6 November, destroying over 400 Japanese aircraft, for the most part on the ground. After a kamikaze hit Lexington during the operation, McCain shifted his flag from that carrier to Wasp and, a short time later, returned in her to Guam to exchange air groups.

Wasp returned to the Philippines a little before mid-month and continued to send strikes against targets in the Philippines, mostly on Luzon, until the 26th, when the Army Air Force assumed responsibility for providing air support for troops on Leyte. TF 38 then retired to Ulithi. There, the carriers received greater complements of fighter planes and, in late November and early December, conducted training exercises to prepare them better to deal with Japan's new threat to the American warships, kamikazes or suicide planes. Task Force 38 sortied from Ulithi on 10 and 11 December 1944 and proceeded to a position east of Luzon for round-the-clock strikes against air bases on that island from the 14th through the 16th to prevent Japanese fighter planes from endangering landings on the southwest coast of Mindoro scheduled for the 15th. Then, while withdrawing to a fueling rendezvous point east of the Philippines, TF 38 was caught in a terribly destructive typhoon which battered its ships and sank three American destroyers. The carriers spent most of the ensuing week repairing storm damage and returned to Ulithi on Christmas Eve.

But the accelerating tempo of the war ruled out long repose in the shelter of the lagoon. Before the year ended, the carriers were back in action against airfields in the Philippines on Sakishima Gunto and on Okinawa. These raids were intended to smooth the way for General MacArthur's invasion of Luzon through the Lingayen Gulf. While the carrier planes were unable to knock out all Japanese air resistance to the Luzon landings, they did succeed in destroying many enemy planes and thus reduced the air threat to manageable proportions.

On the night after the initial landings on Luzon, Halsey took TF 38 into the South China Sea for a week's rampage in which his ships and planes took a heavy toll of Japanese shipping and aircraft before they retransited Luzon Strait on 16 January 1945 and returned to the Philippine Sea. Bad weather prevented Halsey's planes from going aloft for the next few days; but, on the 21st, they bombed Formosa, the Pescadores, and the Sakishimas.

The following day, the aircraft returned to the Sakishimas and the Ryukyus for more bombing and reconnaissance. The overworked Fast Carrier Task Force then headed for Ulithi and entered that lagoon on the 26th.

While the flattops were catching their breath at Ulithi, Adm. Spruance relieved Halsey in command of the Fleet, which was thereby transformed from the 3rd to the 5th Fleet in February. The metamorphosis also entailed Mitscher's replacing McCain and Clark's resuming command of TG 38.1, still Wasp's task group.

The next major operation dictated by Allied strategy was the capture of Iwo Jima in the Volcano Islands. Iwo was needed as a base for Army Air Force fighter planes, which were to protect Mariana-based B-29 bombers during raids against the Japanese home islands and as an emergency landing point for crippled warplanes. Task Force 58 sortied on 10 February, held rehearsals at Tinian, and then headed for Japan. Fighter planes took off from the carriers before dawn on the 16th to clear the skies of Japanese aircraft. They succeeded in this mission, but Wasp lost several of her fighters during the sweep. Bombing sorties, directed primarily at aircraft factories in Tokyo, followed; but clouds hid many of these plants, forcing some planes to drop their bombs on secondary targets. Bad weather, which also hampered Mitscher's fliers during raids the next morning, prompted him to cancel strikes scheduled for the afternoon and head the task force west.

During the night, Mitscher turned the carriers toward the Volcano Islands to be on hand to provide air support for the Marines who would land on the beaches of Iwo Jima on the morning of 19 February 1945.

For the next few days, planes from the American carriers continued to assist the Marines who were engaged in a bloody struggle to wrest the island from its fanatical defenders. On the 23rd, Mitscher led his carriers back to Japan for more raids on Tokyo. Planes took off on the morning of the 25th, but when they reached Tokyo, they again found their targets obscured by clouds. Moreover, visibility was so bad the next day that raids on Nagoya were called off, and the carriers steamed south toward the Ryukyus to bomb and reconnoiter Okinawa, the next prize to be taken from the Japanese Empire. Planes left the carriers at dawn on 1 March; and, throughout the day, they hammered and photographed the islands of the Ryukyu group. Then, after a night bombardment by surface ships, TF 58 set a course for the Carolines and anchored in Ulithi lagoon on the 4th.

Wasp recorded, from 17 to 23 March 1945, what was often referred to as the busiest week in flattop history. In these seven

days, Wasp accounted for 14 enemy planes in the air, destroyed six more on the ground, scored two 500-pound bomb hits on each of two Japanese carriers, dropped two 1,000-pound bombs on a Japanese battleship, put one 1,000-pounder on another battleship, hit a heavy cruiser with three 500-pound missiles, dropped another 1,000-pound bomb on a big cargo ship, and heavily strafed and probably sank a large Japanese submarine. During this week, which also included a bomb hit on the carrier, Wasp was under almost continuous attack by shore-based aircraft and experienced several close kamikaze attacks. The carrier's gunners fired more than 10,000 rounds at the determined Japanese attackers. On 13 April 1945, Wasp returned to the Puget Sound Navy Yard, Bremerton, Wash., and had the damage caused by the bomb hit repaired. Once whole again, she steamed to Hawaii and, after a brief sojourn at Pearl Harbor, headed toward the western Pacific on 12 July 1945. Wasp conducted a strike at Wake Island and paused briefly at Eniwetok before rejoining the rampaging Fast Carrier Task Force. In a series of strikes, unique in the almost complete absence of enemy airborne planes, Wasp pilots struck Yokosuka Naval Base near Tokyo, numerous airfields, and hidden manufacturing centers. On 9 August, a suicide plane swooped down at the carrier, but a Wasp pilot flying above the ship forced the enemy to splash into the sea. Then, on 15 August, when the fighting should have been over, two Japanese planes tried to attack Wasp's task group. Fortunately, Wasp pilots were still flying on combat air patrol and sent both enemies smoking into the sea. This was the last time Wasp pilots and gunners were to tangle with the Japanese.

On 25 August 1945, a severe typhoon, with winds reaching 78 knots, engulfed Wasp and stove in about 30 feet of her bow. The carrier, despite the hazardous job of flying from such a shortened deck, continued to launch her planes on missions of mercy or patrol as they carried food, medicine, and long-deserved luxuries to American prisoners of war at Narumi, near Nagoya. The ship returned to Boston for Navy Day, 27 October 1945. On 30 October, Wasp got underway for the naval shipyard in New York for a period of availability to have additional facilities installed for maximum transportation of troops. This work was completed on 15 November 1945 and enabled her to accommodate the larger, heavier, and faster planes of the jet age. Upon the completion of this conversion, the ship was recommissioned on 10 September 1951.

Wasp reported to the Atlantic Fleet in November 1951 and began a period of shakedown training, which lasted until February 1952. After returning from the shakedown cruise, she spent a month in the New York Naval Shipyard preparing for duty in distant waters. On 26 April 1952, Wasp collided with destroyer minesweeper USS Hobson (DD-464) while conducting night flying operations en route to Gibraltar. Hobson lost 176 of the crew, including her skipper. Rapid rescue operations saved 52 men. Wasp sustained no personnel casualties, but her bow was torn by a 75-foot saw-tooth rip.

The carrier proceeded to Bayonne, N.J., for repairs and, after she entered drydock there, the bow of aircraft carrier USS Hornet (CV 12), then undergoing conversion, was removed and floated by barge from Brooklyn, N.Y., and fitted into position on Wasp, replacing the badly shattered forward end of the ship. This remarkable task was completed in only 10 days, enabling the carrier to get underway to cross the Atlantic. On 2 June 1952, Wasp relieved USS Tarawa (CV-40) at Gibraltar and joined Carrier Division (CarDiv) 6 in the Mediterranean Sea. After conducting strenuous flight operations between goodwill visits to many Mediterranean ports, Wasp was relieved at Gibraltar on 5 September by USS Leyte (CV-32).

After taking part in NATO Exercise "Mainbrace" at Greenock, Scotland, and enjoying a liberty period at Plymouth, England, Wasp headed home and arrived at Norfolk early on the morning of 13 October 1952.

On 7 November 1952, Wasp entered the New York Naval Shipyard to commence a seven-month yard period to prepare her for a world cruise which was to bring her into the Pacific Fleet once more. After refresher training in the Caribbean, Wasp departed Norfolk on 16 September 1953.

After transiting the Panama Canal and crossing the Pacific, the carrier made a brief visit to Japan and then conducted strenuous operations with the famed TF 77. While operating in the western Pacific, she made port calls at Hong Kong, Manila, Yokosuka, and Sasebo.

On 10 January 1954, China's Generalissimo Chiang Kai-shek spent more than four hours on board Wasp watching simulated air war maneuvers in Formosan waters. On 12 March, President Ramon Magsaysay of the Republic of the Philippines came on board to observe air operations as a guest of American Ambassador Raymond A. Spruance. Wasp operated out of Subic Bay, Philippines, for a time, then sailed for Japan where, in April 1954, she was relieved by USS Boxer (CV-21) and sailed for her new home port of San Diego, Calif.

Wasp spent the next few months preparing for another tour of the Orient. She departed the United States in September 1954 and steamed to the Far East visiting Pearl Harbor and Iwo Jima en route. She relieved Boxer in October 1954 and engaged in air operations in the South China Sea with Carrier Task Group 70.2. Wasp visited the Philippine Islands in November and December and proceeded to Japan early in 1955 to join TF 77. While operating with that naval organization, Wasp provided air cover for the evacuation of the Tachen Islands by the Chinese Nationalists.

After the Tachen evacuation, Wasp stopped at Japan before returning to San Diego, Calif., in April 1955. She entered the San Francisco Naval Shipyard in May for a seven-month conversion and overhaul. On 1 December, the carrier returned to duty displaying a new canted flight deck and a hurricane bow. As 1955 ended, Wasp had returned to San Diego and was busily preparing for another Far Eastern tour.

After training during the early months of 1956, Wasp departed San Diego, Calif., on 23 April for another cruise to the Far East with Carrier Air Group 15 embarked. She stopped at Pearl Harbor to undergo inspection and training and then proceeded to Guam where she arrived in time for the Armed Forces Day ceremonies on 14 May. En route to Japan in May, she joined TF 77 for Operation Sea Horse, a five-day period of day and night training for the ship and air group. The ship arrived at Yokosuka on 4 June 1956; visited Iwakuni, Japan, then steamed to Manila for a brief visit. Following a drydock period at Yokosuka, Wasp again steamed south to Cubi Point, Philippine Islands, for the commissioning of the new naval air station there. Carrier Air Group 15 provided an air show for President Ramon Magsaysay of the Philippines and Adm. Arthur Radford.

During the third week of August, Wasp was at Yokosuka enjoying what was scheduled to be a fortnight's stay, but she sailed a week early to aid other ships in searching for survivors of a Navy patrol plane which had been shot down on 23 August 1956 off the coast of communist China. After a futile search, the ship proceeded to Kobe, Japan, and made a final stop at Yokosuka before leaving the Far East.

Wasp returned to San Diego on 15 October and while there was reclassified an antisubmarine warfare aircraft carrier, CVS-18, effective on 1 November 1956. She spent the last days of 1956 in San Diego preparing for her transfer to the east coast. Wasp left San Diego on the last day of January 1957, rounded Cape Horn for operations in the South Atlantic and Caribbean Sea, then proceeded to Boston where she arrived on 21 March. The carrier came into Norfolk, Va., on 6 April to embark members of her crew from the Antisubmarine Warfare School. The carrier spent the next few months in tactics along the eastern seaboard and in the waters off Bermuda before returning to Boston on 16 August.

On 3 September, Wasp got underway to participate in NATO Operations Seaspray and Strikeback, which took her to the coast of Scotland and simulated nuclear attacks and counterattacks on 130 different land bases. The carrier returned to Boston on 23 October 1957 and entered the Boston Naval Shipyard for a major overhaul which was not completed until 10 March 1958 when she sailed for antisubmarine warfare practice at Guantanamo Bay, Cuba. Upon returning to Boston on 29 April and picking up air squadrons at Quonset Point, R.I., on 12 May, she became the hub of TF 66, a special antisubmarine group of the 6th Fleet.

The carrier began her Atlantic crossing on the 12th of May and sailed only a few hundred miles when trouble flared in Lebanon. Wasp arrived at Gibraltar on 21 May 1958 and headed east, making stops at Souda Bay, Crete, Rhodes, and Athens. Wasp next spent 10 days at sea conducting a joint Italian-American antisubmarine warfare exercise in the Tyrrhenian Sea off Sardinia. On 15 July, the carrier put to sea to patrol waters off Lebanon. Her Marine helicopter transport squadron left the ship five days later to set up camp at the Beirut International

Airport. They flew reconnaissance missions and transported the sick and injured from Marine battalions in the hills to the evacuation hospital at the airport. She continued to support forces ashore in Lebanon until 17 September 1958 when she departed Beirut Harbor, bound for home. She reached Norfolk on 7 October, unloaded supplies, and then made a brief stop at Quonset Point before arriving in her home port of Boston on 11 October.

Four days later, Wasp became flagship of Task Group Bravo, one of two new antisubmarine defense groups formed by the Commander in Chief of the Atlantic Fleet. Wasp's air squadrons and seven destroyers were supported by shore-based seaplane patrol aircraft. She sailed from Quonset Point on 26 November for a 17-day cruise in the North Atlantic. This at-sea period marked the first time her force operated together as a team. The operations continued day and night to coordinate and develop the task group's team capabilities until she returned to Boston on 13 December 1958 and remained over the Christmas holiday season. Wasp operated with Task Group Bravo throughout 1959, cruising along the eastern seaboard conducting operations at Norfolk, Va., Bermuda, and Quonset Point, R.I. On 27 February 1960, she entered the Boston Naval Shipyard for overhaul. In mid-July, the carrier was ordered to the South Atlantic where she stood by when civil strife broke out in the newly independent Congo and operated in support of the United Nations airlift. She returned to her home port on 11 August 1960 and spent the remainder of the year operating out of Boston with visits to Guantanamo Bay, Cuba, for refresher training and exercises conducted in the Virginia capes operating areas and the Caribbean operating areas. The carrier returned to Boston on 10 December 1960 and remained in port there into the New Year.

On 9 January 1961, Wasp sailed for the Virginia capes operating area and devoted the first half of 1961 to exercises there, at Narragansett Bay, R.I., and at Nova Scotia. On 9 June, Wasp got underway from Norfolk, Va., for a three-month Mediterranean cruise. The ship conducted exercises at Augusta Bay, Sicily, Barcelona, Spain; San Remo and La Spezia, Italy, Aranci Bay, Sardinia; Genoa, Italy, and Cannes, France, and returned to Boston on 1 September 1961.

The carrier entered the Boston Naval Shipyard for an interim overhaul and resumed operations on 6 November 1961. After loading food, clothing, and equipment, Wasp spent the period from 11 to 18 January 1962 conducting antisubmarine warfare exercises and submarine surveillance off the east coast. After a brief stop at Norfolk, the ship steamed on to further training exercises and anchored off Bermuda from 24 to 31 January. Wasp then returned to her home port.

On 17 February, a delegation from the Plymouth Plantation presented a photograph of the Mayflower II to Captain Brewer, who accepted this gift for Wasp's "People to People" effort in the forthcoming European cruise. On 18 February, Wasp departed Boston, bound for England, and arrived at Portsmouth on 1 March. On 16 March 1962, the carrier arrived at Rotterdam, Netherlands, for a week's goodwill visit.

From 22 to 30 March, Wasp traveled to Greenock, Scotland, and then to Plymouth, England. On 17 April, Capt. Brewer presented Alderman A. Goldberg, Lord Mayor of Plymouth, England, with a large picture of Mayflower II as a gift from the people of Plymouth, Mass. On 5 May, Wasp arrived at Kiel, West Germany, and became the first aircraft carrier to ever visit that port. The ship made calls at Oslo, Norway, Reykjavik, Iceland, and Argentia, Newfoundland, before returning to Boston, Mass., on 16 June 1962.

From August through October, Wasp visited Newport, R.I., New York, and Earle, N.J., then conducted a dependents' cruise, as well as a reserve cruise, and visitors' cruises. The 1st of November gave Wasp a chance to use her capabilities when she responded to a call from President Kennedy and actively participated in the Cuban blockade. After tension relaxed, the carrier returned to Boston on 22 November for upkeep work, and, on 21 December, she sailed to Bermuda with 18 midshipmen from Boston-area universities. Wasp returned to Boston on 29 December and finished out the year there. The early part of 1963 saw Wasp conducting antisubmarine warfare exercises off the Virginia capes and steaming along the Caribbean coast of Costa Rica in support of the presidential visit. On 21 March, President Kennedy arrived at San Jose for a conference with presidents of six Central American nations. After taking part in Fleet exercises off Puerto Rico, the carrier returned to Boston on 4 April. From 11 to 18 May, Wasp took station off Bermuda as a backup recovery ship for Major Gordon Cooper's historic Mercury space capsule recovery. The landing occurred as planned in the mid-Pacific near Midway Atoll, and carrier USS Kearsage (CVA 33) picked up Cooper and his Faith 7 spacecraft. Wasp then resumed antisubmarine warfare exercises along the Atlantic seaboard and in the Caribbean until she underwent overhaul in the fall of 1963 for FRAM (Fleet Rehabilitation and Modernization) overhaul in the Boston Naval Shipyard. In March 1964, the carrier conducted sea trials out of Boston. During April, she operated out of Norfolk and Narragansett Bay, R.I. She returned to Boston on 4 May and remained there until 14 May when she got underway for refresher training in waters between Guantanamo Bay, Cuba, and Kingston, Jamaica, before returning to her home port on 3 June 1964.

On 21 July 1964, Wasp began a round-trip voyage to Norfolk and returned to Boston on 7 August. She remained there through 8 September when she headed, via the Virginia capes operating area, to Valencia, Spain. She then cruised the Mediterranean, visiting ports in Spain, France, and Italy, and returned home on 18 December 1964. The carrier remained in port until 8 February 1965 and sailed for fleet exercises in the Caribbean. Operating along the eastern seaboard, she recovered the Gemini IV astronauts White and McDivitt with their spacecraft on 7 June. During the summer, the ship conducted search and rescue operations for an Air Force C-121 plane that had gone down off Nantucket. Following an orientation cruise for 12 congressmen on 20 to 21 August, Wasp participated in joint training exercises with German and French forces. From 16 to 18 December, the carrier recovered the astronauts of Gemini VI and VII, and then returned to Boston on 22 December to finish out the year.

On 24 January 1966, Wasp departed Boston for fleet exercises off Puerto Rico. En route, heavy seas and high winds caused structural damage to the carrier. She put into Roosevelt Roads, Puerto Rico, on 1 February to determine the extent of her damages and effect as much repair as possible. Engineers were flown from Boston who decided that the ship could cease "Springboard" operations early and return to Boston. The ship conducted limited antisubmarine operations from 6 to 8 February prior to leaving the area. She arrived at Boston on 18 February and was placed in restricted availability until 7 March, when her repair work was completed.

Wasp joined in exercises in the Narragansett Bay operating areas. While the carrier was carrying out this duty, a television film crew from the National Broadcasting Company was flown to Wasp on 21 March and stayed on the ship during the remainder of her period at sea, filming material for a special color television show to be presented on Armed Forces Day.

The carrier returned to Boston on 24 March 1966 and was moored there until 11 April. On 27 March, Doctor Ernst Lemberger, the Austrian Ambassador to the United States, visited the ship. On 18 April, the ship embarked several guests of the Secretary of the Navy and set courses for Guantanamo Bay, Cuba. She returned to Boston on 6 May. A week later, the veteran flattop sailed to take part in the recovery of the Gemini IX spacecraft. Embarked in Wasp were some 66 persons from NASA, the television industry, media personnel, an underwater demolition recovery team, and a Defense Department medical team. On 6 June 1966, she recovered astronauts Lt. Col. Thomas P. Stafford and Lt. Comdr. Eugene Cernan and flew them to Cape Kennedy, Fla. Wasp returned their capsule to Boston.

Wasp participated in "ASWEX III," an antisubmarine exercise that lasted from 20 June through 1 July 1966. She spent the next 25 days in port at Boston for upkeep. On the 25th, the carrier got underway for "ASWEX IV." During this exercise, the Soviet intelligence collection vessel, Agi Traverz, entered the operation area, necessitating a suspension of the operation and eventual repositioning of forces. The exercise was terminated on 5 August. She then conducted a dependents' day cruise on 8 and 9 August, and orientation cruises on 10, 11, and 22 August 1966. After a two-day visit to New York, Wasp arrived in Boston on 1 September and underwent upkeep until the 19th. From that day to 4 October, she conducted hunter/killer operations with the Royal Canadian Navy aircraft embarked.

Following upkeep at Boston, the ship participated in the Gemini XII recovery operation from 5 to 18 November 1966. The recovery took place on 15 November when the space capsule splashdown occurred within three miles of Wasp. Capt. James A. Lovell and Maj. Edwin E. Aldrin were lifted by helicopter hoist to the deck of Wasp and there enjoyed two days of celebration. Wasp arrived at Boston on 18 November with the Gemini XII spacecraft on board. After off-loading the special Gemini support equipment, Wasp spent ten days making ready for her next period at sea. On 28 November, Wasp departed Boston to take part in the Atlantic Fleet's largest exercise of the year, "Lantflex-66," in which more than 100 United States ships took part. The carrier returned to Boston on 16 December where she remained through the end of 1966.

Wasp served as a carrier qualification duty ship for the Naval Air Training Command from 24 January to 26 February 1967 and conducted operations in the Gulf of Mexico and off the east coast of Florida. She called at New Orleans for Mardi Gras from 4 to 8 February, at Pensacola on the 11th and 12th, and at Mayport, Fla., on the 19th and 20th. Returning to Boston a week later, she remained in port until 19 March, when she sailed for "Springboard" operations in the Caribbean. On 24 March, Wasp joined USS Salamonie (AO-26) for an underway replenishment but suffered damage during a collision with the oiler. After making repairs at Roosevelt Roads, she returned to operations on 29 March and visited Charlotte Amalie, St. Thomas, United States Virgin Islands, and participated in the celebration from 30 March to 2 April which marked the 50th anniversary of the purchase of the Virgin Islands by the United States from Denmark. Wasp returned to Boston on 7 April, remained in port four days, then sailed to Earle, N.J., to off-load ammunition prior to overhaul. She visited New York for three days then returned to the Boston Naval Shipyard and began an overhaul on 21 April 1967 which was not completed until early 1968.

Wasp completed her cyclical overhaul and conducted post-repair trials throughout January 1968. Returning to the Boston Naval Shipyard on the 28th, the ship made ready for two months of technical evaluation and training which began early in February.

The 28th of February marked the beginning of almost five weeks of refresher training for Wasp under the operational control of Commander, Fleet Training Group, Guantanamo Bay, Cuba. On 30 March, Wasp steamed north and was in Boston from 6 to 29 April 1968 for routine upkeep and minor repairs. She then departed for operations in the Bahamas and took part in "Fixwex C," an exercise off the Bermuda coast. The carrier set course for home on 20 May but left five days later to conduct carrier qualifications for students of the Naval Air Training Command in the Jacksonville, Fla., operations area.

On 12 June 1968, Wasp and USS Truckee (AO 147) had a minor collision during an underway replenishment. The carrier returned to Norfolk, where an investigation into the circumstances of the collision was conducted. On 20 June, Wasp got underway for Boston, where she remained until 3 August when she moved to Norfolk to take on ammunition.

On 15 June, Wasp's home port was changed to Quonset Point, R.I., and she arrived there on 10 August to prepare for overseas movement. Ten days later, the carrier got underway for a deployment in European waters. The northern European portion of the cruise consisted of several operational periods and port visits to Portsmouth, England; Firth of Clyde, Scotland; Hamburg, Germany; and Lisbon, Portugal. Wasp, as part of TG 87.1, joined in the NATO Exercise Silvertower, the largest combined naval exercise in four years. Silvertower brought together surface, air, and subsurface units of several NATO navies.

On 25 October 1968, the carrier entered the Mediterranean and, the following day, became part of TG 67.6. After a port visit to Naples, Italy, Wasp departed on 7 November to conduct antisubmarine warfare exercises in the Tyrrhenian Sea, Levantine Basin, and Ionian Basin. After loading aircraft in both Taranto and Naples, Italy, Wasp visited Barcelona, Spain, and Gibraltar. On 19 December, the ship returned to Quonset Point, R.I., and spent the remainder of 1968 in port.

Wasp began 1969 in her home port of Quonset Point. Following a yard period which lasted from 10 January through 17 February, the carrier conducted exercises as part of the White Task Group in the Bermuda operating area. The ship returned to Quonset Point on 6 March and began a month of preparations for overseas movement. On 1 April 1969, Wasp sailed for the eastern Atlantic and arrived at Lisbon, Portugal, on 16 April. From 21 to 26 April, she took part in joint Exercise Trilant, which was held with the navies of the United States, Spain, and Portugal. One of the highlights of the cruise occurred on 15 May as Wasp arrived at Portsmouth, England, and served as flagship for TF 87, representing the United States in a NATO review by Queen Elizabeth and Prince Philip in which 64 ships from the 11 NATO countries participated. After conducting exercises and visiting Rotterdam, Oslo, and Copenhagen, Wasp headed home on 30 June 1969 and, but for a one-day United Fund cruise on 12 August, remained at Quonset Point until 24 August. The period from 29 August to 6 October was devoted to alternating operations between Corpus Christi, Tex., for advanced carrier qualifications, and Pensacola for basic qualifications, with in-port periods at Pensacola.

A period of restricted availability began on 10 October and was followed by operations in the Virginia capes area until 22 November. In December, Wasp conducted a carrier qualification mission in the Jacksonville operations area which lasted through 10 December. The ship arrived back at Quonset Point on 13 December and remained there for the holidays.

The carrier welcomed the year 1970 moored in her home port of Quonset Point but traveled over 40,000 miles and was

away from home port 265 days. On 4 January, she proceeded to Earle, N.J., and off-loaded ammunition prior to entering the Boston Naval Shipyard for a six-week overhaul on 9 January.

The carrier began a three-week shakedown cruise on 16 March but returned to her home port on 3 April and began preparing for an eastern Atlantic deployment. Wasp reached Lisbon on 25 May 1970 and dropped anchor in the Tagus River. A week later, the carrier got underway to participate in NATO Exercise Night Patrol with units from Canada, the Netherlands, Portugal, the United Kingdom, and West Germany. On 8 June, Wasp proceeded to the Naval Station, Rota, Spain, to embark a group of midshipmen for a cruise to Copenhagen. During exercises in Scandinavian waters, the carrier was shadowed by Soviet naval craft and aircraft. The ship departed Copenhagen on 26 June and, three days later, crossed the Arctic Circle. On 13 July 1970, Wasp arrived at Hamburg, Germany, and enjoyed the warmest welcome received in any port of the cruise. A Visitors' Day was held, and over 15,000 Germans were recorded as visitors to the carrier. After calls at Edinburgh and Glasgow, Scotland, Wasp got underway on 10 August for operating areas in the Norwegian Sea. The carrier anchored near Plymouth, England, on 28 August and, two days later, sailed for her home port.

Wasp returned to Quonset Point on 8 September and remained there through 11 October when she got underway to off-load ammunition at Earle, N.J., prior to a period of restricted availability at the Boston Naval Shipyard beginning on 15 October. The work ended on 14 December; and, after reloading ammunition at Earle, Wasp returned to Quonset Point on 19 December to finish out the year 1970.

On 14 January 1971, Wasp departed Quonset Point, R.I., with Commander, ASWGRU 2, CVSG-54, and Detachment 18 from Fleet Training Group, Guantanamo Bay, Cuba, embarked. After refresher training at Bermuda, she stopped briefly at Rota, Spain, then proceeded to the Mediterranean for participation in the "National Week VIII" exercises with several destroyers for the investigation of known Soviet submarine operating areas. On 12 February, Secretary of the Navy John Chafee, accompanied by Commander, 6th Fleet, Vice Adm. Isaac C. Kidd, visited the carrier.

Wasp detached early from the "National Week" exercise on 15 February to support USS John F. Kennedy (CVA-67) as she steamed toward Gibraltar. Soviet ships trailed Wasp and John F. Kennedy until they entered the Strait of Sicily when the Soviets departed to the east. After a brief stop at Barcelona, Spain, Wasp began her homeward journey on 24 February and arrived at Quonset Point on 3 March 1971.

After spending March and April in port, Wasp got underway on 27 April and conducted a nuclear technical proficiency inspection and prepared for the forthcoming Exotic Dancer exercise which commenced on 3 May. Having successfully completed the week-long exercise, Wasp was heading home on 8 May when an American Broadcasting Co. television team embarked and filmed a short news report on carrier antisubmarine warfare operations.

On 15 May 1971, the veteran conducted a dependents' day cruise, and one month later, participated in Exercise Rough Ride at Great Sound, Bermuda, which took her to Halifax, Nova Scotia. Wasp returned to Quonset Point on 2 July 1971 and spent the next two months in preparation and execution of Exercise Squeeze Play IX in the Bermuda operating area. During August, the ship conducted exercises with an east coast naval reserve air group while proceeding to Mayport, Fla. She returned to her home port on 26 August and spent the next month there. On 23 September, Wasp got underway for Exercise Lantcortex 1-72 which terminated on 6 October. For the remainder of the month, the carrier joined in a crossdeck operation which took her to Bermuda, Mayport, and Norfolk. She arrived back at Quonset Point on 4 November 1971.

Four days later, the carrier set her course for the Newport News Shipbuilding and Drydock Co. where she was in drydock until 22 November. She then returned to Quonset Point and remained in her home port for the remainder of the year preparing for decommissioning.

On 1 March 1972, it was announced that Wasp would be decommissioned and stricken from the Navy list. Decommissioning ceremonies were held on 1 July 1972. The ship was sold on 21 May 1973 to the Union Minerals and Alloys Corp., of New York City, and subsequently scrapped.

Wasp earned eight battle stars for her World War II service.

CV-19 • USS HANCOCK

(Displacement: 27,100 tons; Length: 888 feet; Beam: 93 feet; extreme width at flight deck: 147½ feet; Draft: 28 feet 7 inches; Speed: 33 knots; Complement: 3,448 crew; Armament: 12 5-inch guns, 44 40mm guns, 59 20mm guns; Aircraft: 80+; Class: Essex.) The fourth Hancock (CV-19) was laid down as Ticonderoga on 26 January 1943 by the Bethlehem Steel Co., Quincy, Mass.; renamed Hancock on 1 May 1943, launched on 24 January 1944; sponsored by Mrs. DeWitt C. Ramsey, wife of Rear Adm. Ramsey, Chief of the Bureau of Aeronautics; and commissioned on 15 April 1944, with Captain Fred C. Dickey in command.

After fitting out in the Boston Navy Yard and shakedown training off Trinidad and Venezuela, Hancock returned to Boston for alterations on 9 July. She departed Boston on 31 July 1944 en route to Pearl Harbor via the Panama Canal and San Diego, and from there sailed on 24 September to join Adm. W. F. Halsey's Third Fleet at Ulithi on 5 October. She was assigned to Rear Adm. Bogan's Carrier Task Group 38.2.

Hancock got underway the following afternoon for a rendezvous point 375 miles west of the Marianas where units of Vice Adm. Mitscher's Fast Carrier Task Force 38 were assembling in preparation for the daring cruise to raid Japanese air and sea bases in the Ryukyus, Formosa, and the Philippines. Thus, enemy air power was paralyzed during General MacArthur's invasion of Leyte. When the armada arrived off the Ryukyu Islands on 10 October 1944, Hancock's planes rose off her deck to wreak destruction upon Okinawan airfields and shipping. Her planes destroyed seven enemy aircraft on the ground and assisted in the destruction of a submarine tender, 12 torpedo boats, two midget submarines, four cargo ships, and several sampans. Next on the agenda were Formosan air bases, where on 12 October Hancock's pilots downed six enemy planes and destroyed nine more on the ground. She also reported one cargo ship definitely sunk, three probably destroyed, and several others damaged.

As they repelled an enemy air raid that evening, Hancock's gunners accounted for a Japanese plane and drove countless others off during seven hours of uninterrupted general quarters. The following morning her planes resumed their assault, knocking out ammunition dumps, hangars, barracks, and industrial plants ashore and damaging an enemy transport.

As Japanese planes again attacked the Americans during their second night off Formosa, Hancock's antiaircraft fire brought down another raider which splashed about 500 yards off her flight deck. On the morning of the third day of operations against this enemy stronghold, Hancock lashed out again at airfields and shipping before retiring to the southeast with her task force. As the American ships withdrew, a heavy force of Japanese aircraft roared in for a parting crack. One dropped a bomb off Hancock's port bow a few seconds before the carrier's guns splashed the attacker into the sea. Another bomb penetrated a gun platform but exploded harmlessly in the water. The surviving attackers then turned tail, and the task force was thereafter unmolested as they sailed toward the Philippines to support the landings at Leyte.

On 18 October 1944, she launched planes against airfields and shipping at Laoag, Aparri, and Camiguin Island in Northern Luzon. Her planes struck the islands of Cebu, Panay, Negros, and Masbate, pounding enemy airfields and shipping. The next day she retired toward Ulithi with Vice Admiral John S. McCain's Carrier Task Group 38.1. She received orders on 23 October to turn back to the area off Samar to assist in the search for units of the Japanese fleet reportedly closing on Leyte to challenge the American fleet and destroy amphibious forces that were struggling to take the island from Japan. Hancock did not reach Samar in time to assist the heroic escort carriers and destroyers of "Taffy 3" during the main action of the Battle off Samar, but her planes did manage to lash the fleeing Japanese Center Force as it passed through the San Bernardino Straits. Hancock then rejoined Rear Adm. Bogan's Task Group with which she struck airfields and shipping in the vicinity of Manila on 29 October 1944. During operations through 19 November, her planes gave direct support to advancing Army troops and attacked Japanese shipping over a 350-mile area. She became the flagship of the Fast Carrier Task Force 38 on 17 November 1944 when Vice Adm. McCain came on board.

Unfavorable weather prevented operations until 25 November, when an enemy aircraft roared toward Hancock in a suicide dive out of the sun. Antiaircraft fire exploded the plane some 300 feet above the ship, but a section of its fuselage landed amidships and a part of the wing hit the flight deck and burst into flames. Prompt and skillful teamwork quickly extinguished the blaze and prevented serious damage. Hancock returned to Ulithi on 27 November 1944 and departed from that island with her task group to maintain air patrol over enemy airfields on Luzon to prevent enemy suicide attacks on amphibious vessels of the landing force in Mindoro. The first strikes were launched on 14 December against Clark and Angeles Airfields as well as

enemy ground targets on Salvador Island. The next day her planes struck installations at Masinloc, San Fernando, and Cabatuan, while fighter patrols kept the Japanese airmen down. Her planes also attacked shipping in Manila Bay.

Hancock encountered a severe typhoon on 17 December and rode out the storm in waves which broke over her flight deck, some 55 feet above her waterline. She put into Ulithi on 24 December and got underway six days later to attack airfields and shipping around the South China Sea. Her planes struck hard blows at Luzon airfields on 7 and 8 January 1945 and turned their attention back to Formosa on 9 January, hitting fiercely at airfields and the Tokyo Seaplane Station. An enemy convoy north of Camranh Bay, Indochina, was the next victim with two ships sunk and 11 damaged. That afternoon Hancock launched strikes against airfields at Saigon and shipping on the northeastern bulge of French Indochina. Strikes by the fast and mobile carrier force continued through 16 January, hitting Hainan Island in the Gulf of Tonkin, the Pescadores Islands, and shipping in the harbor of Hong Kong. Raids against Formosa were resumed on 20 January 1945. The next afternoon one of her planes returning from a sortie made a normal landing, taxied to a point abreast of the island, and disintegrated in a blinding explosion which killed 50 men and injured 75 others. Again, outstanding work quickly brought the fires under control in time to land other planes which were still aloft. She returned to formation and launched strikes against Okinawa the next morning. Hancock reached Ulithi on 25 January 1945 where Vice Adm. McCain left the ship and relinquished command of the 5th Fleet. She sortied with the ships of her task group on 10 February and launched strikes against airfields in the vicinity of Tokyo on 16 February. During that day her air group downed 71 enemy planes, and accounted for 12 more the next. Her planes hit the enemy naval bases at Chichi Jima and Haha Jima on 19 February. These raids were conducted to isolate Iwo Jima from air and sea support when Marines hit the beaches of that island to begin one of the most bloody and fierce campaigns of the war. Hancock took station off this island to provide tactical support through 22 February, hitting enemy airfields and strafing Japanese troops ashore. Returning to waters off the enemy home islands, Hancock launched her planes against targets on northern Honshu, making a diversionary raid on the Nansei-shoto islands on 1 March before returning to Ulithi on 4 March.

Back in Japanese waters, Hancock joined other carriers in strikes against Kyushu airfields, southwestern Honshu, and shipping in the Inland Sea of Japan, on 18 March 1945. Hancock was refueling the destroyer USS Halsey Powell (DD 686) on 20 March when suicide planes attacked the task force. One plane dove for the two ships but was disintegrated by gunfire when about 700 feet overhead. Fragments of the plane hit Hancock's deck while its engine and bomb crashed on the fantail of the destroyer. Hancock's gunners shot down another plane as it neared the release point of its bombing run on the carrier. Hancock was reassigned to Carrier Task Group 58.3 with which she struck the Nansei-shoto islands from 23 through 27 March and Minami Daito Jima and Kyushu at the end of the month.

When the 10th Army landed on the western coast of Okinawa on 1 April, Hancock was on hand to provide close air support. A suicide plane cartwheeled across her flight deck on 7 April and crashed into a group of planes while its bomb hit the port catapult to cause a tremendous explosion. Although 62 men were killed and 71 wounded, heroic efforts doused the fires within half an hour, enabling her to be back in action before an hour had passed.

Hancock was detached from her task group on 9 April 1945 and steamed to Pearl Harbor for repairs. She sailed back into action on 13 June and left lethal calling cards at Wake Island on 20 June en route to the Philippines. Hancock sailed from San Pedro Bay with the other carriers on 1 July and attacked Tokyo airfields on 10 July. She continued to operate in Japanese waters until she received confirmation of Japan's capitulation on 15 August 1945, when she recalled her planes from their deadly missions before they reached their targets. However, planes of her photo division were attacked by seven enemy aircraft over Sagami Wan. Three were shot down, and a fourth escaped in a trail of smoke. Later that afternoon, planes of Hancock's air patrol shot down a Japanese torpedo plane as it dived on a British task force. Her planes flew missions over Japan in search of prison camps, dropping supplies and medicine on 25 August. Information collected during these flights led to landings under the command of Commodore R. W. Simpson, which brought doctors and supplies to all Allied prisoner of war encampments.

When the formal surrender of the Japanese Imperial Government was signed on board the battleship USS Missouri, Hancock's planes flew overhead. The carrier entered Tokyo Bay on 10 September 1945 and sailed on 30 September, embarking 1,500 passengers at Okinawa for transportation to San Pedro, California, where she arrived on 21 October. Hancock was fitted out for "Magic Carpet" duty at San Pedro and sailed for Seeadler Harbor, Manus, Admiralty Islands, on 2 November. On her return voyage, she carried 4,000 passengers who were debarked at San Diego on 4 December. A week later, Hancock departed for her second "Magic Carpet" voyage, embarking 3,773 passengers at Manila for return to Alameda, California, on 20 January 1946. She embarked Air Group 7 at San Diego on 18 February for air operations off the coast of California. She sailed from San Diego on 11 March to embark men of two air groups and aircraft at Pearl Harbor for transportation to Saipan, arriving on 1 April 1946. After receiving two other air groups on board at Saipan, she loaded a cargo of aircraft at Guam and steamed by way of Pearl Harbor to Alameda, California, arriving on 23 April 1946. She then steamed to Seattle, Washington, on 29 April to await inactivation. The proud ship decommissioned and entered the reserve fleet at Bremerton, Washington.

Hancock commenced conversion and modernization to an attack aircraft carrier in Puget Sound on 15 December 1951 and was reclassified CVA-19 on 1 October 1952. She recommissioned on 15 February 1954, with Captain W. S. Butts in command. She was the first carrier of the United States Fleet with steam catapults capable of launching high-performance jets.

She was off San Diego on 7 May 1954 for operations along the coast of California that included the launching on 17 June of the first aircraft to take off from a United States carrier by means of a steam catapult. After a year of operations along the Pacific

coast that included testing of Sparrow I and Regulus missiles and Cutlass jet aircraft, she sailed on 10 August 1955 for 7th Fleet operations ranging from the shores of Japan to the Philippines and Okinawa. She returned to San Diego on 15 March 1956 and decommissioned on 13 April for conversion that included the installation of an angled flight deck.

Hancock recommissioned on 15 November 1956 for training out of San Diego until 6 April 1957 when she again sailed for Hawaii and the Far East. She returned to San Diego on 18 September 1957 and again departed for Japan on 15 February 1958. She was a unit of powerful carrier task groups taking station off Taiwan when the Nationalist Chinese islands of Quemoy and Matsu were threatened with Communist invasion in August 1958. The carrier returned to San Diego on 2 October 1958 for overhaul in the San Francisco Naval Shipyard, followed by rigorous at-sea training out of San Diego. On 1 August 1959, she sailed to reinforce the 7th Fleet as troubles in Laos demanded the watchful presence of powerful American forces in waters off Southeast Asia. She returned to San Francisco on 18 January 1960 and put to sea early in February to participate in a new demonstration of communications by reflecting ultra-high-frequency waves off the moon. She again departed in August to steam with the 7th Fleet in waters off Laos until the lessening of tension in that area permitted operations ranging from Japan to the Philippines.

Hancock returned to San Francisco in March 1961, then entered the Puget Sound Naval Shipyard for an overhaul that gave her new electronics gear and many other improvements. She again set sail for Far Eastern waters on 2 February 1962, patrolling in the South China Sea as crisis and strife mounted both in Laos and in South Vietnam. She again appeared off Quemoy and Matsu in June 1962 to stem a threatened Communist invasion there, then trained along the coast of Japan and in waters reaching to Okinawa. She returned to San Francisco on 7 October 1962, made a brief cruise to the coast of Hawaii while qualifying pilots, then again sailed on 7 June 1963 for the Far East.

Hancock joined in combined defense exercises along the coast of South Korea, then deployed off the coast of South Vietnam after the coup which resulted in the death of President Diem. She entered the Hunter's Point Naval Shipyard on 16 January 1964 for modernization that included the installation of a new ordnance system, hull repairs, and aluminum decking for her flight deck. She celebrated her 20th birthday on 2 June 1964 while visiting San Diego. The carrier made a training cruise to Hawaii, then departed Alameda on 21 October 1964 for another tour of duty with the 7th Fleet in the Far East. Hancock reached Japan on 19 November and soon was on patrol at Yankee Station in the Gulf of Tonkin. She remained active in Vietnamese waters fighting to thwart Communist aggression until heading for home early in the spring of 1965.

November found the carrier steaming back to the war zone. She was on patrol off Vietnam on 16 December 1965; and, but for brief respites at Hong Kong, the Philippines, or Japan, Hancock remained on station launching her planes for strikes at enemy positions ashore until returning to Alameda, California, on 1 August 1966. Her outstanding record during this combat tour won her the Navy Unit Commendation. Following operations off the west coast, Hancock returned to Vietnam early in 1967 and resumed her strikes against Communist positions. After fighting during most of the first half of 1967, she returned to Alameda on 22 July and promptly began preparations for returning to battle.

Aircraft from Hancock, along with those from USS Ranger (CV 61) and USS Oriskany (CV 34), joined with other planes for air strikes against North Vietnamese missile and antiaircraft sites south of the 19th parallel in response to attacks on unarmed U. S. reconnaissance aircraft on 21-22 November 1970. Hancock alternated with Ranger and with USS Kitty Hawk (CV 63) on Yankee Station until 10 May 1971 when she was relieved by USS Midway (CV 41).

Hancock, along with USS Coral Sea (CV 43), was back on Yankee Station by 30 March 1972 when North Vietnam invaded South Vietnam. In response to the invasion, naval aircraft from Hancock and other carriers flew tactical sorties during Operation Freedom Train against military and logistics targets in the southern part of North Vietnam. By the end of April, the strikes covered more areas in North Vietnam throughout the area below 20° 25'N. Between 25 and 30 April, aircraft from Hancock's VA-55, VA-164, and VA-211 struck enemy-held territory around Kontum and Pleiku. Hancock was again deployed to the waters off South Vietnam in 1975. Departing Subic Bay, Philippines, on 23 March, she, along with the carriers Coral Sea, Midway, USS Enterprise (CVN 65), and the amphibious assault ship USS Okinawa (LPH 3), stood by for the possible evacuation of refugees after North Vietnam overran two-thirds of the south. Nearly 9,000 were evacuated: 1,373 U. S. personnel and 6,422 of other nationalities. On 12-14 May, she was alerted, although not utilized, for the recovery of SS Mayaguez, a U. S. merchantman with 39 crew, seized in international waters on 12 May by the Communist Khmer Rouge.

Hancock was decommissioned on 30 January 1976. She was stricken from the Navy list the following day and sold for scrap by the Defense Reutilization and Marketing Service (DRMS) on 1 September 1976.

Hancock was awarded the Navy Unit Commendation and received four battle stars for service in World War II.

CV-20 • USS BENNINGTON

(Displacement: 27,100 tons length: 872 feet beam: 93 feet; extreme width at flight deck: 147½ feet draft: 28 feet 7 inches speed: 32.7 knots complement: 3,448 crew armament: 12 5-inch guns class: Essex.)

The second Bennington (CV-20) was launched on 28 February 1944 by New York Navy Yard, sponsored by Mrs. Melvin J. Maas, wife of Congressman Maas of Minnesota; and commissioned on 6 August 1944, with Captain J. B. Sykes in command.

On 15 December, Bennington got underway from New York and transited the Panama Canal on the 21st. The carrier arrived at Pearl Harbor on 8 January 1945 and then proceeded to Ulithi Atoll, Caroline Islands, where she joined TG 58.1 on 8 February. Operating out of Ulithi, she took part in strikes against the Japanese home islands (16-17 and 25 February), Volcano Islands (18 February-4 March), Okinawa (1 March), and the raids in support of the Okinawa campaign (18 March-11 June). On 7 April, Bennington's planes participated in the attacks on the Japanese task force moving through the East China Sea toward Okinawa, resulting in the sinking of the Japanese battleship Yamato, light cruiser Yahagi, and four destroyers. On 5 June, the carrier was damaged by a typhoon off Okinawa and retired to Leyte for repairs, arriving on 12 June. Her repairs completed, Bennington left Leyte on 1 July and during 10 July-15 August took part in the final raids on the Japanese home islands.

She continued operations in the western Pacific, supporting the occupation of Japan until 21 October 1945. On 2 September, her planes participated in the mass flight over USS Missouri (BB-63) and Tokyo during the surrender ceremonies.

Bennington arrived at San Francisco on 7 November 1945 and early in March 1946 transited the Panama Canal en route to Norfolk. Following a pre-inactivation overhaul, she went out of commission in reserve at Norfolk on 8 November 1946.

The carrier began modernization at New York Naval Shipyard on 30 October 1950 and was recommissioned on 13 November 1952. Her shakedown lasted until May 1953, when she returned to Norfolk for final fleet preparations. Between 14 May 1953 and 27 May 1954, she operated along the eastern seaboard; made a midshipman cruise to Halifax, Nova Scotia; and a cruise in the Mediterranean. At 0811 on 28 May 1954, while cruising off Narragansett Bay, the fluid in one of her catapults exploded, setting off a series of secondary explosions which killed 103 crewmen and injured 201 others. Bennington proceeded under her own power to Quonset Point, R.I., to land her injured.

Moving to New York Naval Shipyard for repairs, she was completely rebuilt from 12 June 1954 to 19 March 1955. On 22 April 1955, the Secretary of the Navy came aboard and presented medals and letters of commendation to 178 of her crew in recognition of their heroism on 26 May 1954. Bennington served as a platform for innovations in Naval Aviation. On 22 August 1955, operational testing of the mirror landing system installed on the ship was begun by VX-3. Commanding officer Cmdr. Robert G. Dose, flying an FJ-3 Fury, made the first landing with the device. Bennington stayed with the Atlantic Fleet until departing Mayport, Fla., on 8 September 1955 for the Pacific. She steamed by way of Cape Horn and arrived at San Diego one month later.

On 18 May 1966, the XC-142A tri-service V/STOL transport made its first carrier takeoffs and landings during tests conducted aboard Bennington at sea off San Diego. The tests, including 44 short and six vertical takeoffs, were made with wind over the deck varying from zero to 32 knots. Lt. Roger L. Rich Jr., and other pilots from the Navy, Marine Corps, and Army took turns at the controls.

Bennington was also involved in the U.S. space program, recovering the unmanned Apollo IV spacecraft about 600 miles northwest of Hawaii, after its 8 1/2-hour orbital flight on 9 November 1967.

Bennington was decommissioned on 15 January 1970. She was stricken from the Navy List in 1989 and sold on 1 December 1994 by the Defense Re-utilization and Marketing Service (DRMS) for scrapping.

CV-21 • USS BOXER

(Displacement: 27,100 tons length: 888 feet beam: width at flight deck: 147½ feet draft: 28 feet 7 inches speed: 33 knots complement: 3,448 crew armament: 12 5-inch guns class: Essex.)

The fifth Boxer (CV-21) was launched on 14 December 1944 by Newport News Shipbuilding and Dry Dock Co., Newport News, Va.; sponsored by Miss Ruth D. Overton, daughter of the Senator from Louisiana, and commissioned on 16 April 1945, with Captain D. F. Smith in command.

Completed too late to take part in World War II, Boxer joined the Pacific Fleet at San Diego in August 1945. From September 1945 to 23 August 1946, she operated out of Guam as the flagship of TF 77 in the western Pacific. During this tour, she visited Japan, Okinawa, the Philippines, and China. She returned to San Francisco on 10 September 1946 and operated off the west coast, engaged in normal peacetime duty until departing for the Far East on 11 January 1950. After service with the 7th Fleet in the Far East during the first half of 1950, she returned to San Diego, arriving on 25 June.

With the outbreak of the Korean conflict, she was pressed into service to carry planes to the fighting. On 23 July 1950, she completed a record crossing of the Pacific from Alameda, Calif., to Yokosuka, Japan, in 8 1/2 days, carrying 145 P-51 Mustang and six L-5 aircraft for the Air Force, 19 Navy planes, 1,012 troops, and 2,000 tons of supplies. On her return trip (27 July-4 August), she cut the record to 7 days, 10 hours, and 36 minutes. After fast repairs, she departed for the Far East on 24 August, this time to join TF 77 in giving air support to the troops. Her planes supported the landing at Inchon (15 September 1950) and other ground action until November, when she departed for the west coast and overhaul.

Boxer departed San Diego for her second Korean tour on 2 March 1951. Again, she operated with TF 77 supporting the ground troops. On 29 March, Carrier Air Wing 101 — composed of Naval Reserve squadrons called to active duty from Dallas, Tex.; Glenview, Ill.; Memphis, Tenn.; and Olathe, Kans. — flew its first combat mission from Boxer, the first carrier strikes by Naval Reserve units against North Korean forces. She returned to San Francisco on 24 October 1951.

Sailing on 8 February 1952 for her third tour in Korea, Boxer again served with TF 77. On 23 June, 35 AD Skyraiders and 35 F9F-2 Panther jets from Boxer, USS Princeton (CV-37), and USS

Philippine Sea (CV-47) joined Air Force Thunderjets in an attack on the heavily defended hydroelectric power plant at Suiho, North Korea, the fourth largest such facility in the world. The plant was completely knocked out. The raid was part of a two-day aerial offensive against North Korea's 13 major power plants. On 5 August 1952, Boxer had nine men killed and two seriously injured in a fire that swept the hangar deck. After emergency repairs at Yokosuka, Japan (11-23 August), Boxer returned to duty off Korea. She arrived at San Francisco on 25 September and underwent repairs until March 1953.

The carrier departed for the Far East on 30 March 1953 and went into action a month later. She took part in the final actions of the Korean conflict and remained in Asiatic waters until November.

Boxer was reclassified CVA-21 on 1 October 1952 and CVS-21 on 1 February 1956.

On 2 October 1958, the Commander in Chief, U.S. Atlantic Fleet, announced the formation of a new amphibious squadron, composed of Boxer and four LSDs equipped with helicopter platforms, which would provide a highly mobile unit capable of employing Marine Corps helicopters squadrons and combat troops in the fast-landing concept of vertical envelopment. The first permanent Marine Aviation Detachment afloat was activated on board Boxer on 10 November to provide supply, maintenance, and flight deck control functions necessary to support the Marine helicopter squadrons and troops. The ship was reclassified as LPH-4 on 30 January 1959.

Boxer and two LSDs arrived off the coast of Hispaniola on 29 August 1964 to provide medical aid and helicopter evacuation services to people in areas of Haiti and the Dominican Republic badly damaged by Hurricane Cleo. Boxer returned to the Dominican Republic on 27 April 1965, sending her Marines ashore while the embarked HMM-264 began an airlift in which over 1,000 U.S. nationals were evacuated to the naval task force offshore as a revolt in the country threatened their safety.

Boxer also participated in the U.S. space program. On 26 February 1966, the first unmanned spacecraft of the Apollo series, fired into suborbital flight by a Saturn 1B rocket from Cape Kennedy, Fla., was recovered in the southeast Atlantic Ocean, 200 miles east of Ascension Island by a helicopter from the ship. Boxer was decommissioned on 1 December 1969 and stricken from the Navy List. She was sold for scrapping on 13 March 1971.

Boxer received eight battle stars for her service off Korea.

CVL-22 • USS INDEPENDENCE

(Displacement: 11,000 tons length: 622½ feet beam: 71½ feet; extreme width at flight deck: 109 feet 2 inches draft: 26 feet speed: 31 knots complement: 1,569 crew armament: 26 40mm guns class: Independence.)

The fourth Independence (CV-22), begun as Amsterdam (CL-59), was launched as CV-22 on 22 August 1942 by New York Shipbuilding Corp., Camden, N.J.; sponsored by Mrs. Rawleigh Warner; and commissioned on 14 January 1943, with Captain G. R. Fairlamb, Jr., in command.

The first of a new class of carriers converted from cruiser hulls, Independence conducted shakedown training in the Caribbean. She then steamed through the Panama Canal to join the Pacific Fleet, arriving in San Francisco on 3 July 1943. Independence got underway for Pearl Harbor on 14 July, and after two weeks of vital training exercises, sailed with carriers USS Essex (CV-9) and USS Yorktown (CV-10) for a devastating raid on Marcus Island. Planes from the carrier force struck on 1 September and destroyed over 70 percent of the installations on the island. The carrier began her next operation, a similar strike against Wake Island on 5 to 6 October, as CVL-22, redesignated on 15 July 1943.

Independence sailed from Pearl Harbor for Espiritu Santo on 21 October, and during an ensuing carrier attack on Rabaul on 11 November 1943, the ship's gunners scored their first success—six Japanese planes shot down. After this operation, the carrier refueled at Espiritu Santo and headed for the Gilberts and pre-landing strikes on Tarawa from 18 to 20 November 1943. During a Japanese counterattack on 20 November, Independence was attacked by a group of planes low on the water. Six were shot down, but the planes managed to launch at least five torpedoes, one of which scored a hit on the carrier's starboard quarter.

Seriously damaged, the ship steamed to Funafuti on 23 November for repairs. With the Gilberts operation, the first step on the mid-Pacific road to Japan underway, Independence returned to San Francisco on 2 January 1944 for more permanent repairs. The veteran carrier returned to Pearl Harbor on 3 July 1944. During her repair period, the ship had been fitted with an additional catapult, and upon her arrival in Hawaiian waters, Independence began training for night carrier operations. She continued this pioneering work from 24 to 29 August out of Eniwetok. The ship sailed with a large task group on 29 August to take part in the Palaus operation, aimed at securing bases for the final assault on the Philippines in October. Independence provided night reconnaissance and night combat air patrol for Task Force 38 during this operation.

In September, the fast carrier task force regularly pounded the Philippines in preparation for the invasion. When no Japanese counterattacks developed in this period, Independence shifted to regular daytime operations, striking targets on Luzon. After replenishment at Ulithi in early October, the great force sortied on 6 October 1944 for Okinawa. In the days that followed, the carriers struck Okinawa, Formosa, and the Philippines in a striking demonstration of the mobility and balance of the fleet. Japanese air counterattacks were repulsed, with Independence providing day strike groups in addition to night fighters and reconnaissance aircraft for defensive protection.

As the carrier groups steamed east of the Philippines on 23 October, it became apparent, as Admiral Carney later recalled, that "something on a grand scale was underfoot." And indeed it was, as the Japanese fleet moved on a three-pronged effort to turn back the American beachhead on Leyte Gulf. Planes from Independence's Task Group 38.2, under Rear Admiral Bogan, spotted Kurita's striking force in the Sibuyan Sea on 24 October 1944, and the carriers launched a series of attacks. Planes from Independence and other ships sank the giant battleship Musashi and disabled a cruiser.

That evening, Admiral Halsey made his fateful decision to turn Task Force 38 northward in search of Admiral Ozawa's carrier group. Independence's night search planes made contact and shadowed the Japanese ships until dawn on 26 October, when the carriers launched a massive attack. In this second part of the great Battle for Leyte Gulf, all four Japanese carriers were sunk. Meanwhile, American heavy ships had won a great victory in Surigao Strait, and a light carrier force had outfought the remainder of Kurita's ships in the Battle of Samar. After the great battle, which virtually spelled the end of the Japanese Navy as a major threat, Independence continued to provide search planes and night fighter protection for Task Force 38 in strikes on the Philippines. In these operations, the ship had contributed to a major development in carrier group operations.

Independence returned to Ulithi for long-delayed rest and replenishment from 9 to 14 November, but soon got underway to operate off the Philippines on night attacks and defensive operations. This phase continued until 30 December 1944, when the great task force sortied from Ulithi once more and moved northward. From 3 to 9 January 1945, the carriers supported the Lingayen landings on Luzon, after which Adm. Halsey took his fleet on a daring foray into the South China Sea. In the days that followed, the aircraft struck at air bases on Formosa and on the coasts of Indo-China and China. These operations in support of the Philippines campaign marked the end of the carrier's night operations, and she sailed on 30 January 1945 for repairs at Pearl Harbor. Independence returned to Ulithi on 13 March 1945 and got underway the next day for operations against Okinawa, the last target in the Pacific before Japan itself. She carried out pre-invasion strikes from 30 to 31 March, and after the assault on 1 April, remained off the island supplying Combat Air Patrol and strike aircraft. Her planes shot down numerous enemy planes during the desperate Japanese attacks on the invasion force. Independence remained off Okinawa until 10 June when she sailed for Leyte.

During July and August, the carrier took part in the final carrier strikes against Japan itself, attacks that lowered enemy morale and had much to do with the eventual surrender. After the end of the war on 15 August, Independence's aircraft continued surveillance flights over the mainland, locating prisoner of war camps, and covered the landings of Allied occupation troops. The ship departed Tokyo on 22 September 1945, arriving in San Francisco via Saipan and Guam on 31 October.

Independence joined the "Magic Carpet" fleet beginning on 15 November 1945, transporting veterans back to the United States until arriving in San Francisco once more on 28 January 1946. Assigned as a target vessel for the Bikini atomic bomb tests, she was placed within one-half mile of ground zero for the 1 July explosion. The veteran ship did not sink, however, and after taking part in another explosion on 25 July, was taken to Kwajalein and decommissioned on 28 August 1946. The highly radioactive hulk was later taken to Pearl Harbor and San Francisco for further tests and was finally sunk in weapons tests off the coast of California on 29 January 1951.

Independence received eight battle stars for World War II service.

CVL-23 • USS PRINCETON

(Displacement: 11,000 tons length: 622½ feet beam: 71½ feet; extreme width at flight deck: 109 feet 2 inches draft: 26 feet speed: 31 knots complement: 1,569 crew armament: 22 40mm guns, 16 20mm guns class: Independence.)

The fourth Princeton was laid down as Tallahassee (CL-61) by the New York Shipbuilding Corp., Camden, N.J., on 2 June 1941. She was reclassified as CV-23 on 16 February 1942, renamed Princeton on 31 March 1942, launched on 18 October 1942, sponsored by Mrs. Harold Dodds, and commissioned at Philadelphia on 25 February 1943, with Capt. George R. Henderson in command.

Following shakedown in the Caribbean, and reclassification to CVL-23 on 15 July 1943, Princeton, with Air Group 23 embarked, got underway for the Pacific. Arriving at Pearl Harbor on 9 August, she sortied with Task Force (TF) 11 on the 25th and headed for Baker Island. There she served as flagship, Task Group (TG) 11.2, and provided air cover during the occupation of the island and the construction of an airfield there from 1-14 September. During that time, her planes downed Japanese Emily reconnaissance planes and, more importantly, furnished the fleet with photographs of them.

Completing that mission, Princeton rendezvoused with TF 15, conducted strikes against enemy installations on Makin and Tarawa, then headed back to Pearl Harbor. In mid-October 1943, she sailed for Espiritu Santo where she joined TF 38 on the 20th. With that force, she sent her planes against airfields at Buka and Bonis on Bougainville (1-2 November) to diminish Japanese aerial resistance during the landings at Empress Augusta Bay. On the 5th and 11th, her planes raided Rabaul, and on the 19th, with TF 50, helped neutralize the airfield at Nauru. Princeton then steamed northeast, covered the garrison groups en route to Makin and Tarawa, and, after exchanging operational aircraft for damaged planes from other carriers, got underway for Pearl Harbor and the West Coast.

An availability at Bremerton followed, and on 3 January 1944, Princeton steamed west. At Pearl Harbor, she rejoined the fast carriers of TF 50, now designated TF 58. On the 19th, she sortied with TG 58.4 for strikes at Wotje and Taroa (29-31 January) to support amphibious operations against Kwajalein and Majuro.

Her planes photographed the next assault target, Eniwetok, on 2 February, and on the 3rd returned on a more destructive assignment—the demolition of the airfield on Engebi. For three days, the atoll was bombed and strafed. On the 7th, Princeton retired to Kwajalein only to return to Eniwetok on the 10th-13th and 16th-28th, when her planes softened the beaches for the invasion force and then provided air cover during the assault and ensuing fight.

From Eniwetok, Princeton retired to Majuro, thence to Espiritu Santo for replenishment. On 23 March 1944, she got underway for strikes against enemy installations and shipping in the Carolines. After striking the Palaus, Woleai, and Yap, the force replenished at Majuro and sortied again on 13 April. Steaming to New Guinea, the carriers provided air cover for the Hollandia operation (21-29 April), then crossed back over the International Date Line to raid Truk (29-30 April) and Ponape (1 May). On 11 May 1944, Princeton returned to Pearl Harbor only to depart again on the 29th for Majuro. There, she rejoined the fast carriers and pointed her bow toward the Marianas to support the assault on Saipan. From 11-18 June, she sent her planes against targets on Guam, Rota, Tinian, Pagan, and Saipan, then steamed west to intercept a Japanese fleet reported to be en route from the Philippines to the Marianas. In the ensuing Battle of the Philippine Sea, Princeton's planes contributed 30 kills, and her guns accounted for another three, plus one assist, to the devastating toll inflicted on Japan's naval air arm.

Returning to the Marianas, Princeton again struck Pagan, Rota, and Guam, then replenished at Eniwetok. On 14 July 1944, she got underway again as the fast carriers returned their squadrons to the Marianas to furnish air cover for the assault and occupation of Guam and Tinian. On 2 August, the force returned to Eniwetok, replenished, then sailed for the Philippines. En route, its planes raided the Palaus, then on 9-10 September, struck airfields on northern Mindanao. On the 11th, they pounded the Visayas. At mid-month, the force moved back over the Pacific chessboard to support the Palau offensive, then returned to the Philippines to hit Luzon, concentrating on Clark and Nichols fields. The force then retired to Ulithi, and in early October, bombed and strafed enemy airfields, installations, and shipping in the Nansei Shoto and Formosa area in preparation for the invasion of the Philippines. On 20 October 1944, landings were made at Dulag and San Pedro Bay, Leyte. Princeton, in TG 38.3, cruised off Luzon and sent her planes against airfields there to prevent Japanese land-based aircraft attacks on Allied ships massed in Leyte Gulf. On the 24th, however, enemy planes from Clark and Nichols fields found TG 38.3 and reciprocated. Shortly before 1000 on 24 October 1944, a lone enemy dive-bomber came out of the clouds above Princeton. At 1500 feet, the pilot released his bomb. It hit between the elevators, crashed through the flight deck and hangar, then exploded. Initial fires soon expanded as further explosions sent black smoke rolling off the flight deck and red flames along the sides from the island to the stern. Covering vessels provided rescue and firefighting assistance and shielded the stricken carrier from further attack. At 1524, another, much heavier explosion, possibly the bomb magazine, blew off the carrier's stern and with it the after flight deck. USS Birmingham (CL-62), alongside to fight fires, suffered heavy damage and casualties.

Efforts to save Princeton continued, but at 1604 the fires won. Boats were requested to take off remaining personnel and shortly after 1706, USS Irwin (DD-794) began to fire torpedoes at the burning hulk. At 1746, USS Reno (CL-96) relieved Irwin and at 1749 the last, and biggest, explosion occurred. Flames and debris shot up 1000 to 2000 feet. Princeton's forward section was gone. Her after section appeared momentarily through the smoke. By 1750 she had disappeared, but 1,361 of her crew survived. Included in that number was Capt. John M. Hoskins, who had been the prospective commanding officer of CVL-23 and lost his right foot with her, but who, despite the loss, would become the first commanding officer of the fifth Princeton (CV-37). Losses and damage to assisting vessels were heavy—Birmingham: 85 killed, 300 wounded, a heavily damaged topside, and the loss of two 5" guns, two 40mm. guns, and two 20mm. guns; USS Morrison (DD-560): foremast lost, portside smashed; Irwin: forward 5" mounts and director out, starboard side smashed; and Reno: one 40mm. gun smashed. Princeton earned nine battle stars during World War II.

CVL-24 • USS BELLEAU WOOD

(Displacement: 11,000; l. 622'6"; b. 109'2"; dr. 26'; S.; 31.6 k.; cpl. 1569; a. 26 40 mm.; cl. Independence.)

New Haven (CL-76) was reclassified CV-24 and renamed Belleau Wood on 16 February 1942. She became CVL-24 on 15 July 1943. Belleau Wood was launched on 6 December 1942 by New York Shipbuilding Corp., Camden, N.J.; sponsored by Mrs. Thomas Holcomb, wife of the Commandant of the Marine Corps; and commissioned on 31 March 1943, with Captain A. M. Pride in command.

After a brief shakedown cruise, Belleau Wood reported to the Pacific Fleet, arriving at Pearl Harbor on 26 July 1943. After supporting the occupation of Baker Island (1 September) and taking part in the Tarawa (18 September) and Wake Island (5-6 October) raids, she joined TF 50 for the invasion of the Gilbert Islands (19 November-4 December 1943).

Belleau Wood operated with TF 58 during the seizure of Kwajalein and Majuro Atolls, Marshall Islands (29 January-3 February 1944); the Truk raid (16-17 February); the Saipan-Tinian-Rota-Guam raids (21-22 February); the Palau-Yap-Ulithi-Woleai raid (30 March-1 April); the Sawar and Wakde Island raids in support of the landings at Hollandia, New Guinea (22-24 April); the Truk-Satawan-Ponape raid (29 April-1 May); the occupation of Saipan (11-24 June); the 1st Bonins raid (15-16 June); the Battle of the Philippine Sea (19-20 June); and the 2nd Bonins raid (24 June). During the Battle of the Philippine Sea, Belleau Wood's planes sank the Japanese carrier Hiyo.

After an overhaul at Pearl Harbor (29 June-31 July 1944), Belleau Wood rejoined TF 58 for the last stages of the occupation of Guam (2-10 August). She joined TF 38 and took part in the strikes in support of the occupation of the southern Palaus (6 September-14 October); the Philippine Islands raids (9-24 September); the Morotai landings (15 September); the Okinawa raid (10 October); the northern Luzon and Formosa raids (11-14 October); the Luzon strikes (15 and 17-19 October); and the Battle of Cape Engano (24-26 October). On 30 October, while Belleau Wood was patrolling with her task group east of Leyte, she shot down a Japanese suicide plane which fell on her flight deck aft causing fires that set off ammunition. Before the holocaust could be brought under control, 92 men were killed or went missing.

After temporary repairs at Ulithi (2-11 November), Belleau Wood steamed to Hunter's Point, Calif., for permanent repairs and an overhaul, arriving on 29 November. She departed San Francisco Bay on 20 January 1945 and joined TF 58 at Ulithi on 7 February. During 15 February-4 March, she took part in the raids on Honshu Island, Japan, and the Nansei Shoto, as well as supporting the landings on Iwo Jima. She also took part in the 5th Fleet strikes against Japan (17 March-26 May) and the 3rd Fleet strikes (27 May-11 June). After embarking a new air group at Leyte (13 June-1 July), she rejoined the 3rd Fleet for the final strikes against the Japanese home islands (10 July-15 August).

Belleau Wood launched her planes on 2 September for the mass flight over Tokyo, Japan, during the surrender ceremonies. She remained in Japanese waters until 13 October. Arriving at Pearl Harbor on 28 October, she departed three days later with 1,248 servicemen for San Diego. She remained on "Magic Carpet" duty, returning servicemen from Guam and Saipan to San Diego, until 31 January 1946. During the next year, Belleau Wood was moored at various docks in the San Francisco area undergoing inactivation. She was placed out of commission in reserve at Alameda Naval Air Station on 13 January 1947. She remained in reserve until transferred to France on 5 September 1953 under the Mutual Defense Assistance Program.

Belleau Wood received the Presidential Unit Citation and 11 battle stars during World War II.

CV-25 • USS COWPENS

The Revolutionary War battle fought on 17 January 1781, 7 miles north of the town of Cowpens, S.C., was an American victory in the campaign that led to the British surrender at Yorktown.

(Displacement: 11,000; length. 622'6"; beam. 71'6"; ew. 109'2"; draft.26'; speed. 32 k.; complement. 1,569; armor. 26 40mm.; cl. Independence.)

The fourth Cowpens (CV-25) was launched on 17 January 1943 by New York Shipbuilding Corp., Camden, N.J., sponsored by Mrs. M. H. Spruance, daughter of Vice Admiral W. F. Halsey; and commissioned on 28 May 1943, with Captain R. P. McConnell in command. She was reclassified CVL-25 on 15 July 1943.

Departing Philadelphia on 29 August 1943, Cowpens arrived at Pearl Harbor on 19 September to begin an active and distinguished war career, which was to earn a Navy Unit Commendation. She sailed with TF 14 for the strike on Wake Island on 5 and 6 October, then returned to Pearl Harbor to prepare for strikes on the Marshall Islands preliminary to invasion. She sortied from Pearl Harbor on 10 November to launch air strikes on Mille and Makin atolls between 19 and 24 November and Kwajalein and Wotje on 4 December, returning to her base on 9 December.

Joining the vast carrier TF 58, Cowpens sailed from Pearl Harbor on 16 January 1944 for the invasion of the Marshalls. Her planes pounded Kwajalein and Eniwetok during the last three days of the month to prepare for the assault landing on the 31st. Using Majuro as a base, the force struck at Truk on 16 and 17 February and the Marianas on 21 and 22 February before putting in at Pearl Harbor on 4 March. Returning to Majuro, TF 58 based there for attacks on the western Carolines; Cowpens supplied air and anti-submarine patrols during the raids on Palau, Yap, Ulithi, and Woleai from 30 March to 1 April. After operating off New Guinea during the invasion of Hollandia from 21 to 28 April, Cowpens took part in the strikes on Truk, Satawan, and Ponape between 29 April and 1 May, returning to Majuro on 14 May for training.

From 6 June to 10 July 1944, Cowpens operated in the Marianas operation. Her planes struck the island of Saipan to aid the assault troops, and made supporting raids on Iwo Jima,

Pagan, Rota, and Guam. They also took part in the Battle of the Philippine Sea on 19 and 20 June, accounting for a number of the huge tally of enemy planes downed. After a brief overhaul at Pearl Harbor, Cowpens rejoined the fast carrier task force at Eniwetok on 17 August. On the 29th, she sailed for the pre-invasion strikes on the Palaus, whose assault was an essential preliminary to the return to the Philippines. From 13 to 17 September, she was detached from the force to cover the landings on Morotai, then rejoined it for sweep, patrol, and attack missions against Luzon from 21 to 24 September. Cowpens, with her task group, flew strikes to neutralize Japanese bases on Okinawa and Formosa from 10 to 14 October, and when Canberra (CA-70) and Houston (CL-81) were hit by torpedoes, Cowpens provided air cover for their safe withdrawal, rejoining her task group on 20 October. En route to Ulithi, she was recalled when the Japanese Fleet threatened the Leyte invasion, and during the Battle of Surigao Strait phase of the decisive Battle for Leyte Gulf on 25 and 26 October, provided combat air patrol for the ships pursuing the fleeing remnant of the Japanese fleet. Continuing her support of the Philippines advance, Cowpens' planes struck Luzon repeatedly during December. During the disastrous typhoon of 18 December, Cowpens lost one man, planes, and equipment, but skillful work by her crew prevented major damage, and she reached Ulithi safely on 21 December to repair her storm damage.

Between 30 December 1944 and 26 January 1945, Cowpens was at sea for the Lingayen Gulf landings. Her planes struck targets on Formosa, Luzon, the Indo-Chinese coast, and the Hong Kong-Canton area, and Okinawa during January. On 10 February, Cowpens sortied from Ulithi for the Iwo Jima operation, striking the Tokyo area, supporting the initial landings from 19 to 22 February, and hitting Okinawa on 1 March.

After overhaul at San Francisco and training at Pearl Harbor, Cowpens sailed on 13 June for San Pedro Bay, Leyte, striking Wake Island on 20 June. Rejoining TF 58, Cowpens sailed from San Pedro Bay on 1 July to join in the final raids on the Japanese mainland. Her planes pounded Tokyo, Kure, and other cities of Hokkaido and Honshu until 15 August. Remaining off Tokyo Bay until the occupation landings began on 30 August, Cowpens launched photographic reconnaissance missions to patrol airfields and shipping movements, and to locate and supply prisoner-of-war camps. Men from Cowpens were largely responsible for the emergency activation of Yokosuka airfield for Allied use. Between 8 November 1945 and 28 January 1946, Cowpens made two voyages to Pearl Harbor, Guam, and Okinawa to return veterans. Placed in commission in reserve at Mare Island on 3 December 1946, Cowpens was decommissioned on 13 January 1947.

In addition to her Navy Unit Commendation, Cowpens received 12 battle stars for World War II service.

CV-26 • USS MONTEREY

(Displacement. 11,000; length. 622'6"; beam. 71'6": ew. 109'2"; draft.26', speed. 31.6 k.; compliment 1,569; a. 26 40mm., 20 20mm., 45 dct; cl. Independence.)

The USS Monterey (CVL-26), initially laid down as Dayton (CL-78) on 29 December 1941 by the New York Shipbuilding Corp., Camden, N.J., was reclassified as CV-26 on 27 March 1942 and renamed Monterey on 31 March 1942. She was launched on 28 February 1943, sponsored by Mrs. P. N. L. Bellinger, and commissioned on 17 June 1943, with Captain Lestor T. Hundt in command.

Monterey was reclassified as CVL-26 on 15 July 1943, shortly after commissioning, and, after shakedown, departed Philadelphia for the western Pacific. She reached the Gilberts on 19 November 1943, in time to help secure Makin Island. She took part in strikes on Kavieng, New Ireland, on 25 December as part of TG 37.2, and supported the landings at Kwajalein and Eniwetok until 8 February 1944. The light carrier then operated with TF 58 during raids in the Carolines, Marianas, northern New Guinea, and the Bonins from February through July 1944. During this time she was also involved in the Battle of the Philippine Sea on 29 and 30 April.

Monterey then sailed to Pearl Harbor for overhaul, departing once again on 29 August. On 3 September, she launched strikes against Wake Island. Then, she joined TF 38 and participated in strikes in the southern Philippines and the Ryukyus. October through December 1944 were spent in the Philippines, supporting first the Leyte and then the Mindoro landings.

Though enemy planes had been unable to damage Monterey, she did not complete her first full year of service unscathed. In December, she steamed into the path of a howling typhoon with winds over 100 knots. At the height of the storm, which lasted two days, several planes tore loose from their cables, causing several fires on the hangar deck. Monterey arrived at Bremerton, Wash., for overhaul in January 1945. She rejoined TF 58 and supported Okinawa operations by launching strikes against Nansei Shoto and Kyushu from 9 May through 1 June. She rejoined TF 38 for the final strike against Honshu and Hokkaido from 1 July to 15 August.

She departed Japanese waters on 7 September, having embarked troops at Tokyo, and steamed home, arriving in New York City on 17 October. Monterey left behind an impressive and enviable war record. Her planes sank five enemy warships and damaged others. She was responsible for the destruction of thousands of tons of Japanese shipping, hundreds of planes, and vital industrial complexes. She was assigned "Magic Carpet" duty and made several voyages between Naples and Norfolk. She decommissioned on 11 February 1947 and was assigned to the Atlantic Reserve Fleet, Philadelphia Group.

With the outbreak of hostilities in Korea, Monterey recommissioned on 15 September 1950. She departed Norfolk on 3 January 1951 and proceeded to Pensacola, Fla., where she operated for the next four years under the Naval Training Command, training thousands of naval aviation cadets, student pilots, and helicopter trainees. Between 1 and 11 October 1954, she took part in a flood rescue mission in Honduras. She departed Pensacola on 9 June 1955 and steamed to rejoin the Atlantic Reserve Fleet, Philadelphia Group. She decommissioned on 16 January 1956. Reclassified as AVT-2 on 15 May 1959, she remained berthed at Philadelphia into 1969.

Monterey received 11 battle stars for World War II service.

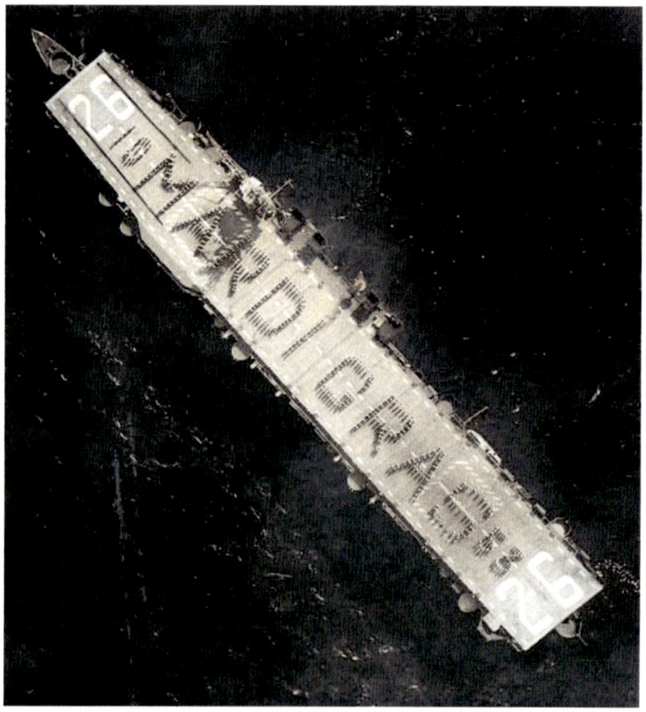

CVL-27 • USS LANGLEY

(Displacement 11,000, length. 622'6", beam. 71'6", ew. 109'2"; draft. 26'; speed. 31 k.;compliment. 1,569; armor. 26 40mm., dct 45; class Independence.)

Langley (CVL-27), originally named Fargo (CL-85), was laid down as Crown Point (CV-27) by New York Shipbuilding Corp., Camden, N.J., on 11 April 1942. She was renamed Langley on 13 November 1942, launched on 22 May 1943, sponsored by Mrs. Harry L. Hopkins, wife of the Special Assistant to President Roosevelt, reclassified as CVL-27 on 15 July 1943, and commissioned on 31 August 1943, with Capt. W. M. Dillon in command.

After shakedown in the Caribbean, Langley departed Philadelphia on 6 December 1943 for Pearl Harbor, where she participated in training operations. On 19 January 1944, she sailed with Rear Adm. Marc Mitscher's Task Force 58 for the attack on the Marshall Islands. From 29 January to 6 February, the carrier's air group conducted raids on Wotje and Taroa to support Allied landings at Kwajalein and repeated the performance from 10 to 28 February at Eniwetok. After a brief respite at Espiritu Santo, New Hebrides Islands, Langley hit Japanese positions on Palau, Yap, and Woleai, Caroline Islands, from 30 March to 1 April. She next proceeded to New Guinea to take part in the capture of Hollandia on 25 April. A mere 4 days later, the tireless carrier engaged in the 2-day strike against the Japanese bastion Truk, rendering the formidable naval base almost useless to the "Sons of Nippon." During the raid, Langley and her aircraft accounted for some 35 enemy planes destroyed or damaged, while losing only one aircraft herself.

Langley next departed Majuro Atoll on 7 June for the Marianas campaign. On 11 June, Admiral Mitscher's carrier groups took over from the land-based Army Air Force bombers. At 1300, the Task Force launched a strike of 208 fighters and eight torpedo bombers against enemy bases and airfields on Saipan and Tinian. From 11 June to 8 August, the battle raged for control of the Marianas. The Allied assault on the key to Japan's inner defenses on 15 June forced the enemy to engage our fleet for the first time since Midway. During the 2-day Battle of the Philippine Sea, 19 to 20 June, the enemy suffered such serious losses that he was not able to again seriously challenge U.S. sea power until the invasion of Leyte. When Adm. Jisaburo Ozawa retreated with his battered Mobile Fleet, he was minus 426 aircraft and three carriers. Langley had added her strength to break this Japanese effort to reinforce the Marianas.

The carrier departed Eniwetok on 29 August and sortied with Task Force 38, under the command of Adm. William F. Halsey, for air assaults on Peleliu and airfields in the Philippines as the preliminary steps in the invasion of the Palaus from 15 to 20 September. During October, she was off Formosa and the Pescadores Islands, attached to Vice Admiral Mitscher's Fast Carrier Force. Later in the month, as the Navy carried General MacArthur back to the Philippines, Langley was with Rear Admiral Sherman's Task Group protecting the Leyte beachheads. In a desperate effort to parry this deadly thrust into her inner defenses, Japan struck back with her entire fleet. On 24 October, Langley's planes helped to blunt the first and most powerful prong of this counteroffensive, Admiral Kurita's Center Force, as it steamed toward the San Bernardino Strait and the American beachhead. The following day, upon word of Japanese carriers north of Leyte, she raced to intercept. In the ensuing battle off Cape Engano, Mitscher's force pulverized the enemy fleet. The Japanese lost four carriers, two battleships, four heavy cruisers, one light cruiser, and five destroyers. Langley's aircraft had assisted in the destruction of the carriers Zuiho and Zuikaku, the latter being the only remaining carrier of the six that had participated in the Pearl Harbor attack. Japan's chances for final victory had been reduced to nil by the great Battle of Leyte Gulf.

During November, Langley was lending her support to the Philippine landings and striking the Manila Bay area, Japanese reinforcement convoys, and Luzon airfields. On 1 December, the flattop withdrew to Ulithi for reprovisioning.

During January 1945, Langley participated in the daring raid into the South China Sea supporting Lingayen Gulf operations. Raids were made against Formosa, Indo-China, and the China coast from 30 December 1944 to 25 January 1945. The thrust into this area, which the enemy had considered a private lake, netted a staggering number of Japanese ships, aircraft, supplies, and destroyed installations.

Langley next joined in the sweeps against Tokyo and Nansei Shoto in support of the conquest and occupation of Iwo Jima from 10 February to 18 March. She next raided airfields on the Japanese homeland and arrived off Okinawa on 23 March. Until 11 May, the ship divided her attention between the Okinawa invasion and strikes on Kyushu, Japan, in an effort to knock out kamikaze bases in southern Japan, which were launching desperate and deadly attacks.

After touching Ulithi and Pearl Harbor, she steamed to San Francisco, arriving on 3 June for repairs and modernization. She departed on 1 August for the forward area and reached Pearl Harbor on 8 August. While there, word arrived that hostilities had ended. She completed two "Magic Carpet" voyages to the Pacific and got underway on 1 October for Philadelphia. She departed from that port on 15 November for the first of two trips to Europe, transporting Army troops returning home from that theater. She returned to Philadelphia on 6 January 1946 and was assigned to the Atlantic Reserve Fleet, Philadelphia Group, on 31 May. She decommissioned on 11 February 1947 and was transferred to France under the Mutual Defense Assistance Program on 8 January 1951. In French service, she was renamed Lafayette (R-96). The carrier was returned to the United States on 20 March 1963 and sold to the Boston Metals Co., Baltimore, Md., for scrapping.

Langley received nine battle stars for World War II service.

CVL-28 • USS CABOT

(Displacement. 11,000; l. 622'6"; b. 71'6"; ew. 109'2"; dr. 26'; s. 32 k.; cpl. 1,569; a. 26 40 mm.; cl. Independence.)

The second Cabot (CVL-28) was laid down as Wilmington (CL-79), redesignated CV-28 on 2 June 1942, renamed Cabot on 23 June 1942, converted while building, and launched on 4 April 1943 by New York Shipbuilding Co., Camden, N.J.; sponsored by Mrs. A. C. Read; reclassified CVL-28 on 16 July 1943; and commissioned on 24 July 1943, Captain M. F. Shoeffel in command.

Cabot sailed from Quonset Point, R.I., on 8 November 1943 for Pearl Harbor, where she arrived on 2 December. Clearing for Majuro on 15 January 1944, she joined TF 58 to begin the consistently high-quality war service which was to win her a Presidential Unit Citation. From 4 February to 4 March 1944, she launched her planes in strikes on Roi, Namur, and the island stronghold of Truk, aiding in the neutralization of these Japanese bases as part of the invasion of the Marshalls.

Cabot returned to Pearl Harbor for a brief repair period but was back in action from Majuro for the pounding raids on the Palaus, Yap, Ulithi, and Woleai at the close of March 1944. She sailed to provide valuable air cover for the Hollandia operation from 22 to 25 April, and 4 days later began to hurl her air power at Truk, Satawan, and Ponape. She cleared Majuro again on 6 June for the preinvasion air strikes in the Marianas, and on 19 and 20 June launched sorties in the key Battle of the Philippine Sea, the famous "Marianas Turkey Shoot," which hopelessly crippled Japanese naval aviation. Cabot's air units pounded Japanese bases on Iwo Jima, Pagan, Rota, Guam, Yap, and Ulithi as the carrier continued her support of the Marianas operation until 9 August.

Preinvasion strikes in the Palaus in September 1944, along with air attacks on Mindanao, the Visayas, and Luzon, paved the way for the long-awaited return to the Philippines. On 6 October, Cabot sailed from Ulithi for raids on Okinawa and to provide air cover for her task group during the heavy enemy attacks off Formosa on 12 and 13 October. Cabot joined the group which screened "Cripple Division 1," the cruisers Canberra (CA-70) and Houston (CL-81), which had been torpedoed off Formosa, to the safety of the Carolines, then rejoined her group for continued air strikes on the Visayas, and the Battle for Leyte Gulf on 24 and 25 October.

Cabot remained on patrol off Luzon, conducting strikes in support of operations ashore and repelling desperate suicide attacks. On 25 November, a particularly vicious one occurred. Cabot had fought off several kamikazes when one, already flaming from hits, crashed into the flight deck on the port side, destroying the still-firing 20 mm gun platform, disabling the 40 mm mounts, and a gun director. Another of Cabot's victims crashed close aboard and showered the port side with shrapnel and burning debris. Cabot lost 62 men killed and wounded, but careful training had produced a crew that handled damage control smoothly and coolly. While she continued to maintain her station in formation and operate effectively, temporary repairs were made. On 28 November, she arrived at Ulithi for permanent repairs.

Cabot returned to action on 11 December 1944, steaming with the force striking Luzon, Formosa, Indo-China, Hong Kong, and the Nansei Shoto in support of the Luzon operations. From 10 February to 1 March 1945, her planes pounded the Japanese homeland and the Bonins to suppress opposition to the invasion of Iwo Jima. Continued strikes against Kyushu and Okinawa in March prepared for the invasion of the latter island. After these prolonged, intensive operations, Cabot was homeward bound for San Francisco for a much-needed overhaul completed in June.

After refresher training at Pearl Harbor, the carrier launched strikes on Wake Island on 1 August while en route to Eniwetok. Here she remained on training duty until the end of the war. Sailing on 21 August, she joined TG 38.3 to support the landings of occupation troops in the Yellow Sea area in September and October. Embarking homeward-bound men at Guam, Cabot arrived at San Diego on 9 November, then sailed for the east coast. Cabot was placed out of commission in reserve at Philadelphia on 11 February 1947.

Recommissioned on 27 October 1948, Cabot was assigned to the Naval Air Reserve training program. She operated out of Pensacola, then Quonset Point, on cruises to the Caribbean and had one tour of duty in European waters from 9 January to 20 March 1952. Cabot was again placed out of commission in reserve on 21 January 1955. She was reclassified AVT-3 on 15 May 1959.

In addition to the Presidential Unit Citation, Cabot received nine battle stars for World War II service.

CVL-29 • USS BATAAN

(Displacement: 11,000 tons; length: 622½ feet; beam: 71½ feet; extreme width at flight deck: 109 feet 2 inches; draft: 26 feet; speed: 32 knots; complement: 1,569 crew; armament: 26 40mm guns; class: Independence.)

The cruiser Buffalo (CL-99) was reclassified as CV-29 and renamed Bataan on 2 June 1942. The ship was reclassified as CVL-29 on 15 July 1943 and launched on 1 August 1943 by New York Shipbuilding Corp., Camden, N.J.; sponsored by Mrs. George D. Murray, wife of Rear Adm. Murray. Bataan was commissioned on 17 November 1943, with Captain V. H. Schaeffer in command, and reported to the Pacific Fleet.

In her initial engagement with the Japanese, Bataan's planes supported the attack on Hollandia, New Guinea, between 21 and 24 April 1944. Following this action were strikes against Truk, Satawan, and Ponape (29 April - 1 May 1944); Saipan, Marianas (11 June - 10 August); 1st Bonins raid (15-16 June); Battle of the Philippine Sea (19-20 June), and the 2nd Bonins raid (24 June).

Bataan then returned to the United States for repairs. With repairs completed, she joined TF 58 and participated in the fleet raids in support of the Okinawa operation (17 March - 30 May 1945), during which her aircraft assisted in the sinking of the Japanese submarine I-56 on 18 April 1945, in 26°42' N., 130°38' E. Retiring to the Philippines, Bataan joined the 3rd Fleet for operations against the Japanese home islands (10 July - 15 August).

Bataan returned to the United States, arriving in New York on 17 October 1945, and was assigned to "Magic Carpet" duty. On 10 January 1946, she arrived in Philadelphia to prepare for inactivation. Bataan went out of commission in reserve on 11 February 1947. Bataan was recommissioned on 13 May 1950 at Philadelphia. In July 1950, she stood out for San Diego, upon arrival loaded Air Force cargo and personnel, and departed on 16 November for Tokyo Bay. She arrived in Korean waters on 15 December and until June 1951, her aircraft flew strikes in support of the ground forces.

Bataan departed for the west coast on 2 June 1951 and, after a brief stop in San Diego, steamed to Bremerton, Wash., on 9 July for overhaul. She returned to San Diego on 20 November and, on 27 January 1952, departed for Yokosuka, Japan, and then to Buckner Bay, Okinawa. She conducted air exercises and other training maneuvers off Okinawa until 29 April, when she sailed for Korean waters. Bataan continued operating between Japan and Korea throughout the summer of 1952, carrying personnel and supplies to the fighting area and launching strikes against the enemy. She left the fighting zone on 11 August for San Diego. On 27 October 1952, the carrier once again stood out for the Far East and operated off Korea until 10 May 1953 when she departed for San Diego.

She remained in the San Diego area undergoing overhaul and training until 31 July. Then she sailed via Pearl Harbor to Kobe and Yokosuka, Japan, and then back to the United States where she reported for inactivation on 26 August 1953. She went out of commission in reserve on 9 April 1954 in San Francisco. She was stricken from the Navy List in September 1959 and sold for scrapping in May 1961. Bataan received six battle stars for her World War II service and seven for her Korean service.

CVL-30 • USS SAN JACINTO

(Displacement: 11,000 tons; length: 622½ feet; beam: 71½ feet; extreme width at flight deck: 109 feet 2 inches; draft: 26 feet; speed: 32 knots; complement: 1,569 crew; armament: 26 40mm guns; class: Independence.)

The second San Jacinto (CVL-30) was laid down as the light cruiser NEWARK (CL-100) on 26 October 1942 by the New York Shipbuilding Co., Camden, N.J.; redesignated CV-30 and renamed REPRISAL on 2 June 1942; renamed San Jacinto on 30 January 1943, converted, while building, to a light aircraft carrier and reclassified as CVL-30; launched on 26 September 1943; sponsored by Mrs. Jesse Jones; and commissioned on 15 November 1943, Capt. Harold M. Martin in command.

After shakedown in the Caribbean, San Jacinto sailed, via the Panama Canal, San Diego, and Pearl Harbor, for the Pacific war zone. Arriving at Majuro, Marshall Islands, she became part of the growing might of Vice Admiral Marc A. Mitscher's Task Force 58/38, the fast carrier striking force of the Pacific Fleet. There, San Jacinto embarked Air Group 51, whose fighters and torpedo planes would be the ship's chief weapons in battle.

After providing search patrols to protect other carriers striking at Wake and Marcus Islands, San Jacinto, by 5 June 1944, was ready to participate in the largest fleet action since the Battle of Midway, almost exactly two years before. On that day, Task Force 58 sortied from Majuro and headed toward the Marianas to conduct air strikes preparatory to American seizure of Saipan and to protect the invasion forces from enemy air and naval attack.

This American thrust triggered a strong Japanese reaction; on 19 June, the Japanese Fleet launched more than 400 planes against the invasion fleet and the covering carrier force. In the ensuing air battle, known to American pilots as the "Marianas Turkey Shoot," more than 300 enemy planes were shot down. While San Jacinto's planes were achieving their most one-sided victory of the war, her gunners helped to down the few attackers able to get near the American ships. Then, at dusk, Admiral Mitscher dispatched an all-carrier attack after the retreating enemy fleet. The night recovery of the returning planes was accomplished amid considerable confusion. Reportedly, a Japanese carrier plane attempted a landing approach on San

Jacinto, only to be waved off by the landing signal officer because its hook wasn't down.

San Jacinto then participated in strikes against Rota and Guam and furnished combat air patrol (CAP) and anti-submarine patrol (ASP) for her task group. During these raids, a San Jacinto fighter pilot was shot down over Guam and spent 17 days in a life raft trying to attract attention and 16 nights hiding on the island.

After a refueling and replenishment stop at Eniwetok Atoll, San Jacinto joined in carrier strikes against the Palaus on 15 July. On 5 August, her targets were Chichi, Haha, and Iwo Jima. A brief stop at Eniwetok preceded dawn-to-dusk CAP and ASP duty while other carriers struck at Yap, Ulithi, Anguar, and Babelthuap, pinning down Japanese air forces while the Palaus were being assaulted on 15 September.

Following a replenishment stop at Manus, Admiralty Islands, San Jacinto joined in strikes against Okinawa and furnished photographic planes to get information necessary for future invasion plans. After refueling at sea, she once again supplied dawn-to-dusk air protection as other carriers sent strikes against Formosa, northern Luzon, and the Manila Bay area from 12 to 19 October. During operations on 17 October, a fighter plane made a very hard landing and inadvertently fired its machine guns into the ship's island structure, killing two men, wounding 24, including her commanding officer, and causing considerable damage to radar. Despite this accident, San Jacinto remained battleworthy.

As American troops landed on Leyte in the central Philippines on 20 October, San Jacinto provided close air support. On 24 October, this mission was interrupted by news of the tri-pronged approach of the Japanese fleet which precipitated the largest fleet battle in naval history.

San Jacinto sent planes against the central force in the Sibuyan Sea, then raced north to launch strikes against the northern force, resulting in heavy damage to the Japanese carriers and surface combatants off Cape Engano. On 30 October, her fighters furnished air protection over Leyte while her guns shot down two planes attempting suicide attacks on the ship. After a pause at Ulithi, the carrier joined in attacks on the Manila Bay area; then took a side trip to Guam to exchange air groups, receiving Air Group 45. She received slight damage during a typhoon in December 1944.

After completing repairs at Ulithi, San Jacinto and the rest of her fast carrier force entered the South China Sea and launched massive air attacks on the airfields of Formosa and

against shipping at Cam Ranh Bay, French Indochina, and at Hong Kong. By refueling and replenishing at sea, Task Force 38 was able to continue its pressure on the enemy and strategic support for the American invasion of Luzon by strikes against the Ryukyu Islands.

Next, San Jacinto joined in the first carrier strikes against the home islands of Japan. During the raids on 16 and 17 February 1945, carrier-based aircraft downed many enemy planes during fierce dogfights over airfields in the Tokyo area. These operations were designed to cover the imminent invasion of Iwo Jima. Next came air support for the landing marines, followed by further strikes against Tokyo and Okinawa before San Jacinto returned to Ulithi.

While conducting operations off Kyushu, Japan, she witnessed the conflagration on Franklin (CV-13); and, on 19 March 1945, narrowly escaped destruction herself when a kamikaze barely missed her. More massive enemy attacks came with Operation "Iceberg" as the carrier force furnished air support for the invasion of Okinawa. On 5 April, more than 500 planes, primarily kamikazes, attacked. Fighter planes and antiaircraft guns shot down about 300, but many got through. San Jacinto's gunners shot the wing off a would-be suicide plane, deflecting its dive, and splashed another only 50 feet off her port bow. Her mission of covering the Okinawa invasion entailed heavy air activity and kept the ship almost constantly at general quarters while supporting ground forces and repelling frequent attacks by suicide planes. On 7 April, San Jacinto's bombers torpedoed the Japanese destroyer Hamakaze, part of a naval suicide attack in which the super battleship Yamato was also sunk. San Jacinto then returned to the dangerous job of defending against the suicide plane attacks, striking at the kamikaze airfields on Kyushu, and providing close air support for ground forces fighting on Okinawa. On 5 June, she successfully rode out another typhoon and after replenishing at Leyte, sortied for her final raids as part of Task Force 58. Her aircraft struck at Hokkaido and Honshu, Japan, on 9 July and continued to operate off the coast of Japan until the end of hostilities on 15 August 1945. Her air missions over Japan then became mercy flights over Allied prisoner-of-war camps, dropping food and medicine until the men could be rescued. Her wartime mission completed, San Jacinto returned home and tied up at Alameda, Calif., on 14 September 1945.

She was decommissioned on 1 March 1947 and joined the Pacific Reserve Fleet berthed at San Diego. Reclassified as an auxiliary aircraft transport (AVT-5) on 15 May 1959; she was struck from the Navy list on 1 June 1970.

San Jacinto earned five battle stars and was awarded the Presidential Unit Citation. Her hull was sold for scrapping on 15 December 1971 to National Metal and Steel Co., Terminal Island, Calif.

CV-31 • USS BON HOMME RICHARD

(Displacement: 27,100 tons; length: 872 feet; beam: 147½ feet; draft: 28 feet 7 inches; speed: 33 knots; complement: 3,448 crew; armament: 12 5-inch guns; class: Essex.)

The second Bon Homme Richard (CV-31) was launched on 29 April 1944 by the New York Navy Yard, sponsored by Mrs. J. S. McCain, wife of Vice Adm. McCain, and commissioned on 26 November 1944, Captain A. O. Rule, Jr., in command.

Bon Homme Richard departed Norfolk on 19 March 1945 to join the Pacific Fleet and arrived at Pearl Harbor on 5 April 1945. Following additional training in Hawaiian waters, the carrier joined TF 38 off Okinawa on 6 June 1945. During 7-10 June, she participated in attacks on Okino Daito Jima and then served with the 3rd Fleet during the air strikes against Japan (2 July-15 August). She remained off Japan until 16 September and, after a short training period off Guam, proceeded to San Francisco, arriving on 20 October. She left San Francisco on 29 October 1945 and steamed to Pearl Harbor to undergo conversion for troop transport duty. From 8 November 1945 to 16 January 1946, she made trans-Pacific voyages, returning servicemen to the United States. Bon Homme Richard then reported to Puget Sound Naval Shipyard for inactivation and was placed out of commission in reserve on 9 January 1947.

Bon Homme Richard was recommissioned on 15 January 1951, and on 10 May departed San Diego for the Far East. She joined TF 77 off Korea on 29 May and launched her first air strikes on 31 May. Bon Homme Richard continued operations with TF 77 until 20 November 1951. The carrier reached San Diego in mid-December and on 20 May 1952 was off again to the Far East.

She joined TF 77 once more on 23 June 1952 and took part in the heavy strikes against the North Korean power complex (24-25 June) and the amphibious feint at Kojo (12-16 October). She continued operations against North Korean targets until 18 December 1952 and then steamed to San Francisco, where she arrived on 8 January 1953. Her classification was changed from CV-31 to CVA-31 on 1 October 1952.

Bon Homme Richard went out of commission on 15 May 1953 in preparation for modernization. When recommissioned on 6 September 1955, she had an angled and strengthened flight deck, enclosed bow, enlarged elevators, and steam catapults. She completed her conversion period on 31 October 1955 and commenced sea trials in the Alameda-San Diego area.

Recommissioned in September 1955, she began the first of a long series of Seventh Fleet deployments. The initial west coast deployment of a squadron equipped with the new Sidewinder missile was with Fighter Squadron 211, equipped with FJ-3s, aboard Bon Homme Richard in September 1956.

On 6 June 1957, two F8U Crusaders and two A3D Skywarriors flew non-stop from Bon Homme Richard off the California coast to USS Saratoga (CVA 60) off the east coast of Florida. This, the first carrier-to-carrier transcontinental flight, was completed by the F8Us in 3 hours 28 minutes and by the A3Ds in 4 hours 1 minute.

Bon Homme Richard made additional western Pacific cruises in 1957, 1958-1959, 1959-60, 1961, 1962-63, and 1964, with the last including a voyage into the Indian Ocean. The ship entered the Indian Ocean on 4 April 1964 with the "Concord Squadron," composed of Bon Homme Richard, USS Shelton (DD 790), USS Blue (DD 744), USS Frank Knox (DD 742), and the fleet oiler USS Hassayampa (AO 145). The cruise lasted six weeks and went near Iran, the Arabian Peninsula, down the African coast, and into many ports along the way for goodwill visits.

The Vietnam War escalation in early 1965 brought Bon Homme Richard into a third armed conflict, and she deployed on five Southeast Asia combat tours over the next six years. Her aircraft battled North Vietnamese MiGs on many occasions, downing several, as well as striking transportation and infrastructure targets. Occasional excursions to other Asian areas provided some variety to her operations.

Bon Homme Richard was ordered inactivated at the end of her 1970 deployment. She decommissioned in July 1971, becoming part of the Reserve Fleet at Bremerton, Washington. The ship was stricken from the Navy List in 1989 and was sold for scrapping on 4 February 1992.

Bon Homme Richard received one battle star for her World War II service and five battle stars for participation in the Korean conflict.

CV-32 • USS LEYTE

(Displacement: 27,100 tons; length: 888 feet; beam: 147½ feet; draft: 28 feet 7 inches; speed: 33 knots; complement: 3,448 crew; armament: 12 5-inch guns, 72 40mm guns; aircraft: 80+; class: Essex.)

The third Leyte was laid down as Crown Point (CV-32) by Newport News Shipbuilding and Dry Dock Co., Newport News, Va., on 21 February 1944; renamed Leyte on 8 May 1945; launched on 23 August 1945, sponsored by Mrs. James M. Mead; and commissioned on 11 April 1946, Capt. Henry F. MacComsey in command.

Leyte joined USS Wisconsin (BB-64) on a goodwill cruise down the western seaboard of South America in the fall of 1946 before returning to the Caribbean on 18 November to resume shakedown operations. The following three years were spent in numerous fleet exercises in the Atlantic and Caribbean, training naval reservists, and four deployments in the Mediterranean: April to June 1947, July to November 1947, September 1949 to January 1950, and May to August 1950. The latter included a demonstration of air power over Beirut, Lebanon, on 13 August, supporting the Middle East against Communist pressure. Leyte returned to Norfolk on 24 August, and after two weeks of preparation, departed on 6 September 1950 to join TF 77 in the Far East to support United Nations Forces in Korea.

Leyte arrived at Sasebo, Japan, on 8 October 1950 and made final preparations for combat operations. From 9 October through 19 January 1951, the ship and her aircraft spent 92 days at sea and flew 3,933 sorties against the North Korean aggressors.

On 4 December 1950, Ensign Jesse L. Brown, the Navy's first black pilot, flying a close support mission from Leyte, was forced to make a crash landing near Hagaru-Ri when his plane was hit by enemy ground fire. Observing that Ens. Brown was unable to get out of his cockpit, one of his squadron mates, Lt. (j.g.) Thomas J. Hudner, fearlessly landed to assist. Ens. Brown died before he could be removed from the wreckage. Lt. (j.g.) Hudner was rescued by helicopter and later was awarded the Medal of Honor. Ens. Brown was posthumously decorated with the Distinguished Flying Cross.

All told, Leyte's pilots accumulated 11,000 hours in the air while inflicting massive damage upon enemy positions, supplies, transportation, and communications. Leyte returned to Norfolk for overhaul on 25 February 1951.

After fleet training exercises in the Caribbean terminated on 21 August 1951, the carrier departed for her fifth tour of duty with the 6th Fleet on 3 September. She returned to Norfolk on 21 December for operations out of Hampton Roads, and again steamed for the Mediterranean on 29 August 1952. Reclassified CVA-32 on 1 October 1952, she returned to Boston on 16 February 1953 for deactivation. On 8 August, however, she was ordered to be retained in the active fleet, and, redesignated CVS-32 on the same day, work was begun converting her to an ASW support carrier.

On 16 October 1953, at 1515, while still under conversion to an antisubmarine carrier, Leyte suffered an explosion in her port catapult machinery room. Within minutes, naval base and city fire trucks were on the scene. After a hard and gallant fight, the fire was extinguished at 1957. As a result of the fire, 37 men died and 28 were injured.

Conversion completed on 4 January 1954, Leyte departed Boston for Quonset Point, R.I., as the flagship of CarDiv 18. She conducted anti-submarine operations in the Atlantic and Caribbean over the next five years. She also served briefly as an interim amphibious assault ship in 1957, with her normal air group replaced with Marine Corps transport helicopters.

Leyte departed Quonset Point in January 1959 for the New York Navy Yard where she commenced pre-inactivation overhaul. She was redesignated AVT-10 and decommissioned both on 15 May 1959, and was assigned to the Philadelphia group of the Atlantic Reserve Fleet with a berth in New York. The ship was stricken from the Navy List on 1 June 1969, and was sold for scrapping in September 1970. Leyte received two battle stars for Korean service.

CV-33 • USS KEARSARGE

(Displacement: 30,800 tons; length: 888 feet; beam: 93 feet; extreme width at flight deck: 136 feet; draft: 28 feet 7 inches; speed: 33 knots; complement: 2,100 crew; armament: 12 5-inch guns, 44 40mm guns, 36 20mm guns; class: Essex.)

The third Kearsarge (CV-33) was launched on 5 May 1945 by the New York Naval Shipyard, New York, sponsored by Mrs. Aubrey W. Fitch, and commissioned on 2 March 1946, Captain Francis J. McKenna in command.

Kearsarge arrived at her homeport, Norfolk, on 21 April 1946, and for the next year engaged in training operations along the east coast and Caribbean. She cleared Norfolk on 7 June 1947 on a midshipmen training cruise to the United Kingdom. Upon her return to the United States in August, the carrier engaged in maneuvers for 10 months before departing Hampton Roads on 1 June 1948 for duty with the 6th Fleet. During her tour in the Mediterranean, units of the 6th Fleet were placed on alert to ensure peace in the Arab-Israeli area. Kearsarge returned to Quonset Point, R.I., on 2 October 1948, and operated along the Atlantic Coast and the Caribbean until 27 January 1950 when she sailed for the west coast. The carrier arrived at Puget Sound Navy Yard on 23 February and decommissioned there on 16 June 1950 for a modernization overhaul that would enable her to handle new jet aircraft.

Kearsarge recommissioned on 15 February 1952, Captain Louis B. French in command. Following shakedown, the carrier cleared San Diego on 11 August for intensive flight training in the Hawaiian Islands. Her readiness complete, she sailed for the Far East to engage in combat missions in the Korean War. Arriving at Yokosuka on 8 September 1952, Kearsarge joined the fast carrier Task Force 77 off the east coast of Korea six days later. For the next five months, the carrier's planes flew nearly 6,000 sorties against Communist forces in North Korea, unleashing considerable damage on enemy positions. She completed her tour in late February 1953, returning to her homeport, San Diego, on 17 March. While serving in Korea her classification was changed to CVA-33.

Kearsarge sailed again for the Far East on 1 July 1953 and operated with the 7th Fleet fast carrier force during the uneasy truce in Korea. The "Mighty Kay" also kept watch over the Formosa Straits to prevent the Communists from interfering with the Chinese Nationalists on Taiwan. Kearsarge returned to San Diego on 18 January 1954 to resume training operations off California.

Clearing San Diego on 7 October 1954, she steamed toward her third deployment to the Far East. While operating with the 7th Fleet, the carrier stood by to assist the Nationalist Chinese in the evacuation of the Tachen Islands. From 6 to 13 February 1955, Kearsarge supported units of the fleet in the successful evacuation of 18,000 civilians and 20,000 military personnel from the islands. Her cruise ended at San Diego on 12 May. The carrier was again modernized in 1956-57, receiving an enclosed "hurricane" bow and an angled flight deck to better equip her to operate high-performance aircraft.

Kearsarge was assigned a new role in October 1958, becoming an antisubmarine warfare (ASW) support aircraft carrier, with the new designation CVS-33. Following intensive training in her new role, the carrier sailed on 5 September 1959 for 7th Fleet operations in the Far East. Early in her tour, Japan was hit with a violent typhoon, and Kearsarge played an important role in providing relief to the victims. Her planes landed parties of medical and supply units, while her crew and air group donated clothing and money to the distressed people. After participating in SEATO exercises and 7th Fleet operations, she cleared Yokosuka on 3 March 1960 for her homeward voyage. Three days later in stormy waters 1,200 miles off Wake Island, four Russians were rescued after drifting 49 days in a disabled landing craft. They were flown back to their country after Kearsarge arrived at Alameda, Calif., on 15 March; and the carrier received thanks from the Soviet Union for this gesture.

A year of training operations preceded her next deployment from San Diego which began on 3 March 1961. The antisubmarine carrier steamed to Southeast Asian waters as the Communists intensified their effort to overthrow the government in Laos. The power and determination of the 7th Fleet was observed by the enemy and the crisis eased. Peace prevailed. After six months in the Far East, Kearsarge arrived at Puget Sound on 1 November for the second phase of her modernization. Upon completion of repairs and training, Kearsarge departed Long Beach on 1 August 1962 to station herself in the Pacific missile range as a recovery ship in the Mercury orbital space flight of astronaut Walter Schirra. In October, after a flawless flight, the carrier played her role in the space age by retrieving Schirra and his capsule and returning him to Honolulu for a flight back to the States.

Kearsarge resumed training exercises, continuing these for six months before arriving at Pearl Harbor on 29 April 1963 to once again take part in the space program. The carrier repeated her earlier recovery by plucking astronaut Gordon Cooper on 18 May 1963, after he orbited the earth 22 times in his capsule Faith 7. She returned the space hero to Pearl Harbor, then departed on 4 June for her eighth cruise to the Far East. Operations with the 7th Fleet included keeping watch on the unsettled problems in Southeast Asia. Kearsarge returned to Long Beach on 3 December for training exercises off California. Six months later, on 19 June 1964, the antisubmarine carrier was deployed on her ninth Far Eastern cruise. Arriving at Yokosuka on 30 July, Kearsarge was dispatched to the South China Sea following the North Vietnamese patrol boat attack on U.S. destroyers in the Gulf of Tonkin. While U.S. Navy planes destroyed North Vietnamese oil and supply depots, Kearsarge provided antisubmarine protection for the 7th Fleet. The decisiveness of American action persuaded the Communists to delay their objectives for the time being; and Kearsarge returned to Long Beach on 16 December 1964.

After an overhaul during the first half of 1965, Kearsarge operated off the West Coast until sailing for the Far East on 9 June 1966. Steaming via Hawaii and Japan, she reached "Yankee Station" on 8 August and operated off Vietnam through 24 October. The next day she headed for the Kuala Lumpur area and anchored in the Strait of Malacca on the 30th. She returned via Subic Bay to "Yankee Station" on 5 November and operated there through the 23rd. The next day, the carrier started home via Hong Kong and Japan, arriving at San Diego on 20 December 1966. She operated on the west coast until departing San Diego on 18 August 1967 and reached Pearl Harbor 10 days later to prepare for future action.

The general fleet drawdown of the late 1960s and early 1970s made the USS Kearsarge redundant. She was decommissioned on 13 February 1970. Following three years in the Reserve Fleet, she was stricken from the Naval Vessel Register in May 1973 and sold for scrapping in February 1974.

Kearsarge received two battle stars for Korean War service.

CVA-34 • USS ORISKANY

(Displacement: 27,100 tons; length: 888 feet; beam: 147'6" (ew); draft: 31 feet; speed: 33 knots; complement: 3,460 crew; armament: 8 5-inch guns, 14 3-inch guns, 80 aircraft; class: Oriskany.)

Oriskany (CVA-34), an attack aircraft carrier, was laid down on 1 May 1944 by the New York Naval Shipyard, launched on 13 October 1945, and sponsored by Mrs. Clarence Cannon. While still incomplete, her construction was suspended on 12 August 1947. She remained in a state of preservation until after the outbreak of hostilities in Korea in June 1950, then was rushed to completion. She was commissioned at the New York Naval Shipyard on 25 September 1950, with Captain Percy H. Lyon in command.

Oriskany departed New York on 6 December 1950 for carrier qualification operations off Jacksonville, Fla., followed by a Christmas call at Newport, R.I. She resumed operations off Jacksonville through 11 January 1951, when she embarked Carrier Air Group 1 for shakedown out of Guantanamo Bay, Cuba.

After major modifications at New York Naval Shipyard from 6 March through 2 April, she embarked Carrier Air Group 4 for training off Jacksonville, then departed Newport on 15 May 1951 for a Mediterranean deployment with the 6th Fleet.

For the next few months, she added her far-reaching air arm to the strength of the 6th Fleet, the silent, flexible, and controlling weapon of deterrence to overt Soviet aggression in the Mediterranean and the Near East. The mobile 7th Fleet was deeply committed to a shooting war to help restore the independence and freedom of South Korea. Oriskany was part of the affirmative answer to the crucial question of whether the Korean War would affect the Navy's ability to maintain the status quo in the Mediterranean.

Having swept from ports of Italy and France to those of Greece and Turkey, thence to the shores of Tripoli, Oriskany returned to Quonset Point, R.I., on 4 October 1951. She entered Gravesend Bay, New York, on 6 November 1951 to offload ammunition and to have her masts removed to allow passage under the East River Bridges to the New York Naval Shipyard. Overhaul included the installation of a new flight deck, steering system, and bridge. Work was completed by 15 May 1952, and the carrier steamed the next day to take on ammunition at Norfolk from 19-22 May. She then got underway to join the Pacific Fleet, steaming via Guantanamo Bay, Rio de Janeiro, Cape Horn, Valparaiso, and Lima, arriving in San Diego, Calif., on 21 July.

Following carrier qualifications for Air Group 102, Oriskany departed San Diego on 15 September 1952 to aid UN forces in Korea. She arrived at Yokosuka on 17 October and joined the Fast Carrier Task Force 77 off the Korean Coast on 31 October. Her aircraft struck hard with bombing and strafing attacks against enemy supply lines and coordinated bombing missions with surface gun strikes along the coast. Her pilots downed two Soviet-built MiG-15 jets and damaged a third on 18 November.

Strikes continued through 11 February, heaping destruction upon enemy artillery positions, troop emplacements, and supply dumps along the main battle front. Following a brief upkeep period in Japan, Oriskany returned to combat on 1 March 1953. She continued in action until 29 March, called at Hong Kong, then resumed air strikes on 8 April. She departed the Korean Coast on 22 April, touched at Yokosuka, and then departed for San Diego on 2 May, arriving there on 18 May.

Following readiness training along the California coast, Oriskany departed San Francisco on 14 September to aid the 7th Fleet watching over the uneasy truce in Korea, arriving at Yokosuka on 15 October. Thereafter she cruised the Sea of Japan, the East China Sea, and the area of the Philippines. After providing air support for Marine amphibious assault exercises at Iwo Jima, the carrier returned to San Diego on 22 April 1954. She entered San Francisco Naval Shipyard for a modernization overhaul completed on 22 October, when she stood out to sea for the first of a series of coastal operations.

Oriskany arrived at Yokosuka from San Francisco on 2 April 1955 and operated with the Fast Carrier Task Force ranging from Japan and Okinawa to the Philippines. This deployment ended on 7 September, and the carrier arrived at Alameda, Calif., on 21 September.

The attack carrier cruised the California Coast while qualifying pilots of Air Group 9, then put to sea from Alameda on 11 February 1956 for another rigorous Westpac deployment. She returned to San Francisco on 13 June and entered the shipyard for overhaul on 1 October. She decommissioned there on 2 January 1957 for modernization work that included a new angled flight deck and enclosed hurricane bow. New, powerful steam catapults were installed by the Puget Sound Naval Shipyard, Bremerton, Wash.

Oriskany recommissioned at the San Francisco Naval Shipyard on 7 March 1959, Capt. James Mahan Wright in command. Four days later, she departed for shakedown out of San Diego with Carrier Air Group 14 embarked. Operations along the west coast continued until 14 May 1960, when she again deployed to Westpac, returning to San Diego on 15 December. She entered San Francisco Naval Shipyard on 30 March 1961 for a five-month overhaul that included the first aircraft carrier installation of the Naval Tactical Data System (NTDS).

Oriskany departed the shipyard on 9 September for underway training out of San Diego until 7 June 1962 when she again deployed to the Far East with Carrier Air Group 16 embarked. She returned to San Diego on 17 December 1962 for operational readiness training off the west coast.

The carrier again stood out of San Diego on 1 August 1963 for Far Eastern waters, with Carrier Air Group 16 embarked. She arrived at Subic Bay on 31 August 1963; thence to Japan. She stood out of Iwakuni, Japan, the morning of 31 October en route to the coast of South Vietnam. There she stood by for any eventuality as word was received of the coup d'etat taking place in Saigon. When the crisis abated, the carrier resumed operations from Japanese ports.

Oriskany returned to San Diego on 10 March 1964. After overhaul at Puget Sound Naval Shipyard, she steamed for refresher training out of San Diego, followed by qualifications for Carrier Wing 16. During this period, her flight deck was used to test the E-2A "Hawkeye," the Navy's new airborne early warning aircraft. She also provided orientation to senior officers of eight allied nations.

Oriskany departed San Diego on 5 April 1965 for Westpac, arriving at Subic Bay on 27 April. By this time, more United States troops had landed in South Vietnam to support Vietnamese troops against increased Viet Cong pressure to destroy the independence of that nation. Oriskany added her weight to the massive American naval strength supporting the freedom of South Vietnam. In combat operations that brought her and embarked Carrier Wing 16 the Navy Unit Commendation for exceptionally meritorious service between 10 May and 6 December 1965, she carried out over 12,000 combat sorties and delivered nearly 10,000 tons of ordnance against enemy forces. She departed Subic Bay on 30 November and returned to San Diego on 16 December.

Oriskany again stood out of San Diego for the Far East on 26 May 1966, arriving at Yokosuka on 14 June. She steamed for "Dixie Station" off South Vietnam on 27 June. Wearisome days and nights of combat shifted to "Yankee Station" in the Gulf of Tonkin on 8 July. In the following months, there were brief respites for replenishment in Subic Bay, then back into the action that saw her launch 7,794 combat sorties.

The carrier was on station the morning of 27 October 1966 when a fire erupted on the starboard side of the ship's forward hangar bay and raced through five decks, claiming the lives of 44 men. Many who lost their lives were veteran combat pilots who had flown raids over Vietnam a few hours earlier. Oriskany had been put in danger when a magnesium parachute flare exploded in the forward flare locker of Hangar Bay 1 beneath the carrier's flight deck. Her crewmen performed fantastic feats in jettisoning heavy bombs which lay within reach of the flames. Other men wheeled planes out of danger, rescued pilots, and helped quell the blaze through three hours of prompt and daring actions. Medical assistance was rushed to the carrier from sister aircraft carriers Constellation and Franklin D. Roosevelt.

Oriskany steamed to Subic Bay on 28 October, where victims of the fire were transferred to waiting aircraft for transportation to the United States. A week later the carrier departed for San Diego, arriving 16 November. San Francisco Bay Naval Shipyard completed repairs on 23 March 1967, and Oriskany, with Carrier Air Wing 16 embarked, underwent training. She then stood out of San Francisco Bay on 16 June to take station in waters off Vietnam. Designated flagship of Carrier Division 9 in Subic Bay on 9 July, she commenced "YankeeStation" operations on 14 July. While on the line 26 July she provided medical assistance to the fire-ravaged attack carrier Forrestal.

Oriskany returned to the Naval Air Station pier at Alameda, California, on 31 January 1968 and entered San Francisco Bay Naomi Shipyard on 7 February for an eight-month overhaul. Upon completion of work, the carrier underwent refresher training and flight qualifications before deploying to the Far East in April 1969. Into 1970, she continued to serve her nation in the Pacific.

Oriskany received two battle stars for Korean Service and five for Vietnamese service.

CV-36 • USS ANTIETAM

(Displacement: 27,100 tons; length: 872 feet; beam: 93 feet; draft: 28 feet 7 inches; speed: 32.7 knots; complement: 3,448 crew; armament: 12 five-inch guns; class: Essex.)

The second Antietam (CV-36) was laid down on 15 March 1943 by the Philadelphia Navy Yard; launched on 20 August 1944; sponsored by Mrs. Millard E. Tydings, the wife of Senator Tydings of Maryland; and commissioned on 28 January 1945, Capt. James R. Tague in command.

The aircraft carrier completed fitting out at Philadelphia until 2 March when she got underway for her shakedown cruise. The ship arrived in Hampton Roads on the 5th and conducted operations from Norfolk until 22 March when she stood out of Chesapeake Bay bound for Trinidad in the British West Indies. At the conclusion of her shakedown cruise, Antietam returned to Philadelphia on 28 April to begin post-shakedown availability. She completed repairs on 19 May and departed Philadelphia that same day. After a three-day stop at Norfolk, the warship resumed her voyage to the Panama Canal in company with Higbee (DD-806), George W. Ingram (APD-43), and Ira Jeffery (APD-44). She arrived at Cristobal on 31 May, transited the canal the next day, and continued her voyage up the coast to San Diego. She stopped at San Diego from 10 to 13 June before beginning the first leg of her transpacific voyage. Antietam arrived in Pearl Harbor on the 19th and remained in the Hawaiian Islands conducting training missions until 12 August. On that day, she shaped a course for the western Pacific.

Three days out of Oahu, she received word of the Japanese capitulation and the consequent cessation of hostilities. Thus, by the time of her arrival in Eniwetok Atoll on the 19th, her mission had changed from combat to occupation support duty. On the 21st, she exited the lagoon in company with Cabot (CVL-28) and a screen of destroyers bound for Japan. En route, she suffered some internal damage which forced her into port at Apra Harbor, Guam, for inspections. The inspection party deemed the damage minimal; and the carrier remained operational, resuming her course on the 27th. By that time, however, her destination had been changed to the coast of the Asian mainland. She stopped at Okinawa between 30 August and 1 September and arrived in Chinese waters near Shanghai the following day.

The aircraft carrier remained in the Far East for a little more than three years. The Yellow Sea constituted her primary theater of operations while her air group provided support for the Allied occupation of North China, Manchuria, and Korea. During the latter stages of that assignment, her airmen conducted surveillance missions in that area as a result of the civil war in China between communist and nationalist factions, which later resulted in the expulsion of Chiang Kai-shek's forces from mainland China and the establishment of Mao Tse-tung's communist People's Republic of China. Throughout the period, however, she did depart the Yellow Sea on occasion for visits to Japan, the Philippines, Okinawa, and the Marianas. Early in 1949, she completed her mission in the Orient and headed back to the United States for deactivation.

Antietam remained in reserve at Alameda, Calif., until communist forces from the north invaded South Korea in the summer of 1950. She began reactivation preparations on 6 December and went back into commission on 17 January 1951, Capt. George J. Dufek in command. Initially, the carrier conducted shakedown training and carrier qualifications along the California coast, first out of Alameda and-after 14 May-out of San Diego. She made one voyage to Pearl Harbor and back to San Diego in July and August before departing the latter port on 8 September and heading for the Far East. Antietam arrived in the Far East later that fall and, by late November, began the only combat deployment of her career. During that tour, she made four cruises with Task Force (TF) 77, in the combat zone off the coast of Korea. In between fighting assignments, she returned to Yokosuka, Japan. During each of those periods, her air group carried out a variety of missions in support of United Nations forces combatting North Korean aggression. Those missions included combat air patrol, logistics interdiction-particularly against railroad and highway traffic-reconnaissance, antisubmarine patrols, and night heckler missions. Between late November 1951 and mid-March 1952, Antietam's air group flew nearly 6,000 sorties of all types. She returned to Yokosuka on 21 March 1952 at the conclusion of her fourth cruise with TF 77 to begin preparations for her voyage back to the United States.

The aircraft carrier returned home in April and rejoined the Pacific Reserve Fleet briefly. She was reactivated later that summer and, in August, transited the Panama Canal to join the Atlantic Fleet. In September, the warship entered the New York Naval Shipyard for major alterations. In October, she was redesignated an attack aircraft carrier, CVA-36. In December, Antietam emerged from the yard as America's first angled-deck aircraft carrier. She operated out of Quonset Point, R.I., until the beginning of 1955. During the intervening years, she participated in numerous fleet and independent ship's exercises. After August 1953-at which time she was redesignated an antisubmarine warfare (ASW) carrier, CVS-36-Antietam concentrated upon honing her hunter/killer skills. In January 1955, she embarked upon a voyage to the Mediterranean Sea where she served with the 6th Fleet until March. Resuming duty with the Atlantic Fleet ASW forces, she operated along the eastern seaboard until the fall of 1956. In October of that year, she cruised to the waters of the eastern Atlantic for NATO ASW exercises and goodwill visits to ports in Allied countries. While the carrier was in Rotterdam, the Suez crisis broke out in the eastern Mediterranean. Antietam cut short her visit to the Netherlands and headed for the "middle sea" to bolster the 6th Fleet during the evacuation of American citizens from Alexandria, Egypt. At the end of that assignment, she conducted ASW training exercises with Italian naval officers embarked before returning to Quonset Point on 22 December.

After resuming operations along the eastern seaboard early in 1957, Antietam was assigned on 21 April 1957 to training duty with the Naval Air Training Station, Pensacola, Fla. Mayport, however, served as her home port because ships of her draft could not then enter port at Pensacola. For almost two years, the aircraft carrier operated out of Mayport, training new

Navy pilots and conducting tests on new aviation equipment-most noteworthy, on the Bell automatic landing system during August of 1957. She also participated in annual Naval Academy midshipmen cruises each summer.

In January 1959, after the deepening of the channel into Pensacola had been completed, Antietam's home port was changed from Mayport to Pensacola. For the remainder of her active career, the carrier operated out of Pensacola as an aviation training ship. On two occasions, she provided humanitarian services to victims of hurricane damage. The first came in September of 1961 when she rushed to the Texas coast to provide supplies and medical assistance to the victims of hurricane Carla. The second came just over a month later when she carried medical supplies, doctors, nurses, and other medical personnel to British Honduras to help the victims of hurricane Hattie. Otherwise, she spent the final four years of her naval career in routine naval aviation training duty out of Pensacola. On 23 October 1962, Antietam was relieved by Lexington (CVS-16) as aviation training ship at Pensacola and was placed in commission, in reserve, on 7 January 1963. She remained in that status until she was decommissioned on 8 May 1963. Berthed at Philadelphia, Pa., she remained in reserve until May of 1973 when her name was struck from the Navy list. On 28 February 1974, she was sold to the Union Minerals & Alloys Corp. for scrapping.

Antietam (CV-36) earned two battle stars for service in the Korean conflict.

CV-37 • USS PRINCETON

(Displacement: 33,000 tons length: 888 feet beam: 93 feet; extreme width at flight deck: 147½ feet draft: 28 feet 7 inches speed: 33 knots complement: 3,448 crew armament: 12 5-inch guns, 44 40mm guns, 59 20mm guns aircraft: 80+ class: Essex.)

The fifth Princeton was laid down as Valley Forge at the Philadelphia Navy Yard on 14 September 1943, renamed Princeton on 21 November 1944, launched on 8 July 1945, sponsored by Mrs. Harold Dodds, and commissioned on 18 November 1945, Capt. John M. Hoskins in command.

Following shakedown off Cuba, Princeton, with Air Group 81 embarked, remained in the Atlantic and operated with the 8th Fleet until June 1946. She was then transferred to the Pacific Fleet, arriving at San Diego on 30 June and departing again on 3 July to carry the body of Philippine President Manuel Quezon back to Luzon for burial. From Manila, Princeton joined the 7th Fleet in the Marianas, becoming the flagship of TF 77.

During September and October 1946, she operated in Japanese and Chinese waters, then returned to the Marianas where she remained until February 1947. Maneuvers in Hawaiian waters preceded her return to San Diego on 15 March. She cruised the west coast, Hawaiian waters, and the western Pacific (1 October - 23 December) during 1948. She then prepared for inactivation and on 20 June 1948, decommissioned and joined other capital ships in the Pacific Reserve Fleet.

Reactivated with the outbreak of hostilities in Korea fifteen months later, Princeton recommissioned on 28 August 1950. Intensive training refreshed her Naval Reserve crew and on 5 December she joined TF 77 off the Korean coast, her planes and pilots (Air Group 19) making possible the reinstitution of jet combat air patrols over the battle zone. She launched 248 sorties against targets in the Hagaru area to announce her arrival, and for the next six days continued the pace to support Marines fighting their way down the long, cold road from the Chosin Reservoir to Hungnam. By the 11th, all units had reached the staging area on the coast. Princeton's planes, with other Navy, Marine, and Air Force squadrons, then covered the evacuation from Hungnam through its completion on 24 December 1950.

Interdiction missions followed, and by 4 April 1951, Princeton's planes had rendered 54 rail and 37 highway bridges inoperable and damaged 44 more. In May, they flew against the railroad bridges connecting Pyongyang with Sunchon, Sinanju, Kachon, and the trans-peninsula line. Next, they combined close air support with raids on power sources in the Hwachon Reservoir area and, with the stabilization of the front there, resumed interdiction. For much of the summer, they pounded supply arteries, concentrating on highways, and in August, Princeton got underway for the United States, arriving at San Diego on 21 August 1951.

Eight months later, on 30 April 1952, Princeton rejoined TF 77 in the combat zone. For 138 days, her planes flew against the enemy. They sank small craft to prevent the recapture of offshore islands; blasted concentrations of supplies, facilities, and equipment behind enemy lines; participated in air-gun strikes on coastal cities; pounded the enemy's hydroelectric complex at Suiho on the Yalu to turn off power on both sides of that river; destroyed gun positions and supply areas in Pyongyang; and closed mineral processing plants and munitions factories at Sindok, Musan, Aoji, and Najin. Reclassified CVA-37 on 1 October 1952, Princeton returned to California on 3 November for a two-month respite from the western Pacific. In February 1953, she was back off the Korean coast and until the end of the conflict launched planes for close air support, "Cherokee" strikes against supply, artillery, and troop concentrations in enemy territory, and against road traffic. She remained in the area after the truce on 27 July 1953 and on 7 September got underway for San Diego.

In January 1954, Princeton was reclassified CVS-37 and, after conversion at Bremerton, took up antisubmarine/Hunter-Killer (HUK) training operations in the eastern Pacific. For the next five years, she alternated HUK exercises off the west coast with similar operations in the western Pacific and, in late 1957 to early 1958, in the Indian Ocean-Persian Gulf area.

Reclassified again on 2 March 1959, she emerged from conversion as an amphibious assault carrier, LPH-5. Capable of transporting a battalion landing team and carrying helicopters in place of planes, Princeton's mission became that of vertical envelopment — the landing of Marines behind enemy beach fortifications and providing logistics and medical support as they attack from the rear to seize critical points, cut enemy supplies, sever communications, and link up with assault forces landed on the beaches.

From May until January 1960, Princeton trained with Marine units from Camp Pendleton, then deployed to WestPac to train in Okinawan waters. For the next three years, she followed a similar schedule, gaining experience in her primary mission. Interruptions came in October 1961 when she rescued survivors

of the merchantmen Pioneer, Muse, and Sheik grounded on Daito Shima and in April 1962 when she delivered Marine Corps advisers and helicopters to Soc Trang in the Mekong Delta area of the Republic of South Viet Nam.

In October 1964, Princeton exchanged WestPac training for the real thing as she returned to Viet Nam and joined the Pacific Fleet's Ready Group in operations against North Vietnamese and Viet Cong forces. Combat operations, interrupted in November for flood relief work, continued into the new year, 1965, and culminated in May off Chu Lai as she carried out her primary mission, vertical envelopment, for the first time in combat.

Returning to her homeport of Long Beach after that operation, she transported Marine Air Group 36 to Viet Nam in August, and in February 1966 got underway for another tour in the combat zone. Relieving USS Okinawa (LPH-3) as flagship for the Amphibious Ready Group, she engaged the enemy in operations Jackstay, 26 March-6 April, to clear the Rung Sat Special Zone of Viet Cong guerrillas, and Osage, 27 April-4 May, to protect Vietnamese in the Phu Loc area from Viet Cong "harassment."

Search and destroy missions against Viet Cong and North Vietnamese Army units followed as Princeton provided transportation, medical evacuation, logistics, and communication support for the amphibious Operation Deckhouse I, 18-27 June 1966, in the Song Cau district and the Song Cai river valley, then supported 1st Air Cavalry and 101st Airborne units engaged in Operation Nathan Hale to the south of the Deckhouse I area. Deckhouse II and support for Operation Hastings followed as Navy, Marine, and Army units again combined, this time to impede enemy infiltration from the DMZ.

After Hastings, Princeton sailed for home, arriving on 2 September 1966. She deployed again to Viet Nam, 30 January-19 June 1967, and again ranged along that long embattled, highly indented coast. In March 1967, she assisted in countering an enemy threat to the Marine artillery base at Gio Linh and evacuated wounded from Con Thien mountain. In April, she participated in Operation Beacon Star, in the Khe Sanh area, and supported search and destroy operations in conjunction with Operation Shawnee. In May, her helicopters lifted Marines to the DMZ to block enemy forces withdrawing across the Ben Hai river.

A much-needed overhaul followed Princeton's return to the west coast and in May 1968 she again sailed west to Viet Nam. There, as flagship for Amphibious Ready Group Alpha, she provided amphibious assault carrier services for operations Fortress Attack III and IV, Proud Hunter, Swift Pursuit, and Eager Hunter.

In December 1968, she returned to the United States and in April 1969 she was designated the prime recovery ship for Apollo 10, the lunar mission which paved the way for Apollo 11 and the first landing on the moon. On 26 May, Apollo 10 astronauts Thomas P. Stafford, USAF; John W. Young, USN; and Eugene A. Cernan, USN, were recovered by a helicopter from HS-4 off Princeton after their eight-day orbit of the earth. Completing that mission successfully, Princeton resumed exercises off the west coast. After two and a half decades of service, Princeton was decommissioned and stricken from the Naval Vessel Register on 30 January 1970. She was sold for scrapping in May 1971. Princeton received 8 battle stars for service during the Korean Conflict.

CV-38 • USS SHANGRI-LA

(Displacement: 27,100 tons; length: 888 feet; beam: 93 feet; extreme width at flight deck: 147.5 feet; draft: 28 feet 7 inches; speed: 32.7 knots; complement: 3,448 crew; armament: 12 5-inch guns, 44 40mm guns, 60 20mm guns; aircraft: 80+; class: Essex.)

Shangri-La, an aircraft carrier, was laid down by the Norfolk Navy Yard in Portsmouth, Va., on 15 January 1943, launched on 24 February 1944, sponsored by Mrs. James H. Doolittle, and commissioned on 15 September 1944, with Capt. James D. Barner in command.

Shangri-La completed fitting out at Norfolk and took her shakedown cruise to Trinidad, B.W.I., between 15 September and 21 December 1944, at which time she returned to Norfolk. On 17 January 1945, she stood out of Hampton Roads, formed up with the large cruiser USS Guam (CB-2) and USS Harry E. Hubbard (DD-748), and sailed for Panama. The three ships arrived at Cristobal, C.Z., on the 23rd and transited the canal on the 24th. Shangri-La departed from Balboa, C.Z., on 25 January and arrived at San Diego, Calif., on 4 February. There, she loaded passengers, stores, and extra planes for transit to Hawaii and got underway on 7 February. Upon her arrival at Pearl Harbor on 15 February, she commenced two months of duty, qualifying land-based Navy pilots in carrier landings.

On 10 April 1945, she weighed anchor for Ulithi Atoll, where she arrived ten days later. After an overnight stay in the lagoon, Shangri-La departed Ulithi in the company of USS Haggard (DD-555) and USS Stembel (DD-644) to report for duty with Vice Adm. Marc A. Mitscher's Fast Carrier Task Force. On 24 April, she joined Task Group 58.4 while it was conducting a fueling rendezvous with TG 50.8. The next day, Shangri-La and her air group, CVG-85, launched their first strike against the Japanese. The target was Okino Daito Jima, a group of islands several hundred miles to the southeast of Okinawa. Her planes successfully destroyed radar and radio installations there and, upon their recovery, the task group sailed for Okinawa. Shangri-La supplied combat air patrols for the task group and close air support for the 10th Army on Okinawa before returning to Ulithi on 14 May.

While at Ulithi, Shangri-La became the flagship of the 2nd Carrier Task Force. Vice Adm. John S. McCain hoisted his flag on

Shangri-La on 18 May 1945. Six days later, TG 58.4, with Shangri-La in company, sortied from the lagoon. On 28 May, TG 58.4 became TG 38.4, and Vice Adm. McCain relieved Vice Adm. Mitscher as Commander, Task Force 38, retaining Shangri-La as his flagship. On 2 and 3 June, the task force launched air strikes on the Japanese home islands, aimed particularly at Kyushu, the southernmost of the major islands. Facing the stiffest airborne resistance to date, Shangri-La's airmen suffered their heaviest casualties.

On 4 and 5 June, she moved off to the northwest to avoid a typhoon; then, on the 6th, her planes returned to close air support duty over Okinawa. On the 8th, her air group hit Kyushu again, and, on the following day, they came back to Okinawa. On 10 June 1945, the task force cleared Okinawa for Leyte, conducting drills en route. Shangri-La entered Leyte Gulf and anchored in San Pedro Bay on 13 June. She remained at anchor there for the rest of June, engaged in upkeep and recreation. On 1 July, Shangri-La got underway from Leyte to return to the combat zone. On 2 July, the oath of office of Assistant Secretary of the Navy for Air was administered to John L. Sullivan on board Shangri-La, the first ceremony of its type ever undertaken in a combat zone. Eight days later, her air group commenced a series of air strikes against Japan, which lasted until the capitulation on 15 August.

Shangri-La's planes ranged the length of the island chain during these raids. On 10 July, they attacked Tokyo, the first raid there since the strikes of the previous February. On 14 and 15 July, they pounded Honshu and Hokkaido and, on the 18th, returned to Tokyo, also bombing the battleship Nagato, moored close to shore at Yokosuka. From 20 to 22 July, Shangri-La joined the logistics group for fuel, replacement aircraft, and mail. By the 24th, her pilots were attacking shipping in the vicinity of Kure. They returned the next day for a repeat performance, before departing for a two-day replenishment period on the 26th and 27th. On the following day, Shangri-La's aircraft damaged the cruiser Oyoda and the battleship Haruna, the latter so badly that she beached and flooded. She later had to be abandoned. They pummeled Tokyo again on 30 July, then cleared the area to replenish on 31 July and 1 August.

Shangri-La spent the next four days in the retirement area waiting for a typhoon to pass. On 9 August, after heavy fog had caused the cancellation of the previous day's missions, the carrier sent her planes aloft to bomb Honshu and Hokkaido once again. The next day, they raided Tokyo and central Honshu, then retired from the area for logistics. She evaded another typhoon on 11 and 12 August, then hit Tokyo again on the 13th. After replenishing on the 14th, she sent planes to strike the airfields around Tokyo on the

morning of 15 August 1945. Soon thereafter, Japan's capitulation was announced, and the fleet was ordered to cease hostilities. Shangri-La steamed around in the strike area from 15 to 23 August, patrolling the Honshu area on the latter date. Between 23 August and 16 September, her planes sortied on missions of mercy, air-dropping supplies to Allied prisoners of war in Japan.

Shangri-La entered Tokyo Bay on 16 September, almost two weeks after the surrender ceremony on board USS Missouri (BB-63), and remained there until 1 October. Departing Japan, she arrived at Okinawa on 4 October, stayed until the 6th, and then headed for the United States in company with Task Unit 38.1.1. She sailed into San Pedro Bay, Calif., on 21 October and stayed at Long Beach for three weeks. On 5 November, she shifted to San Diego, departing that port a month later for Bremerton, Wash. She entered Puget Sound on 9 December, underwent availability until the 30th, and then returned to San Diego.

Upon her return, Shangri-La began normal operations out of San Diego, primarily engaged in pilot carrier landing qualifications. In May 1946, she sailed for the Central Pacific to participate in Operation Crossroads, the atomic bomb tests conducted at Bikini Atoll. Following this, she made a brief training cruise to Pearl Harbor, then wintered at Puget Sound Naval Shipyard. In March 1947, she deployed again, calling at Pearl Harbor and Sydney, Australia. When she returned to the United States, Shangri-La was decommissioned and placed in the Reserve Fleet at San Francisco on 7 November 1947.

Shangri-La recommissioned on 10 May 1951, with Capt. Francis L. Busey in command. For the next year, she conducted training and readiness operations out of Boston, Mass. Reclassified as an attack aircraft carrier, CVA-38, in 1952, she returned to Puget

Sound that fall and decommissioned again on 14 November, this time for modernization at Puget Sound Naval Shipyard.

During the next two years, she received an angled flight deck, twin steam catapults, and overhauled her aircraft elevators and arresting gear. At a cost of approximately $7 million, she was virtually a new ship when she was commissioned for the third time on 10 January 1955, with Capt. Roscoe L. Newman commanding.

She conducted intensive fleet training for the remainder of 1955, then deployed to the Far East on 5 January 1956. On 2 September 1956, the second day of the National Air Show, Lt. (j. g.) R. Carson, flying an F3H-2N Demon of VF-124, captured the McDonnell Trophy with a non-stop, non-refueling flight from Shangri-La off the coast of San Francisco to Oklahoma City. Lt. (j. g.) Carson covered the 1,436 miles in two hours, 32 minutes, and 13.45 seconds for an average speed of 566.007 mph.

On 16 March 1960, she put to sea from San Diego en route to her new home port, Mayport, Fla. She entered Mayport after visits to Callao, Peru; Valparaiso, Chile; Port of Spain, Trinidad; Bayonne, N.J.; and Norfolk, Va.

After six weeks of underway training in the local operating area around Guantanamo Bay, Cuba, she embarked upon her first Atlantic deployment, a NATO exercise followed by liberty in Southampton, England. Almost immediately after her return to Mayport, Shangri-La was ordered back to sea, this time to the Caribbean in response to trouble in Guatemala and Nicaragua. She returned to Mayport on 25 November 1960 and remained in port for more than two months.

Between 1961 and 1970, Shangri-La alternated between deployments to the Mediterranean and operations in the western Atlantic out of Mayport. She sailed east for her first tour of duty with the 6th Fleet on 2 February 1961. On 1 June 1961, Shangri-La, along with USS Intrepid (CV 11) and USS Randolph (CV 15), was ordered to stand by off southern Hispaniola when a general uprising seemed about to follow the assassination of President Trujillo of the Dominican Republic.

She returned to the United States that fall and entered the New York Naval Shipyard. Back in Mayport by the beginning of 1962, Shangri-La stood out again for the Mediterranean on 7 February 1962. After about six months of cruising with the 6th Fleet, she departed the Mediterranean in mid-August and arrived in Mayport on the 28th.

Following a month's stay at her home port, the aircraft carrier headed for New York and a major overhaul. Shangri-La was modified extensively during her stay in the yard. Four of her 5-inch mounts were removed, but she received a new air search and height-finding radar and a new arrester system. In addition, much of her electrical and engineering equipment was renovated. After sea trials and visits to Bayonne, N.J., and Norfolk, Va., Shangri-La returned to Mayport for a week in late March 1963, then put to sea for operations in the Caribbean. Eight months of similar duty followed before Shangri-La weighed anchor for another deployment. On 1 October 1963, she headed back to the 6th Fleet for a seven-month tour.

Shangri-La continued her 2nd and 6th Fleet assignments for the next six years. During the winter and spring of 1964, she underwent another extensive overhaul, this time in Philadelphia, and then resumed operations as before. On 30 June 1969, she was redesignated an antisubmarine warfare aircraft carrier, CVS-38.

In 1970, Shangri-La returned to the western Pacific after an absence of ten years. She got underway from Mayport on 5 March, stopped at Rio de Janeiro, Brazil, from the 13th to the 16th, and headed east through the Atlantic and Indian oceans. She arrived in Subic Bay, R.P., on 4 April and, during the next seven months, launched combat sorties from Yankee station. Her tours of duty on Yankee station were punctuated by frequent logistics trips to Subic Bay, visits to Manila, R.P., and Hong Kong, B.C.C., in October, and 12 days in drydock at Yokosuka, Japan, in July.

On 9 November 1970, Shangri-La stood out of Subic Bay to return home. En route to Mayport, she visited Sydney, Australia; Wellington, N.Z.; and Rio de Janeiro, Brazil. She arrived in Mayport on 16 December and began preparations for inactivation. After pre-inactivation overhaul at the Boston Naval Shipyard, South Annex, Shangri-La decommissioned on 30 July 1971. She was placed in the Atlantic Reserve Fleet and berthed in Philadelphia.

Shangri-La remained in the reserve fleet for the next 11 years and was stricken from the Navy List on 15 July 1982. On 9 August 1988, the Maritime Administration disposed of her.

CV-39 • USS CHAMPLAIN

(Displacement: 27,100 tons length: 888 feet beam: 93 feet; extreme width at flight deck: 147½ feet draft: 28 feet 7 inches speed: 33 knots complement: 3,448 crew armament: 12 5-inch guns, 72 40mm guns class: Essex.)

The second Lake Champlain (CV-39) was laid down in drydock by the Norfolk Navy Yard, Portsmouth, Va., on 15 March 1943. It was launched by float on 2 November 1944, sponsored on 3 June 1945 by Mrs. Warren Austin, wife of Senator Austin of Vermont, and commissioned the same day, Capt. Logan C. Ramsey in command.

After shakedown and visits to New York and Philadelphia, Lake Champlain was assigned to "Magic Carpet" duty. She departed Norfolk for England on 14 October 1945 and arrived at Southampton on the 19th, where she embarked on veterans and returned them to New York.

She set a speed record for crossing the Atlantic on 26 November 1945 when she arrived at Hampton Roads, Va., having completed a run from Cape Spartel, Africa, in 4 days, 8 hours, and 51 minutes. This record stood until surpassed by SS United States in the summer of 1952.

Lake Champlain retired to the "Mothball Fleet" at Norfolk, Va., on 17 February 1947. After the United States had allowed her active military strength to shrink to the danger point, the Communists struck in Korea. Fortunately, the U.S. had ships in reserve, though it took time to obtain and train crews and provide material. Lake Champlain was reactivated and modernized at Newport News Shipbuilding and Drydock Co., and recommissioned on 19 September 1952, Capt. G. T. Mundroff in command. After shakedown in Cuban and Haitian waters, from 25 November through 25 December 1952, the carrier departed Mayport, Fla., for Korea on 26 April 1953 via the Red Sea, Indian Ocean, and China Sea, becoming the largest ship to transit the Suez Canal up to that time. She moored at Yokosuka, Japan, on 9 June 1953.

As flagship of Carrier Task Force 77, she sailed from Yokosuka on 11 June 1953 and arrived off western Korea on 14 June. The carrier's air group immediately launched sorties cratering runways; assaulting enemy troops; attacking trenches, bunkers, and gun positions; and providing close air support to hard-pressed ground forces. Her planes also escorted B-29 bombers on their way to enemy targets. Lake Champlain continued to strike at the enemy until the truce was signed on 27 July. Relieved by USS Kearsarge (CVA-33) on 11 October 1953, Lake Champlain headed toward the South China Sea, arriving at Singapore on 24 October. Bidding farewell to the Pacific Ocean on 27 October, she steamed toward home, touching at Colombo, Port Said, Cannes, and Lisbon before arriving at Mayport, Fla., on 4 December 1953.

In the years that followed, Lake Champlain made several cruises to the Mediterranean, participating with NATO forces. On 25 April 1957, she joined elements of the fleet in a high-speed run to the scene of tension in the Middle East, cruising in the vicinity of Lebanon and backing Jordan's stand against the threat of Communism. The swift and firm reaction averted a near catastrophe in the Middle East. Tension eased and Lake Champlain returned to Mayport on 27 July.

Converted to an antisubmarine carrier and reclassified (CVS-39) on 1 August 1957, Lake Champlain trained off the eastern seaboard to master her new role. She departed Bayonne, N.J., on 8 February 1958 for a Mediterranean cruise. While in the Mediterranean, she arrived on 16 October 1957 at Valencia, Spain, and provided aid to thousands made homeless by a flood. Lake Champlain returned on 30 October to Mayport, Fla. After yard overhaul, she again departed for the Mediterranean on 10 June 1958 and visited Spain, Denmark, and Scotland, before returning to Mayport on 9 August.

The carrier operated off Florida and in the Caribbean until 15 June 1958 when she sailed on another Mediterranean cruise, returning to her newly assigned homeport, Quonset Point, R.I., on 4 September.

The carrier operated out of Quonset Point, R.I., until 29 June 1960 when she made a midshipmen cruise to Halifax, returning on 12 August. Beginning on 7 February 1961, she made a cruise to the Caribbean, returning on 2 March.

Lake Champlain was selected as the prime recovery ship for the first manned space flight. She sailed for the recovery area on 1 May 1961, and was on station on 5 May when Cmdr. Alan Shepard splashed down in spacecraft Freedom 7, some 300 miles downrange from Cape Kennedy. Helicopters from the carrier visually followed the descent of the capsule and were over the astronaut two minutes after the impact. They skillfully recovered Astronaut Shepard and Freedom 7 and carried them safely to Lake Champlain's flight deck.

For the next year, the ship operated along the Atlantic coast and in the Caribbean. In June 1962, she embarked Naval Academy midshipmen for a summer cruise to Halifax, Nova Scotia, and

Kingston, Jamaica, where she represented the United States at the island's celebration of its independence on 3 August.

On 24 October 1962, Lake Champlain joined in a classic exercise of sea power — the quarantine of Cuba, where the Soviet Union was constructing bases for offensive missiles. To block this grave threat, U.S. warships deployed throughout the western Atlantic, choking off the flow of military supplies to Cuba and enforcing American demands for the withdrawal of the Russian offensive missiles.

After the American demands were substantially complied with, Lake Champlain sailed for home on 23 November, via St. Thomas, Virgin Islands, and arrived at Quonset Point on 4 December 1962. For the next few months, the carrier was in New England waters for operations and overhaul. In mid-October 1963, four Navy ships, including Lake Champlain and the amphibious assault ship USS Thetis Bay (LPH-6), aided by Navy and Marine Corps cargo aircraft from east coast stations, delivered nearly 375 tons of food, clothing, and medical supplies donated by relief agencies to the people of Haiti after that country was devastated by Hurricane Flora.

Lake Champlain returned to Quonset Point on 9 November 1963 for operations in New England waters. She visited Bermuda briefly in the spring of 1964 and steamed to Spain in the fall for landings near Huelva. She sailed on 6 November from Barcelona for the United States, touched at Gibraltar, and arrived at Quonset Point on 25 November 1964.

On 19 January 1965, Lake Champlain recovered an unmanned Project Gemini space capsule launched from Cape Kennedy, Fla., after a suborbital flight 1,879 miles down the Atlantic Missile Range and to within 16 miles of the carrier.

The last major duty of her career occurred on 5 August 1965 when she served as the primary recovery ship for Gemini 5.

Gemini 5 splashed down into the Atlantic 90 miles off target after a record-breaking eight-day space flight, and 45 minutes later, Navy frogmen helped astronauts Gordon Cooper and Charles Conrad out of their space capsule and aboard a helicopter for the ride back to Lake Champlain. Soon after this duty was completed, she sailed to Philadelphia, where she commenced inactivation. She was decommissioned on 2 May 1966.

The 24-year-old Lake Champlain was stricken from the Navy List on 1 December 1969 and sold by the Defense Reutilization and Marketing Service (DRMS) for scrapping on 28 April 1972.

CV-40 • USS TARAWA

Construction and Commissioning: The first USS Tarawa (CV-40) was laid down on 1 March 1944 at the Norfolk Navy Yard. She was launched on 12 May 1945, sponsored by Mrs. Julian C. Smith, the wife of Lieutenant General Julian C. Smith, USMC, who commanded the 2nd Marine Division at Tarawa. The ship was commissioned on 8 December 1945, with Captain Alvin Ingersoll Malstrom in command.

Early Service: Tarawa remained in the Norfolk area until 15 February 1946, when she sailed for shakedown training near Guantanamo Bay, Cuba. She briefly returned to Norfolk on 16 April, visited New York later that month, and came back to Norfolk on 30 April. After completing her post-shakedown overhaul in late June, she departed Hampton Roads on 28 June for the west coast, transiting the Panama Canal in early July and arriving in San Diego on 15 July 1946.

Western Pacific Deployment: Following training and upkeep, Tarawa left San Diego for the western Pacific, reaching Pearl Harbor on 7 August. After arriving at Saipan on 20 August, she operated in the Mariana Islands until late September, then headed to Japan, stopping at Yokosuka and Sasebo. She then proceeded to the northern coast of China, operating near Tsingtao until the end of October before returning to the Marianas.

Post-War Operations: From 7 November 1946, Tarawa operated in the Marianas, with a brief trip to Okinawa in January 1947. She then returned to Pearl Harbor and participated in fleet exercises near Kwajalein. After more training in the Hawaiian Islands, she headed for the west coast, arriving in San Francisco on 29 April 1947. Over the next 16 months, she conducted air operations out of San Francisco and San Diego.

Around-the-World Cruise: On 28 September 1948, Tarawa embarked on an extensive cruise, visiting Tsingtao, China; Hong Kong; Singapore; and Colombo, Ceylon, among other ports. She transited the Suez Canal in January 1949, visited Greece, Turkey, and Crete, and returned to Norfolk on 21 February 1949. After a period of normal operations along the east coast and in the

Caribbean, she was placed out of commission on 30 June 1949 and berthed with the Atlantic Reserve Fleet in New York.

Korean War Reactivation: Tarawa was reactivated on 30 November 1950 due to the Korean War. She was recommissioned on 3 February 1951 but did not see combat in Korea. Instead, she served with the 6th and 2nd Fleets as a replacement for carriers sent to the war zone. On 1 October 1952, she was redesignated as an attack aircraft carrier (CVA-40).

Cold War Service: In the spring of 1954, Tarawa operated in the Far East, including a deployment long after the Korean War armistice. She returned to the east coast later that year and underwent conversion to an antisubmarine warfare (ASW) aircraft carrier, redesignated CVS-40 on 10 January 1955. She operated mainly in the Atlantic, conducting barrier patrols and training missions until her decommissioning in May 1960.

Final Years and Decommissioning: Tarawa was placed in reserve in Philadelphia and redesignated AVT-12 in May 1961. She was struck from the Navy list on 1 June 1967 and sold for scrapping on 3 October 1968.

Throughout her service, USS Tarawa played a significant role in various naval operations, adapting to changing military needs and serving both in combat and training capacities.

CVB-41 • USS MIDWAY

(Dp: 45,000 tons length: 968 feet beam: 113 feet draft: 35 feet speed: 33 knots complement: 4,104 crew armament: 18 5-inch guns, 84 40mm guns, 68 20mm guns class: Midway).

The third Midway (CVB-41) was laid down on 27 October 1943 by Newport News Shipbuilding Co., Newport News, Va.; launched on 20 March 1945, sponsored by Mrs. Bradford William Ripley, Jr.; and commissioned on 10 September 1945, with Capt. Joseph F. Bolger in command.

After shakedown in the Caribbean, Midway joined the Atlantic Fleet training schedule, with Norfolk as her homeport. From 20 February 1946, she served as flagship for CarDiv 1. In March, she tested equipment and techniques for cold weather operations in the North Atlantic. A highlight of her east coast and Caribbean training was Operation Sandy in September 1947, where she test-fired a captured German V-2 rocket from her flight deck, marking the first such launch from a moving platform.

On 29 October 1947, Midway sailed for her first annual deployment with the 6th Fleet in the Mediterranean, acting as a powerful extension of sea/air power. Between deployments, she trained and received alterations necessary to accommodate heavier aircraft.

From 26 to 29 May 1952, the feasibility of the angled deck concept was demonstrated on a simulated angled deck aboard Midway by pilots from the Naval Air Test Center and the Atlantic Fleet. Midway also participated in North Sea maneuvers with NATO forces and was redesignated CVA-41 on 1 October 1952.

Midway cleared Norfolk on 27 December 1954 for a world cruise, sailing via the Cape of Good Hope to Taiwan, where she joined the 7th Fleet on 6 February 1955 for operations in the western Pacific. This was the first deployment of her class in the western Pacific. Midway remained with the 7th Fleet until 28 June 1955 when she sailed for overhaul at Puget Sound Naval Shipyard, where she was out of commission until 30 September 1957 for modernization, including an enclosed bow and an angled flight deck.

Homeported at Alameda, Midway began annual deployments with the 7th Fleet in 1958. On 8 December 1958, VF-64 based aboard Midway conducted the first firing of a Sparrow III air-to-air missile by a squadron deployed outside the U.S.

She was also on duty in the South China Sea during the Laotian crisis in the spring of 1961. During her 1962 deployment, her aircraft tested the air defense systems of Japan, Korea, Okinawa, the Philippines, and Taiwan.

The carrier continued as a research and development platform. On 13 June 1963, Lt. Cmdr. Randall K. Billins and Lt. Cmdr. Robert S. Chew Jr., piloting an F-4A Phantom II and an F-8D Crusader, respectively, made the first fully automatic carrier landings with production equipment on board Midway off the California coast. These landings, made "hands off" with both flight controls and throttles operated automatically by signals from the ship, highlighted almost ten years of research and development.

When Midway sailed for the Far East on 6 March 1965, her aircraft were prepared for combat operations, and from mid-April, they flew strikes against military and logistics installations in North and South Vietnam. On 17 June 1965, while escorting a strike on the barracks at Gen Phu, North Vietnam, Cmdr. Louis C. Page and Lt. Jack E. D. Batson, flying F-4B Phantoms of VF-21, intercepted four MiG-17s and each shot down one, scoring the first U.S. victories over MiGs in Vietnam.

Returning to Alameda on 23 November 1965, she entered the San Francisco Bay Naval Shipyard on 11 February 1966 for extensive modernization, for which she was placed in Reserve, in commission special, on 15 February 1966. She was recommissioned on 31 January 1970 following a four-year conversion-modernization at the shipyard. Midway returned to Vietnam and, on 18 May 1971, after relieving USS Hancock (CVA 19) on Yankee Station, began single carrier operations which continued until the end of the month. She departed Yankee Station on 5 June and completed her final line period on 31 October, returning to her homeport on 6 November.

Midway, with embarked Carrier Air Wing Five (CVW 5), again departed Alameda for operations off Vietnam on 10 April 1972. On 11 May, aircraft from Midway and other carriers continued laying minefields in ports significant to the North Vietnamese. Ships in Haiphong port were advised that the mining would take place and that the mines would be armed 72 hours later. Midway continued Vietnam operations throughout the summer of 1972. On 7 August 1972, an HC-7 Det 110 helicopter from Midway conducted a search and rescue mission for a downed aviator in North Vietnam. Despite heavy ground fire, the helicopter successfully retrieved the pilot, marking the deepest penetration of a rescue helicopter into North Vietnam since 1968. By the end of 1972, HC-7 Det 110 had successfully

accomplished 48 rescues, 35 of which were under combat conditions.

On 5 October 1973, Midway, with CVW 5, put into Yokosuka, Japan, marking the first forward deployment of a complete carrier task group in a Japanese port. This move had strategic significance, facilitating the continuous positioning of three carriers in the Far East.

Midway, along with other carriers, responded on 19 April 1975 to the waters off South Vietnam when North Vietnam overran two-thirds of South Vietnam. Ten days later, Operation Frequent Wind was carried out by U.S. Seventh Fleet forces, evacuating hundreds of U.S. personnel and Vietnamese to waiting ships after the fall of Saigon. One South Vietnamese pilot landed a small aircraft aboard Midway, bringing himself and his family to safety.

On 21 August 1976, a Navy task force headed by Midway made a show of force off the coast of Korea in response to an unprovoked attack on two U.S. Army officers by North Korean guards on 18 August. Midway's response was in support of a U.S. demonstration of military concern vis-à-vis North Korea.

Midway relieved USS Constellation (CV 64) as the Indian Ocean contingency carrier on 16 April 1979, maintaining a significant American naval presence in the Arabian Sea and Persian Gulf. On 18 November, she arrived in the northern part of the Arabian Sea in connection with the ongoing hostage crisis in Iran. Militant followers of Ayatollah Khomeini seized the U.S. Embassy in Tehran on 4 November and held 63 U.S. citizens hostage. Midway was joined by USS Kitty Hawk (CV 63) and USS Nimitz (CVN 68) and their escorts. Midway was relieved by USS Coral Sea (CV 43) on 5 February.

Following a period in Yokosuka, Midway relieved USS Coral Sea on 30 May 1980 on standby south of the Cheju-Do Islands in the Sea of Japan. On 17 August, Midway relieved USS Constellation to begin another Indian Ocean deployment and complement the USS Dwight D. Eisenhower (CVN 69) task group. Midway spent a total of 118 days in the Indian Ocean during 1980.

On 16 March 1981, an A-6 Intruder from VA-115 aboard Midway sighted a downed civilian helicopter in the South China Sea. Midway immediately dispatched HC-1 Det 2 helicopters to the scene, rescuing all 17 people aboard and bringing them safely to the carrier.

Midway continued serving in the western Pacific throughout the 1980s. On 25 March 1986, the final carrier launching of a Navy fleet F-4S Phantom II took place off Midway during flight operations in the East China Sea. The aircraft was manned by pilot Lt. Alan S. Cosgrove and radar intercept officer Lt. Greg Blankenship of VF-151. Phantoms were being replaced by the new F/A-18 Hornets.

On 2 August 1990, Iraq invaded its neighbor Kuwait, and U.S. forces moved into Saudi Arabia as part of Operation Desert Shield to protect against further invasion. On 1 November 1990, Midway was again on station in the North Arabian Sea, relieving USS Independence (CV 62). On 15 November, she participated in Operation Imminent Thunder, an eight-day combined amphibious landing exercise in northeastern Saudi Arabia. As the United Nations set an ultimatum deadline for Iraq to withdraw from Kuwait, Operation Desert Storm began on 16 January 1991. Midway and other carriers launched sorties and Tomahawk missiles against Iraqi forces. President Bush declared Kuwait liberated on 27 February, and Operation Desert Storm ended at midnight.

Midway departed the Persian Gulf on 11 March 1991 and returned to Yokosuka. In August 1991, she returned to Pearl Harbor, turning over duties to USS Independence (CV 62) and sailing to San Diego. Midway was decommissioned at North Island Naval Air Station on 11 April 1992 and was stricken from the Navy List on 17 March 1997. On 30 September 2003, Midway began On 30 September 2003, Midway began her journey from the Navy Inactive Ship Maintenance Facility, Bremerton, Wash., to San Diego where she will be a museum and memorial. She was docked at the Charles P. Howard Terminal in Oakland, Calif., during the first week in October while the construction of her pier in San Diego was completed. The carrier was towed from Oakland to San Diego, and arrived on 5 January 2004. She docked at the Naval Air Station North Island to load historic aircraft for display. She will be part of a major museum devoted to carriers and naval aviation.

CVB-42 • USS FRANKLIN D. ROOSEVELT

(Displacement: 45,000; length. 968'; bow. 113'; ew. 136'; dr. 35'; speed. 33 k.; complement. 4,104; armor. 18 5"; class. Midway.)

The USS Franklin D. Roosevelt (CVB-42) was initially launched as Coral Sea (CVB-42) on 29 April 1945 by the New York Naval Shipyard. It was sponsored by Mrs. John H. Towers, wife of the Deputy Commander in Chief, Pacific Fleet, and was renamed Franklin D. Roosevelt on 8 May 1945. The ship was commissioned on 27 October 1945 under the command of Captain A. Soucek. It was reclassified as CVA-42 on 1 October 1952.

During her shakedown cruise, Franklin D. Roosevelt called at Rio de Janeiro from 1 to 11 February 1946 to represent the United States at the inauguration of Brazilian President Eurico G. Dutra, who came aboard for a short cruise. Fleet maneuvers and training operations in the Caribbean preceded her first deployment to the Mediterranean from 8 August to 4 October. During this time, she visited Athens to bolster the Greek government during its fight against communism and received thousands of visitors at various Mediterranean ports. The ship's first public visit in the United States occurred during Navy Day celebrations in New York City in late October.

Franklin D. Roosevelt operated off the east coast until July 1947 when she entered Norfolk Naval Shipyard for an extensive overhaul, receiving various improvements. On 13 September 1948, she sailed from Norfolk for a second Mediterranean tour, returning on 23 January 1949. Over the next five years, Franklin D. Roosevelt participated in intensive operations off the Virginia Capes, along the east coast, and in the Caribbean, making four more tours of duty in the Mediterranean.

Assigned to extensive conversion at Puget Sound Naval Shipyard, the carrier sailed from Norfolk on 7 January 1954. Too large to pass through the Panama Canal, she rounded Cape Horn and arrived at the shipyard on 5 March. She was decommissioned on 23 April 1954. The carrier was then fitted with an angled flight deck, steam catapults, and a hurricane bow. Franklin D. Roosevelt was recommissioned on 6 April 1956 and arrived in San Francisco on 16 June to load stores for her voyage around the Horn to Mayport, Florida, arriving at her new home port on 8 August. She trained pilots in the use of the mirror landing system and general carrier operations until her emergency deployment to the eastern Atlantic during the Suez Crisis between 7 November and 9 December.

In February 1957, Franklin D. Roosevelt sailed to the Gulf of Maine for cold weather tests of catapults, aircraft, and other carrier equipment, including the Regulus guided missile. In July, she began the first of three post-conversion cruises to the Mediterranean, completed by 1960. Her Mediterranean assignments included NATO exercises and major fleet operations, entertaining distinguished guests annually.

On 24 October 1958, Franklin D. Roosevelt supported the evacuation of 56 U.S. citizens and three foreign nationals from Nicaro, Cuba, during the Cuban revolution.

On 6 March 1965, a Sikorsky SH-3A Sea King helicopter piloted by Cmdr. James R. Williford took off from USS Hornet (CVS-12) in San Diego and landed 15 hours and 51 minutes later on the deck of Franklin D. Roosevelt off Mayport, Florida, surpassing the existing distance record for helicopters by more than 700 miles.

A significant development in carrier fire prevention occurred on 26 May 1969 when Franklin D. Roosevelt put to sea after an 11-month overhaul at Norfolk Naval Shipyard. This overhaul included the installation of a deck edge spray system using the new seawater-compatible fire-fighting chemical, Light Water.

During the Yom Kippur War in October 1973, Franklin D. Roosevelt, along with USS Independence (CV-62) and USS Guadalcanal (LPH-7), stood by for possible evacuation contingencies.

On 4 October 1976, the first overseas operational deployment of the AV-8A Harrier on a carrier began when VMA-231 embarked aboard Franklin D. Roosevelt for a Sixth Fleet deployment. On 13 January 1977, two other Harriers made the first bow-on, downwind landing aboard a carrier at sea.

Franklin D. Roosevelt was decommissioned on 30 September 1977 and stricken from the Navy List the following day. The Defense Reutilization and Marketing Service (DRMS) sold her for scrapping on 1 April 1978.

CV-43 • USS CORAL SEA

(Displacement: 45,000 tons length: 968 feet beam: 113 feet, extreme width at flight deck: 136 feet draft: 35 feet speed: 33 knots complement: 4,104 crew armament: 18 5-inch guns class: Midway.)

The USS Coral Sea (CVB-43) was launched on 2 April 1946 by Newport News Shipbuilding and Drydock Co., Newport News, Virginia. Sponsored by Mrs. T. C. Kinkaid, the ship was commissioned on 1 October 1947 under the command of Captain A. P. Storrs III and reported to the Atlantic Fleet.

The ship achieved several milestones early in her career. On 27 April 1948, two P2V-2 Neptunes, piloted by Cmdr. Thomas D. Davies and Lt. Cmdr. John P. Wheatley, made JATO take-offs from the carrier off Norfolk, Virginia. This marked the first carrier launch of planes of this size and weight. Coral Sea sailed from Norfolk on 7 June 1948 for a midshipmen cruise to the Mediterranean and Caribbean, returning to Norfolk on 11 August.

Following an overhaul period, Coral Sea operated off the Virginia Capes. On 7 March 1949, a P2V-3C Neptune, piloted by Capt. John T. Hayward of VC 5, was launched from the carrier with a 10,000-pound load of dummy bombs. The aircraft flew across the continent, dropped its load on the west coast, and returned nonstop to land at Naval Air Station Patuxent River, Maryland. After training in the Caribbean, Coral Sea sailed on 3 May 1949 for her first Mediterranean tour with the 6th Fleet, returning on 28 September.

On 21 April 1950, Capt. John T. Hayward of VC 5 made the first carrier takeoff of an AJ-1 Savage heavy attack bomber from Coral Sea. The remaining squadron pilots completed carrier qualifications on 31 August. Coral Sea returned to the Mediterranean from 9 September 1950 to 1 February 1951, strengthening the 6th Fleet's role as a peacekeeper in the region.

Following an overhaul and local operations, as well as training with Air Group 17, Coral Sea returned to the Mediterranean on 20 March 1951 as the flagship for Commander, Carrier Division 6. She participated in NATO Exercise Beehive I and returned to Norfolk on 6 October for local and Caribbean operations. She sailed again for the Mediterranean on 19 April 1952, visiting Yugoslavia and carrying Marshal Tito on a one-day cruise to observe carrier operations. Coral Sea was reclassified as CVA-43 on 1 October 1952 while still at sea and returned to Norfolk for an overhaul on 12 October.

In 1953, Coral Sea trained pilots in carrier operations off the Virginia Capes and Mayport, Florida. She embarked the Judiciary Committee of the House of Representatives for a three-day cruise in April. On 26 April, the carrier sailed for a Mediterranean tour, participating in NATO Exercise Black Wave and hosting Deputy Secretary of Defense R. M. Kyes as an observer. She returned to Norfolk on 21 October, conducted tests for the Bureau of Aeronautics, and trained Naval Reserve members at Mayport and Guantanamo Bay.

Coral Sea returned to the Mediterranean from 7 July to 20 December 1954, during which Generalissimo Franco visited her while she was off Valencia, Spain. On her next Mediterranean tour from 23 March to 29 September 1955, she visited Istanbul and participated in NATO exercises.

Sailing from Norfolk on 23 July 1956 for Mayport, Coral Sea embarked Carrier Air Group 10 and continued to the Mediterranean. She participated in NATO exercises and hosted the King and Queen of Greece in October. During the Suez Crisis, she evacuated American citizens and stood by off Egypt until November. She returned to Norfolk on 11 February 1957.

Coral Sea departed Norfolk on 26 February 1957 and visited Santos, Brazil; Valparaiso, Chile; and Balboa, C.Z., before arriving at Bremerton, Washington, on 15 April. She was decommissioned for conversion on 24 May 1957 and recommissioned on 25 January 1960. In September 1960, she conducted training with her new air group along the west coast and then sailed for a tour of duty with the 7th Fleet in the Far East.

Coral Sea continued WestPac/Vietnam deployments until 1975. Significant operations included retaliatory raids in response to Viet Cong attacks, participation in Operation

Rolling Thunder, and the mining campaign against North Vietnamese ports during Operation Pocket Money. In 1975, Coral Sea participated in Operation Frequent Wind, evacuating U.S. personnel and Vietnamese after the fall of Saigon.

From 12 to 14 May 1975, Coral Sea participated in the recovery of the U.S. merchant ship SS Mayaguez, which had been seized by Cambodian forces. The carrier's aircraft provided protective air strikes, and Marines rescued the ship's crew. Coral Sea was awarded the Meritorious Unit Commendation for her actions.

Coral Sea relieved USS Midway in the northern part of the Arabian Sea on 5 February 1980 during the Iranian hostage crisis. The hostages were released on 20 January 1981. Coral Sea returned to the Mediterranean in 1985, deploying the F/A-18 Hornet for the first time. She continued deployments in the Mediterranean and Indian Ocean throughout the 1980s.

Coral Sea was decommissioned on 26 April 1990 and stricken from the Navy List. She was sold for scrapping on 7 May 1993.

CV-45 • USS VALLEY FORGE

(Displacement: 36,380 tons length: 888 feet beam: 93 feet draft: 28 feet 7 inches speed: 32.7 knots complement: 3,448 crew armament: 12 5-inch guns, 72 40mm guns aircraft: 80 class: Ticonderoga.)

USS Valley Forge (CV-45), funded by the citizens of Philadelphia through a special war bond drive, was laid down on 7 September 1944 at the Philadelphia Navy Yard. She was launched on 18 November 1944, sponsored by Mrs. A. A. Vandegrift, the wife of the Commandant of the Marine Corps, and commissioned on 3 November 1946 under the command of Capt. John W. Harris.

Following her fitting out, the new carrier got underway on 24 January 1947 for shakedown training via Norfolk to Guantanamo Bay, Cuba, and the Canal Zone. Completing the cruise on 18 March, she returned to Philadelphia for a post-shakedown overhaul. Valley Forge left Philadelphia on 14 July, headed south, and transited the Panama Canal on 5 August. She arrived at her home port, San Diego, on 14 August and joined the Pacific Fleet. After embarking Air Group 11 and undergoing intensive air and gunnery training in coastal waters, the carrier, flying the flag of Rear Admiral Harold L. Martin, Commander of Task Force 38, got underway for Hawaii on 9 October. The task force spent nearly three months in training operations out of Pearl Harbor before sailing for Australia on 16 January 1948. Following a visit to Sydney, the American warships conducted exercises with units of the Royal Australian Navy and then steamed to Hong Kong.

During a voyage from Hong Kong to Tsingtao, China, orders arrived directing the task force to return home via the Atlantic. The ship continued the round-the-world trip with calls at Hong Kong, Manila, Singapore, Trincomalee, Ceylon, and Ras Tanura, Saudi Arabia. After operating for a time in the Persian Gulf, Valley Forge became the largest aircraft carrier to transit the Suez Canal. She finally arrived at San Diego via the Mediterranean Sea, the Atlantic Ocean, and the Panama Canal.

Valley Forge deployed to the Far East, departing the west coast on 1 May 1950. While anchored in Hong Kong harbor on 25 June, the warship received news that North Korean forces had crossed the 38th parallel into South Korea. Departing Hong Kong the next day, the carrier steamed to Subic Bay for provisioning and fueling before setting a course for Korea.

The first carrier air strike of the Korean conflict was launched from Valley Forge's flight deck on 3 July 1950. South Korean troops, outnumbered and outgunned, fought desperately against the invading North Korean forces. Waves of Douglas AD Skyraiders and Vought F4U Corsairs struck the North Korean airfield at Pyongyang, while Grumman F9F-2 Panthers flew top cover. Tons of bombs pounded hangars, fuel storages, parked Russian-built aircraft, and railroad marshaling yards. Escorting Panthers downed two Yak-9s and damaged another.

Despite efforts by United Nations forces to interdict the flow of Communist infantry and armor, the North Koreans pushed the defending South Koreans back to a tenuous perimeter around Pusan. On 18 September 1950, the American landing at Inchon outflanked the Communist forces, allowing United Nations forces to break out from the Pusan perimeter.

During this period, Valley Forge's Air Group 5 conducted numerous strikes against North Korean targets. Troop concentrations, defensive positions, and supply and communications lines were repeatedly targeted by Skyraiders and Corsairs. Over 5,000 combat sorties delivered 2,000 tons of bombs and rockets between 3 July and 19 November 1950.

Valley Forge returned to San Diego for overhaul, arriving on 1 December 1950, only to be urgently redeployed to Korea. The ship embarked a new air group, replenished, and sailed on 6 December for the Far East. Rejoining Task Force 77 three days before Christmas, Valley Forge recommenced air strikes on 23 December, launching 2,580 sorties and delivering 1,500 tons of bombs during her second deployment.

On 11 December 1951, Valley Forge launched her first air strikes in railway interdiction. Rockets, cannon fire, and bombs hammered North Korean railway targets, severing Communist rail lines in over 5,346 places by June.

Valley Forge returned to the United States in the summer of 1952 but was again deployed to the Far East late in the year. Reclassified as an attack carrier (CVA-45) on 1 October 1952, she began the new year with strikes against Communist supply dumps and troop billeting areas on 2 January 1953. Her aircraft, including propeller-driven Skyraiders and Corsairs, delivered tons of bombs, while jet Panthers conducted flak-suppression missions. This teamwork allowed regular strikes against Korea's eastern coastlines and close-support missions. Valley Forge air groups dropped 3,700 tons of bombs on the enemy before returning to San Diego on 25 June 1953.

After a west coast overhaul, Valley Forge was transferred to the Atlantic Fleet and reclassified as an antisubmarine warfare support

carrier (CVS-45). She was refitted at the Norfolk Naval Shipyard and rejoined the fleet in January 1954. Conducting local operations and antisubmarine warfare exercises, Valley Forge operated off the east coast through late 1956, including midshipman and reservist training cruises and visits to the Caribbean.

In 1957, Valley Forge accomplished a naval "first" by embarking a Marine detachment and HR2S-1 Mojave helicopters, experimenting with "vertical envelopment" during a ship-based air assault exercise. Rear Admiral John S. Thach, the pilot who devised the "Thach Weave" fighter tactic, became the ship's commander on 1 April 1958. Task Group Alpha, built around Valley Forge, included destroyers, submarines, and antisubmarine aircraft, focusing on countering enemy submarines.

Valley Forge remained engaged in operations with TG Alpha until fall 1959 when she entered the New York Naval Shipyard for repairs. She returned to sea on 21 January 1960 for maneuvers in the Caribbean. During Operation Skyhook, the carrier launched balloons for measuring cosmic ray emissions.

After a deployment in the eastern Mediterranean, Valley Forge resumed local operations in Norfolk on 30 August 1960. She continued antisubmarine exercises and participated in Project Mercury operations, retrieving a space capsule in December 1960. The carrier also responded to an SOS from SS Pine Ridge, rescuing 28 survivors from the tanker in a storm off Cape Hatteras.

Entering the Norfolk Naval Shipyard on 6 March 1961 for overhaul and modification to an amphibious assault ship, Valley Forge was reclassified as LPH-8 on 1 July 1961. Following refresher training in the Caribbean, she trained in the Virginia Capes area with newly embarked troop-carrying helicopters. As part of the Atlantic Fleet's ready amphibious force, Valley Forge stood by to evacuate American nationals from the Dominican

Republic during the power struggle following Generalissimo Rafael Trujillo's assassination.

Sailing from Norfolk on 6 January 1962, Valley Forge was assigned to the Pacific Fleet. After three months of training off the west coast, she deployed to the Far East with the 7th Fleet. The ship airlifted Marines into Laos on 17 May 1962 in response to the Communist Pathet Lao assault on the Royal Laotian Government. The Marines were withdrawn in July, and the ship operated in the Far East before returning to the west coast in early 1963.

Valley Forge underwent a Fleet Rehabilitation and Modernization overhaul at Long Beach Naval Shipyard, completed on 27 January 1964. She conducted local operations and training before deploying to the Western Pacific. The ship participated in SEATO exercises and was awarded the Battle Efficiency "E" in July 1964.

During the Gulf of Tonkin incident in August 1964, Valley Forge remained on station off Vietnam, ready to land her Marines. She returned to Long Beach on 5 November, made two round-trip voyages to Okinawa, and deployed to the South China Sea in the fall of 1965. In early 1966, she took part in Operation Double Eagle, providing logistic and medical support for Marines ashore.

Valley Forge returned to the west coast for overhaul and local training, deploying again to WestPac in November 1967. She participated in Operation Fortress Ridge and Operation Beaver Tooth in late 1967 and early 1968, providing resupply and MedEvac support. The ship continued operations off Dong Hoi, took part in Operation Badger Catch, and served as "Hero Haven" for Marine helicopter units.

Returning to the United States on 3 August 1968, Valley Forge underwent an overhaul and conducted training before deploying to WestPac in January 1969. The ship operated off Danang, supporting Operation Defiant Measure and Operation Brave Armada. She returned to Long Beach on 22 September 1969 and was decommissioned on 16 January 1970. After efforts to preserve her as a museum failed, Valley Forge was sold for scrap on 29 October 1971.

CV-47 • USS PHILIPPINE SEA

(Displacement: 27,100 tons length: 855 feet 10 inches beam: 93 feet draft: 39 feet speed: 30 knots complement: 3,310 crew armament: 12 5-inch guns, 44 40mm guns class: Essex.)

USS Philippine Sea (CV-47) was laid down by Bethlehem Steel Co., Quincy, Mass., on 19 August 1944. She was launched on 5 September 1945, sponsored by Mrs. Albert B. Chandler, and commissioned on 11 May 1946 under the command of Capt. D.S. Cornwell. In June 1946, the ship moved to Quonset Point, R.I., for initial crew training. By September 1946, she began her shakedown cruise in the Caribbean with Air Group 20 embarked.

Upon returning from shakedown exercises, Philippine Sea was ordered back to Boston to prepare for the Navy's Antarctic Expedition, Operation Highjump. On 29 January 1947, from a position 660 miles off the Antarctic continent, Philippine Sea launched the first of six R4D transport aircraft she had ferried from Norfolk, Va., as part of Operation Highjump. The first plane off, piloted by Cmdr. William M. Hawkes carried Rear Adm. Richard E. Byrd and his party from the ship to begin their polar explorations from Little America.

Throughout 1947, Philippine Sea operated in the Atlantic and Caribbean. In the spring of 1948, the ship was deployed to the Mediterranean to join Vice Adm. Forrest Sherman's 6th Fleet. With Air Group 9 on board, Philippine Sea displayed the American ensign in France, Greece, Tunisia, and Sicily. In June 1948, the carrier returned to the United States.

During the summer of 1948, the Philippine Sea developed a doctrine for carrier control approach landings, the seagoing equivalent of the ground control approach. By November 1948, she was exploring the lower rim of the Arctic Circle in a cold-weather operation to test planes, ships, and equipment.

In January 1949, the ship was ordered to the Mediterranean with Air Group 7 embarked. Returning at the end of May, she went into overhaul at the Boston Naval Shipyard. Early autumn found the ship back in the Caribbean, this time with Air Group 1. Operational development projects with jet fighters and task force exercises in the North Atlantic kept the ship and her air group busy until the end of the year.

Operating again from her base at Quonset Point, Philippine Sea qualified carrier pilots during the winter of 1950 and participated in extensive fleet exercises in the Atlantic and Caribbean in February and March. In April and May, she conducted demonstration cruises for guests of the Secretary of the Navy, the Armed Forces Industrial College, Air War College, and the Armed Forces Staff College.

On 24 May 1950, Philippine Sea sailed from Norfolk, Va., passed through the Panama Canal, and arrived at her new homeport of San Diego, Calif., becoming a welcome addition to the Pacific Fleet.

With the outbreak of the Korean War, Philippine Sea was ordered to Pearl Harbor. She sailed for Hawaiian waters on 5 July 1950 with Air Group 11 embarked. Departing for the forward area on 24 July, she reached Okinawa on 1 August. Responding to an urgent request from Lt. Gen. Walton H. Walker, commanding general of the U.S. Eighth Army, the carriers of Task Force 77 began providing tactical air support for Army forces holding the hard-pressed Pusan perimeter. Philippine Sea joined the combat operations on 5 August, along with the escort carrier USS Badoeng Strait (CVE 116), marking the beginning of extensive carrier deployments to the combat area.

Philippine Sea sailed into action off Korea as flagship of Task Force 77 on 5 August 1950, launching air strikes that delivered thousands of tons of bombs, rockets, and napalm on strategic targets. The carrier launched as many as 140 sorties a day. Except for brief periods for re-arming, refueling, or repairs, Philippine Sea was continuously in action for almost three years.

Operating with other carriers of Task Force 77, she targeted North Korean rail and communication centers from Seoul to Wonsan in September. In the Yellow Sea, she supported the Inchon invasion coast. On D-Day, 15 September 1950, Philippine Sea's planes ranged far inland to prevent enemy reinforcements. Following the initial assault, she provided close and deep support for the advance inland to Seoul. On 9 November 1950, Lt. Cmdr. W.T. Amen, flying an F9F-2 Panther from Philippine Sea, became the first Navy pilot to destroy an enemy jet fighter, downing a MiG-15.

During the long retreat from the Yalu River, Philippine Sea's planes provided crucial support for the trapped Marines. The carrier's Panther jets, Skyraider attack bombers, and Corsair fighter bombers cleared the way to Hungnam, where Task Force 77's carriers created an aerial umbrella for the evacuation. From 10 through 24 December, hundreds of carrier planes covered the evacuation perimeter, aiding in the rescue of 150,000 troops and civilians.

After rest and repairs at Yokosuka Naval Base, Japan, Philippine Sea exchanged Air Group 11 for Air Group 2 (CVG 2) from USS Valley Forge (CV-45) on 26 March 1951. On 28 March

1951, Philippine Sea became the flagship of Vice Adm. H.M. Martin, Commander 7th Fleet, and immediately departed with CVG 2 for another tour in Korea.

In April 1951, Philippine Sea led Task Force 77 and other 7th Fleet elements from the Sea of Japan through the Strait of Formosa to the South China Sea. The fleet demonstrated strength over Formosa before returning to Korea to support ground forces. Every Chinese offensive in spring 1951 suffered heavy losses as Philippine Sea's planes conducted repeated air strikes.

Returning to San Francisco on 9 June 1951, Philippine Sea underwent yard availability and west coast operations until departing from San Diego on 31 December 1951. On 19 December 1951, Philippine Sea tested emergency assembly capabilities for nuclear weapons, successfully introducing special weapons into the Pacific fleet. She arrived at Pearl Harbor on 8 January 1952 and continued to Yokosuka, Japan, on 20 January. Philippine Sea remained in action in Korea until relieved in late July 1952. On 23 June, Skyraiders and Panther jets from Philippine Sea, USS Boxer (CV 21), and USS Princeton (CV 37) joined Air Force Thunderjets in a successful attack on the heavily defended Suiho hydroelectric power plant in North Korea.

Returning to San Diego on 8 August 1952, Philippine Sea was reclassified as CVA in October. With Carrier Air Group 9 embarked, she departed for the Far East on 15 December 1952. The carrier conducted air strikes that cut Communist supply and transportation arteries, supporting frontline UN troops.

Arriving at Alameda Naval Air Station on 14 August 1953, Philippine Sea underwent overhaul at Hunter's Point and resumed training off San Diego on 9 January 1954. Heading west on 12 March for her fourth tour in the Far East, the carrier operated out of Manila.

On 26 July 1954, while searching for survivors of a Cathay-Pacific Airways passenger liner shot down by Communist planes near Hainan Island, two AD Skyraiders from Philippine Sea were attacked by Communist fighters. Under orders to fire only if attacked, the Skyraiders returned fire and shot down the attackers. This incident became known as the "Hainan Incident."

Returning to San Diego in November 1954, Philippine Sea conducted extensive training operations off the California coast. She began her fifth Far East cruise on 1 April 1955, operating in Japan, Okinawa, and Taiwan waters. Redesignated as CVS on 15 November, she returned to San Diego on 23 November 1955.

Following leave, upkeep, and operations off the southern California coast and in Hawaiian waters, Philippine Sea embarked on another WestPac tour in March 1957. After a brief deployment, she returned to San Diego for local operations. In January 1958, Philippine Sea steamed west for her final 7th Fleet deployment.

Returning to San Diego on 15 July 1958, Philippine Sea commenced inactivation. She was decommissioned on 28 December 1958 and berthed with the Reserve Fleet at Long Beach. Redesignated as AVT-11 on 15 May 1959, she was struck from the Navy List on 1 December 1969 and sold for scrapping in March 1971.

CVL-48 • USS SAIPAN

(Displacement: 14,500 tons length: 684 feet beam: 76 feet 9 inches; extreme width: 115 feet draft: 28 feet speed: 33 knots complement: 1,721 crew armament: 40 40mm guns class: Saipan.)

The first Saipan (CVL-48) was laid down on 10 July 1944 by the New York Shipbuilding Corp., Camden, N.J.; launched on 8 July 1945, sponsored by Mrs. John W. McCormack; and commissioned on 14 July 1946, with Capt. John G. Crommelin in command.

Commissioned eleven months after the close of World War II, Saipan trained student pilots out of Pensacola from September 1946 to April 1947. She was then reassigned to Norfolk as her homeport. After participating in exercises in the Caribbean, she proceeded to Philadelphia for overhaul. In November, she returned to Pensacola, but in late December, after training midshipmen, she steamed back to the east coast to serve with the Operational Development Force.

In February 1948, her work in jet operational techniques, carrier support tactics, and electronic instrument evaluation was briefly interrupted. From 7 to 24 February, she transported the United States delegation to the Venezuelan Presidential inauguration and back. Upon her return, she conducted local operations off the Virginia Capes, and in April, after a visit to Portsmouth, N.H., she resumed work for the Operational Development Force. On the 18th, she also relieved USS Mindoro (CVE-120) as flagship of Carrier Division 17 (CarDiv 17).

On 19 April 1948, she departed Norfolk for Quonset Point, R.I., where, on 3 May, she embarked Fighter Squadron 17A. Three days later, all squadron pilots had qualified in FH-1 Phantom jets, marking the first carrier-based jet squadron. By the end of the month, Saipan returned to Norfolk and was relieved of flagship duties. In June, she returned to New England waters, and in July, she commenced an overhaul at Norfolk. In December, she resumed local operations. On the 24th, she was ordered to embark two of the Navy's latest type helicopters, the XHJS-1, and three Marine Corps HRP-1 helicopters, and proceeded north to Greenland to assist in the rescue of eleven airmen downed on the ice cap. Departing Norfolk on Christmas day, the CVL arrived off Cape Farewell on 28 December 1948 and prepared to launch the helicopters as soon as weather allowed. On the 29th, however, a C-47, equipped with jet assist takeoff and skis, landed on the ice, rescued the marooned airmen, and made it out again.

Saipan returned to Norfolk on 31 December and sailed again on 28 January 1949. Steaming south, she conducted exercises out of Guantanamo Bay until March, then returned to Hampton Roads on the 10th. From the 11th to the 19th, she conducted operations for the development force, followed by a reserve training cruise to Canada. At the end of May, she again commenced work for the Operational Development Force. Three months later, she conducted her second reservist cruise of the year, then qualified Royal Canadian Navy pilots in carrier landings.

From November 1949 to March 1951, Saipan operated from the Virginia Capes to the south. On 6 March 1951, she got underway as flagship, CarDiv 14, and sailed for duty with the 6th Fleet. Deployed for three months, she operated in the western Mediterranean until the end of May, then headed for home. On 8 June, she returned to Norfolk and resumed operations in the western Atlantic, from Greenland to the Caribbean.

For over two years, Saipan continued Second Fleet operations, interrupting them for midshipman cruises during the summers of 1952 and 1953 and for an overhaul. In October 1953, she departed the east coast, transited the Panama Canal, and steamed for the Pacific. On 30 October, she arrived at San Diego, then continued to Pearl Harbor, Yokosuka, and Korea for support of the uneasy truce agreement.

Assigned to TF 95, she conducted surveillance and reconnaissance missions along the coast and patrolled islands south of the 38th Parallel. In January 1954, she provided air support for Japanese-manned LSTs ferrying former Chinese POWs from Inchon to Formosa. In early February, she participated in amphibious exercises in the Ryukyus, then returned to Inchon to stand by for a possible evacuation of Indian troops from Panmunjom. In March, she conducted amphibious exercises in the Bonins, then returned to Japan. Instead of resuming truce patrols, she took on 25 AU-type aircraft and five H19A helicopters at Yokosuka, which she delivered to French forces in Tourane (later Danang) Air Base, French Indo-China. She then proceeded to Manila, delivered the helicopters to Air Force personnel, and resumed operations off Korea at the end of April.

On 8 May, she put into Sasebo, then operated in Japanese waters until returning to Norfolk via the Suez Canal. On 20 July, she completed her round-the-world cruise. In October 1954, she sailed south to the Caribbean, arriving as hurricane "Hazel" hit the Greater Antilles. From 13 to 20 October, she delivered relief supplies and personnel to isolated areas of Haiti, then returned to Norfolk.

After an overhaul, Saipan resumed operations in April 1955 with a cruise to the Caribbean. In June 1955, she was reassigned to the aviation training center at Pensacola, conducting qualification exercises. In late September, she assisted in hurricane relief efforts in Mexico, evacuating 5,439 people and distributing supplies in the flooded Tampico area. She returned to Pensacola on 12 October, remaining there until April 1957, when she sailed to Bayonne, N.J., for inactivation and decommissioning on 30 September 1957.

Reclassified AVT-6 on 15 May 1959, Saipan remained in the Atlantic Reserve Fleet until March 1963. She then entered the Alabama Dry Dock and Shipbuilding Co. yard in Mobile for conversion to a command ship. Briefly designated CC-3, she was reclassified as a Communications Major Relay ship (AGMR-2) on 1 September 1964 while still undergoing conversion. On 8 April 1965, she was renamed Arlington in honor of Arlington County, Va., and on 12 August 1966, she completed her conversion. As Arlington (AGMR-2), she sailed for Norfolk and was recommissioned on 27 August 1966.

Fitting out occupied the remainder of the year. In January 1967, she conducted shakedown exercises in the Caribbean, and in February, she sailed for the Bay of Biscay and exercises off northern Europe. At the end of March, she returned to Norfolk. In April, she again steamed to the Caribbean. On her return to the Hampton Roads area, she prepared for deployment to the western Pacific.

Departing Norfolk on 7 July 1967, the communications ship transited the Panama Canal and proceeded to Pearl Harbor, Yokosuka, and Subic Bay. With USS Annapolis (AGMR-1), she rotated on station off Vietnam. During her first patrol in Tonkin Gulf from 21 August to 18 September, she provided reliable message handling facilities for ships of the Seventh Fleet in support of combat operations and assisted in repairing electronic equipment. After receiving a new satellite communications terminal, she departed Subic for Taiwan on 2 October, continued to Tonkin Gulf, and resumed her communications relay duties. After a brief visit to Hong Kong, she returned to Subic Bay, then to Tonkin Gulf in early December for her third "Yankee Station" communications patrol. She departed the area on 27 December and headed north, arriving at Yokosuka on 4 January 1968, and returned to Vietnam on 24 January.

Participating in exercises in the Sea of Japan, she returned to "Yankee Station" from 13 February to 10 March 1968, then returned to Yokosuka on 14 March. She resumed operations in Tonkin Gulf on 10 April, visited Sydney, and then returned to station in mid-June. She completed two more tours on "Yankee Station" between the end of August and mid-November and headed for Pearl Harbor in early December. Acting as primary landing area communications relay ship, she participated in the recovery of Apollo 8, then returned to Pearl Harbor and sailed for the Philippines. On 17 January 1969, she resumed direct communications support in Tonkin Gulf.

Arlington participated in further Apollo mission recoveries and various exercises until returning to her homeport of Long Beach on 21 August 1969. Decommissioned on 14 January 1970, she was berthed with the Inactive Fleet at San Diego. Stricken from the Navy List on 15 August 1975, she was sold for scrapping on 1 June 1976. Arlington (AGMR-2) earned seven campaign stars for service off Vietnam.

CVL-49 • USS WRIGHT

(Displacement: 14,500 tons length: 684 feet beam: 76 feet 9 inches; extreme width: 115 feet; draft: 28 feet; speed: 33 knots. Complement: 1,787 crew armament: 40 40mm guns aircraft: 50+ class: Saipan).

The second Wright (CVL-49) was laid down on 21 August 1944 at Camden, N.J., by the New York Shipbuilding Corp.; launched on 1 September 1945 (the day before the formal Japanese surrender ceremony on board the battleship USS Missouri (BB-63) in Tokyo Bay); sponsored by Mrs. Harold S. Miller, a niece of the Wright brothers, and commissioned at the Philadelphia Naval Shipyard on 9 February 1947, with Capt. Frank T. Ward in command.

Wright departed Philadelphia on 18 March 1947 and stopped briefly at Norfolk, Va., en route to the Naval Air Training Base at Pensacola, Fla. After her arrival there on 31 March, Wright soon commenced a rigorous schedule of air defense drills and gunnery practice while acting as a qualification carrier for hundreds of student pilots at the Naval Air Training Base, conducting 40 operational cruises—each between one and four days' duration off the Florida coast. Additionally, the carrier embarked a total of 1,081 naval reservists and trained them in a series of three two-week duty tours. On 3 September 1947, Wright embarked 48 midshipmen for temporary training duty and later welcomed 62 Army officers when she set out to sea on 15 October in company with USS Forrest Royal (DD-872) to let her guests observe flight operations in the Pensacola area. The exercises included the catapulting of a Grumman F6F aircraft for rocket-firing operations.

That exercise was her last prior to her departure from Pensacola on 24 October 1947 to return north. She arrived at the Philadelphia Naval Shipyard soon thereafter and, from 1 November to 17 December, underwent post-shakedown repairs and alterations before returning to Pensacola two days before Christmas. She then resumed her regular schedule of pilot qualification training under the operational control of the Chief of Naval Air Training, Commander Air Atlantic. Wright spent 1948 engaged in those pilot carrier qualification operations, before putting into the Norfolk Naval Shipyard on 26 January 1949 to commence a four-month overhaul. Following refresher training in Cuban waters, Wright returned to Norfolk on 1 August 1949 and, four days later, shifted to Newport, R.I., for two weeks of antisubmarine warfare (ASW) training in the Narragansett Bay area with submarines and destroyers. She also visited New York City before taking up a steady schedule of carrier qualifications, air defense tactics, and exercises out of Quonset Point, R.I.; Key West and Pensacola, Fla. Except for 10 days of maneuvers with the Second Task Fleet from 21 to 31 October 1949, she continued that duty until 7 January 1951, when she embarked the last increment of personnel from Fighter Squadron (VF) 14 for temporary duty. Wright then sailed from Norfolk on 11 January 1951 with a fast carrier task group and reached Gibraltar on the 21st for her first tour of duty with the Sixth Fleet in the Mediterranean. Wright's first Mediterranean deployment took her from Gibraltar to Oran, Algeria. She then proceeded to Augusta Bay, Sicily; Souda Bay, Crete; Beirut, Lebanon; and Golfe Juan, France, her replenishment and liberty ports during the never-ending cycle of fleet training and readiness exercises with the Sixth Fleet. Departing Golfe Juan on 19 March 1951, Wright made port at Newport on the 31st. The carrier later entered the Norfolk Naval Shipyard and underwent an overhaul there before she took part in Atlantic Fleet maneuvers out of Guantanamo Bay, Cuba; engaged in ASW tactics and carrier operations in Narragansett Bay, received further repairs at the Boston Naval Shipyard, and participated in a convoy exercise that ran from 25 February to 21 March 1952; ranging from Newport to waters of the Panama Canal Zone and Trinidad in the British West Indies.

As flagship for Carrier Division (CarDiv) 14, Wright sailed on 9 June 1952 in company with four destroyers forming Task Group (TG) 81.4 for ASW operations along the Atlantic seaboard until the 27th, when the ships arrived at New York City. Returning to Quonset Point on 1 July 1952, Wright trained units of the organized naval reserves concurrently with hunter-killer tactics and pilot training in operations out of Narragansett Bay until 26 August. On that day, she set course from Quonset Point and later rendezvoused with Vice Admiral Felix B. Stump's Second Task Fleet en route to northern Europe for combined defense exercises and maneuvers with naval units of other North Atlantic Treaty Organization (NATO) navies.

En route, Wright, escorted by Forrest Royal, was detached to ferry men and gear of Marine Night Fighter Squadron (VMF(N)) 114 to Port Lyautey, French Morocco, an operation she completed on 4 September. Two days later, Wright and her escort rejoined the task force, reaching the Firth of Clyde, Scotland, on 10 September.

Three days later, Wright put to sea with two British destroyers acting as her plane guard for NATO Operation Mainbrace. She conducted air defense maneuvers and tactics

evolutions with the British carriers HMS Illustrious (R-87) and HMS Eagle (R-05) en route to Rotterdam, Holland, where the force arrived on 25 September. On 29 September, Wright departed Rotterdam, bound for the United States, and arrived at Newport on 9 October 1952.

That day, she embarked Rear Admiral W. L. Erdman, Commander, Carrier Division 4, and spent the next few months engaged in carrier qualification duties in waters ranging from Newport to the Virginia Capes, before beginning her second deployment to the Mediterranean. She reached Golfe Juan on 21 February 1953 and operated with the Sixth Fleet until 31 March, sailing for home via the Azores.

Wright returned to Newport and, after a rigorous schedule of training in Narragansett Bay, sailed on 5 May 1953 for the Gulf of Mexico. During that training cruise, she visited Houston, Tex., where she hosted some 14,000 visitors on 16 and 17 May. Returning to Quonset Point on 28 May, Wright operated locally for another month before shifting south for operations out of Mayport, Fla.

Wright was overhauled at the Philadelphia Naval Shipyard from 31 July to 21 November 1953, then conducted refresher training in Cuban waters from 4 January to 16 February 1954. Next, after departing Davisville, R.I., on 5 April, Wright sailed for the Far East—via the Panama Canal, San Diego, Calif., and Pearl Harbor—reaching Yokosuka, Japan, on 28 May. The carrier, with Marine Attack Squadron 211 embarked, operated with the Seventh Fleet off both coasts of Korea and Okinawa before visiting Hong Kong from 24 to 30 September. Departing Yokosuka on 15 October 1954, Wright arrived at San Diego on 31

October and entered the Long Beach Naval Shipyard, where she remained until 23 February 1955.

At that point, Wright was attached to CarDiv 17, Pacific Fleet, and operated locally out of San Diego until 3 May 1955, when she put to sea as part of TG 7.3, formed around the flagship USS Mount McKinley (AGC-7), for the atomic test, Operation Wigwam, carried out in Pacific waters. Returning to the West Coast on 20 May, Wright subsequently cruised briefly to Pearl Harbor before entering the Mare Island Naval Shipyard on 14 July 1955 to commence preparation for inactivation. After shifting to the Puget Sound Naval Shipyard, Bremerton, Wash., on 17 October, for the final phase of preservation for inactivation, Wright was decommissioned at Puget Sound on 15 March 1956 and assigned to the Bremerton group of the Pacific Reserve Fleet. During her time in reserve, Wright was reclassified on 15 May 1959 as an auxiliary aircraft transport, AVT-7. However, she never served in that role, but remained inactive until 15 March 1962, when she was taken to the Puget Sound Naval Shipyard for conversion to a command ship and reclassified as CC-2. The conversion, which lasted a year, included extensive alterations to enable the ship to function as a fully equipped mobile command post afloat for top-echelon commands and staff for the strategic direction of the area or worldwide military operations. Facilities were built into the ship for worldwide communications and rapid, automatic exchange, processing, storage, and display of command data. A portion of the former hangar deck space was utilized for special command spaces and the extensive electronics equipment required, while a major portion of the flight deck was utilized for specially designed communications antenna arrays. In addition, facilities were provided to enable the ship to operate three helicopters.

Recommissioned at Puget Sound on 11 May 1963, with Capt. John L. Arrington, II, in command, Wright (CC-2) operated locally on trials and training evolutions in the waters off the Pacific Northwest until 3 September, when she departed Seattle and proceeded to San Diego, which she reached three days later. For the next three weeks, the ship trained in nearby waters before returning to Puget Sound on 30 September to commence her post-shakedown availability.

Following those repairs and alterations—which took up all of October and most of November—Wright prepared to shift to her new home port of Norfolk, Va. She departed Seattle on 26 November 1963, stopped briefly at San Diego three days later to embark civilian engineers and personnel who were to conduct surveys of communications and air conditioning equipment, and

was steaming south off the coast of northern Mexico when she picked up a distress message from the Israeli merchantman, SS Velos, on 1 December. Wright altered course and rendezvoused with Velos later that same day. The command ship's medical officer was flown across to the Israeli ship and treated a seaman suffering from kidney stones. Upon completion of that mission of mercy, Wright resumed her voyage to Balboa.

Transiting the Panama Canal on 7 and 8 December 1963, Wright steamed via St. Thomas, Virgin Islands, and moored at the Hampton Roads Army Terminal on 18 December. After a subsequent brief operational period off the Virginia Capes, Wright entered port on 21 December and remained there through Christmas and New Year's. For the next six years, Wright operated out of Norfolk, training to perform her assigned mission as an emergency command post afloat. Regular overhauls performed at the Norfolk Naval Shipyard saw the ship receiving the repairs and alterations that continually improved her capabilities to carry out her task. She operated primarily off the Virginia Capes, but ranged as far north as Bar Harbor, Maine, and as far south as Rio de Janeiro, Brazil, and Punta del Este, Uruguay. Her other ports of call included Newport, Fort Lauderdale and Port Everglades, Fla., Boston, New York City, Annapolis, Philadelphia,

Norfolk, and Guantanamo Bay, Cuba. On occasion, she alternated on "alert" status with USS Northampton (CC-1).

There were highlights and breaks from the cycle of periods in port and at sea. From 11 to 14 April 1967, Wright lay at anchor off the coast of Uruguay, providing a worldwide communications capability in support of President Lyndon B. Johnson as he attended the Latin American summit conference at Punta del Este. On 8 May 1968, Wright went to the aid of USS Guadalcanal (LPH-7) after that amphibious assault ship had suffered a machinery failure and had gone dead in the water, 180 miles south of Norfolk. She towed the helpless assault ship 84 miles before other ships arrived on the scene to help out. Later that same year, Wright received the coveted Ney award in the large mess afloat category. That award is given annually to the ship that maintains the highest food standards. During the USS Pueblo (AGER-2) crisis in February 1969, Wright—while en route to Port Everglades, Fla.—was hurriedly recalled to Norfolk and, upon her arrival there, stood by on alert.

Ultimately decommissioned on 27 May 1970, Wright was placed in reserve at the Philadelphia Naval Shipyard. The ship was stricken from the Navy List on 1 December 1977 and sold by the Defense Reutilization and Marketing Service (DRMS) for scrapping on 1 August 1980.

CVA-59 • USS FORRESTAL

(Displacement: 59,900 tons length: 1,046 feet beam: 129 feet 4 inches; extreme width: 252 feet draft: 28 feet speed: 33 knots complement: 4,000+ crew armament: 8 5-inch guns class: Forrestal).

The first of the "supercarriers," Forrestal (CVA-59) was launched on 11 December 1954 by Newport News Shipbuilding and Drydock Co., Newport News, Va.; sponsored by Mrs. James V. Forrestal, widow of Secretary of Defense Forrestal; and commissioned on 1 October 1955, Capt. Roy L. Johnson in command.

Forrestal represented more than one step in the evolutionary chain of modern carrier aviation. Besides her sheer size and weight, she was the first built with an angled flight deck, which allowed simultaneous takeoffs and landings. She also featured four catapults and four deck edge elevators to move aircraft from the hangar bays to the flight deck.

From her homeport, Norfolk, Va., Forrestal spent the first year of her commissioned service in intensive training operations off the Virginia Capes and in the Caribbean. Cmdr. Ralph L. Werner made the first arrested landing on January 3, 1956. An important assignment was training aviators in the use of her advanced facilities, a duty on which she often operated out of Mayport, Fla. On 7 November 1956, she put to sea from Mayport to operate in the eastern Atlantic during the Suez Crisis, ready to enter the Mediterranean should her great strength be necessary. She returned to Norfolk on 12 December to prepare for her first deployment with the Sixth Fleet in the Mediterranean, for which she sailed on 15 January 1957.

On this, as on her succeeding tours of duty in the Mediterranean, Forrestal visited many ports to allow dignitaries and the general public to come aboard and view the tremendous power for peace she represented. For military observers, she staged underway demonstrations to illustrate her capacity to bring air power to and from the sea in military operations on any scale. She returned to Norfolk on 22 July 1957 for exercises off the North Carolina coast in preparation for her first NATO operation, Operation Strikeback, in the North Sea. This deployment, between 3 September and 22 October, found her visiting Southampton, England, as well as drilling in the highly important task of coordinating United States naval power with that of other NATO nations.

The next year found Forrestal participating in a series of major fleet exercises, as well as taking part in experimental flight operations. During the Lebanon Crisis of the summer of 1958, the great carrier was again called upon to operate in the eastern Atlantic to back up naval operations in the Mediterranean. She sailed from Norfolk on 11 July to embark an air group at Mayport two days later, then patrolled the Atlantic until returning to Norfolk on 17 July 1958.

On her second tour of duty in the Mediterranean, from 2 September 1958 to 12 March 1959, Forrestal again combined a program of training, patrol, and participation in major exercises with ceremonial, hospitality, and public visiting. Her guest list during this cruise was headed by Secretary of Defense N. H. McElroy. Returning to Norfolk, she continued the never-ending task of training new aviators, constantly maintaining her readiness for instant reaction to any demand for her services brought on by international events. Visitors during the year included King Hussein of Jordan.

From 1958 through 1966, Forrestal alternated between the Second Fleet in the Atlantic and the Sixth Fleet in the Mediterranean. She again brought her imposing presence to the Sixth Fleet between 28 January 1960 and 31 August, visiting the ports usual to a Mediterranean deployment as well as Split, Yugoslavia. Again she was open for visitors at many ports, as well as taking part in the patrol and training schedule of the Sixth Fleet. Upon her return to the United States, she resumed her schedule of east coast and Caribbean operations for the remainder of 1960.

Forrestal made history in November 1963 when, on the 8th, 21st, and 22nd, Lt. James H. Flatley III and his crew members, Lt. Cmdr. "Smokey" Stovall and Aviation Machinist's Mate (Jets) 1st Class Ed Brennan, made 21 full-stop landings and takeoffs in a C-130F Hercules aboard the ship. The tests were conducted 500 miles out in the North Atlantic off the coast of Massachusetts. In so doing, Forrestal and the C-130 set a record for the largest and heaviest airplane landing on a Navy aircraft carrier. The Navy was trying to determine if the big Hercules could serve as a "Super-COD" — a "Carrier On-board Delivery" aircraft. The problem was there was no aircraft that could provide resupply to a carrier in mid-ocean. The Hercules was stable, reliable, and had a long cruising range and high payload.

The tests were more than successful. At 85,000 pounds, the C-130F came to a complete stop within 267 feet, and at the maximum load, the plane used only 745 feet for takeoff. The Navy concluded that with the C-130 Hercules, it would be possible to lift 25,000 pounds of cargo 2,500 miles and land it on a carrier. However, the idea was considered a bit too risky for routine

COD operations. The C-2A Greyhound program was developed and the first of these planes became operational in 1965. For his effort, the Navy awarded Lt. Flatley the Distinguished Flying Cross.

In June 1967, Forrestal departed Norfolk for duty in waters off Vietnam. As the huge ship cut a wake through the calm waters of the Gulf of Tonkin on 29 July 1967, the hot, tropical sun beat down from a clear sky. Forrestal had been launching aircraft from her flight deck on strikes against an enemy whose coastline was only a few miles over the horizon. For four days, the planes of Attack Carrier Air Wing 17 had been launched on, and recovered from, about 150 missions against targets in North Vietnam. On the ship's four-acre flight deck, her crewmen went about the business at hand, the business of accomplishing the second launch of the fifth day in combat.

It was just about 10:50 a.m. (local time). The launch that was scheduled for a short time later was never made. Lt. Cmdr. John S. McCain III, later a prisoner of war in Vietnam and still later a U.S. Senator from Arizona, said later he heard a "whooshy" sound then a "low-order explosion" in front of him. Suddenly, two A-4s ahead of his plane were engulfed in flaming jet fuel — JP-5 — spewed from them. A bomb dropped to the deck, rolled about six feet, and came to rest in a pool of burning fuel. The awful conflagration, which was to leave 132 Forrestal crewmen dead, 62 more injured, and two missing and presumed dead, had begun. The entire nation felt the tragedy, and Life magazine reported that "in five minutes, everyone became a man." The ship returned to Norfolk for extensive repairs.

Forrestal deployed to Mediterranean waters four times between 1968 and 1973. She sped to Tunisia for rescue operations in the flooded Medjerda River Valley near Tunis. The ship logged three more Mediterranean deployments between 1973 and 1975. On 22 July 1974, as a result of a conflict between Turkish and Greek Cypriot forces on Cyprus, the U.S. Ambassador to Cyprus, Roger Davies, requested the evacuation of U.S. citizens from that island nation. In a joint Navy-Marine Corps effort, HMM-162 from the Sixth Fleet amphibious assault ship USS Inchon (LPH 12) evacuated 466 people, 384 of them U.S. citizens, in only five hours. Forrestal provided air cover for that operation.

In 1975, Forrestal was selected to be the host ship for the International Naval Review in New York City on the nation's Bicentennial. On July 4, 1976, on Forrestal's flight deck, President Gerald Ford rang in the Bicentennial and reviewed over 40 "Tall Ships" from countries around the world.

Shortly after the review, Forrestal participated in a special shock test. It involved the detonation of high explosives near the hull to determine if a capital ship could withstand the strain of close-quarter combat and still remain operational. In September 1977, following a nine-month overhaul, Forrestal departed Norfolk and shifted homeport to Mayport, Fla. The carrier left Mayport on 13 January 1978 for a three-week at-sea period in the Atlantic Fleet Weapons Range of the Roosevelt Roads Operating Area to complete the third phase of Type Commander's Training (TYT-3), and to undergo the Operational Readiness Evaluation (ORE). Tragedy struck Forrestal on the evening of 15 January as an A-7E Corsair II from VA-81 crashed on the flight deck, killing two deck crewmen and injuring 10 others. The pilot ejected safely and was recovered, suffering only minor injuries. The plane crashed as the pilot attempted to land while the aft portion of the flight deck was crowded with aircraft. The Corsair struck a parked A-7 and an EA-6B before careening across the deck in a ball of flames. A small fire on the aft portion of the deck, caused by fuel spilled during the crash, was extinguished within seconds. At the time of the accident, Forrestal was operating about 49 miles off St. Augustine, Fla. A memorial service for the dead was held on board on 19 January. The ship returned to Mayport on 3 February.

Forrestal left Mayport for the Mediterranean on 4 April 1978. At 2200 on 8 April, just minutes after the ship had finished a general quarters drill, the crew was called to G.Q. again, but this time it was not a drill; a fire had broken out in the Number Three Main Machinery Room. Freshly painted lagging in Three Main engine room had been set smoldering by hot steam lines. Watch-standers within the space activated an extinguishing system and had the fire out within seconds.

Three days later, the crew was again called to respond to another emergency G.Q. At midnight on 11 April, a fire was discovered in a catapult steam trunk in the forward part of the ship at about the 01 level, and another fire was found in an adjoining storeroom minutes later. The at-sea fire brigade, working with area repair lockers, had the fires out within the hour.

Forrestal recorded her 227,000th arrested landing on 22 April 1978 while in the Mediterranean. Pilot Lt. j.g. Erick Hitchcock and Radar Intercept Officer (RIO) Lt. j.g. Al Barnet of VF-74 were the crew of the F-4 Phantom that marked the milestone trap. On 10 May 1978, flooding which began in a pump room in the aft portion of the ship rose to a height of 20 feet before it was controlled and spread into food storage rooms, destroying most of the ship's stocks of fresh milk and produce. Divers from the ship's Explosive Ordnance Disposal (EOD) team dropped into the pump room to plug the leak. Total damage from the flooding was estimated at $30,000.

From 19 to 29 May 1978, Forrestal participated in Dawn Patrol, the first of three NATO exercises the ship would be involved in during the deployment. Dawn Patrol involved air and ground forces and over 80 ships from six NATO countries. Forrestal's role during the exercise included protecting a Turkish amphibious task group and working with USS Nimitz (CVN 68) and the French carrier Foch to defend against simulated "enemy" ships and aircraft.

During this sea period, two separate air crashes on successive days left one pilot dead and another injured. On 24 June 1978, Lt. Cmdr. T. P. Anderson, Operations Officer for Carrier Air Wing Seventeen, was killed when his A-7E Corsair II crashed into the sea during a practice bombing mission. On 25 June, a

pilot from VA-83, also flying an A-7E, ejected shortly after takeoff, suffering minor injuries. A rescue crew aboard an SH-3D Sea King helicopter from HS-3 recovered the pilot and returned to the ship within eight minutes after the crash. Both accidents occurred as the ship was operating in the Ionian Sea, east of Sicily.

From 4 to 19 September 1978, Forrestal participated in the massive NATO exercise Northern Wedding, which included over 40,000 men, 22 submarines, and 800 rotary and fixed-wing aircraft from nine NATO countries. Northern Wedding, which takes place every four years, practices NATO's ability to reinforce and resupply Europe in times of tension or war. During the exercise, Forrestal and the British aircraft carrier HMS Ark Royal headed separate task groups, steaming in a two-carrier formation to gain sea control and deploying their aircraft in support of mock amphibious landings in the Shetland Islands and Jutland, Denmark. From 28 September to 10 October, Forrestal participated in Display Determination, the third and final NATO exercise of the deployment. The operation, involving ships, aircraft, and personnel from eight NATO countries, was designed to practice rapid reinforcement and resupply of the southern European region in times of tension or war. Forrestal arrived in Rota on 11 October for the last overseas port stop of the deployment.

On 13 October 1978, the ship put to sea to conduct a one-day exercise with a task group of deploying U.S. ships headed by the aircraft carrier USS Saratoga (CV-60). Air Wing Seventeen's planes conducted mock attacks on the task group to allow the ships to practice anti-air warfare. Forrestal returned to Rota late in the evening on the 13th. Before dawn on 15 October, Forrestal departed Rota and out-chopped from the Sixth Fleet, having been relieved by Saratoga. On the homeward transit, Forrestal took an extreme northerly course as part of a special operation code-named Windbreak. Commander Second Fleet, Vice Adm. Wesley L. McDonald, embarked in Forrestal for the exercise. Windbreak was designed to introduce U.S. sailors and equipment to relatively unfamiliar waters and conditions, and to gauge Soviet interest in U.S. ships in transit to and from the Mediterranean. During the exercise, Forrestal traveled as far north as 62 degrees latitude, 150 miles south of Iceland, encountering seas up to 34 feet, winds in excess of 70 knots, and a wind chill factor that drove the temperature as far down as 0 degrees. Also participating in Windbreak were the guided missile cruiser USS Harry E. Yarnell (CG-17) and the destroyer USS Arthur W. Radford (DD-968). Forrestal returned to Mayport on 26 October 1978. On 13 November, Forrestal commenced a four-month period of upkeep and repair known as an Extended Selected Restricted Availability (ESRA), conducted while the ship was moored alongside the carrier pier in Mayport. Forrestal ended 1978 as she had started it, moored to the carrier pier in Mayport.

After completing two more Mediterranean cruises, she celebrated her silver anniversary in October 1980. On 2 March 1981, Forrestal began her 16th Mediterranean deployment and second quarter century of naval service. During the Syria/Israel missile crisis, Forrestal maintained a high state of readiness for 53 consecutive days at sea. In a Gulf of Sidra exercise, two Libyan aircraft were shot down after firing upon F-14s from USS Nimitz (CVN 68) over international waters. Forrestal aircraft made more than 60% of all the intercepts of Libyan planes. After departing the Mediterranean, she operated above the Arctic Circle as part of NATO Ocean Venture '81.

After a repair period, Forrestal deployed for her 17th Mediterranean cruise on 8 June 1982 and operated in the eastern Mediterranean in support of the Lebanon Contingency Force of 800 U.S. Marines in Beirut. On 12 September 1982, after transiting the Suez Canal for the first time in her 28-year history, she entered the Indian Ocean. This marked the first time that Forrestal had operated with Seventh Fleet since the 1967 Vietnam cruise.

Forrestal completed the five and one-half month deployment with a nighttime arrival at Mayport on 16 November and immediately began preparing for the Service Life Extension Program (SLEP). The ship shifted homeport to Philadelphia, Penn., on 18 January 1983, and embarked on the 28-month, $550 million SLEP, designed to extend the life of U.S. aircraft carriers another 15 to 20 years.

During Forrestal's SLEP, the ship was completely emptied, and most major equipment was removed for rework or replacement. Forrestal's successful SLEP period was completed on time when the ship left Philadelphia on 20 May 1985. After completing a four-day transit to her homeport of Mayport, Fla., Forrestal immediately began a workup cycle in preparation for her first deployment in over four years.

Forrestal departed Mayport on 2 June 1986 on her 18th deployment. During this cruise, Forrestal aircraft frequently operated in the international airspace of the Tripoli Flight region, the international air traffic control sector of Libya. Forrestal also participated in Operation Sea Wind, a joint U.S.-Egyptian training exercise, and Display Determination, which featured low-level coordinated strikes and air combat maneuvering training over Turkey.

In 1987, Forrestal went through yet another period of pre-deployment workups. This included refresher training, carrier qualifications, and a six-week deployment to the North Atlantic to participate in Ocean Safari '87. In this exercise, Forrestal operated with NATO forces in the fjords of Norway.

Forrestal departed on her 19th major deployment on 25 April 1988. She steamed directly to the North Arabian Sea via the Suez Canal in support of America's Earnest Will operations in the region. She spent 108 consecutive days at sea before her first liberty port. During the five and one-half month deployment, Forrestal operated in three ocean areas and spent only 15 days in port. She returned on 7 October 1988 and received the Meritorious Unit Citation for her superior operational performance during the deployment.

After a brief stand down period followed by local operations, Forrestal participated in New York City's Fleet Week in May 1989 and then commenced preparations for her next deployment.

Forrestal's departure for her 20th major deployment was delayed when a fire caused major damage to a primary command and control trunk space. Through the efforts of the ship's crew and civilian contractors, Forrestal was able to depart for her deployment on 6 November 1989, completing the necessary repairs well ahead of projections. The final two months of 1989 proved exciting. Beyond the "routine" exercises and training initiatives, Forrestal's crew became part of history, as they provided support to President George Bush during his Malta Summit. The support included a three-hour Presidential visit to the ship. Forrestal participated in numerous exercises during this deployment, including Harmonie Sud, Tunisian Amphibious, and National Week. She returned to Mayport on 12 April 1990, ending a deployment that had included eight port visits in five different countries.

The year 1991 was a year of anticipation and change for Forrestal and its crew, as she spent the first five months maintaining combat readiness as the East Coast-ready carrier. Maintaining a hectic and challenging period of at-sea operations, Forrestal's anticipated deployment in support of Operation Desert Storm was not to be, and orders to deploy were canceled twice during the conflict.

The call to deploy finally came, and Forrestal commenced its 21st and final operational deployment on 30 May 1991.

No less challenging than the months of maintaining readiness for combat, Forrestal's deployment was repeatedly referred to as "transitional." During the ensuing seven months, Forrestal was called upon to provide air power presence and airborne intelligence support for Operation Provide Comfort, and to initiate, test, and evaluate a wide range of innovative COMSIXTHFLT battle group tactics and new carrier roles. The year ended with Forrestal making advanced preparations for its change of homeport to Pensacola, Fla., and the transition into a new role as the Navy's training carrier, replacing USS Lexington (AVT 16). Forrestal arrived in Philadelphia on 14 September 1992 to begin a 14-month, $157 million complex overhaul prior to assuming the duties as a training carrier. In early 1993, however, the Navy decided to decommission Forrestal and leave the Navy without a dedicated training carrier. Forrestal was decommissioned on 11 September 1993 at Pier 6E in Philadelphia and was stricken from the Navy List the same day. In 2014, the ship was sold for scrapping and was subsequently towed to Brownsville, Texas, where it was dismantled.

CVA-60 • USS SARATOGA

(Displacement: 56,300 tons, length: 1,046 feet, beam: 130 feet 4 inches; extreme width: 249½ feet, draft: 37 feet, speed: 34 knots, complement: 3,950 crew, armament: 8 5-inch guns, class: Forrestal).

The fifth Saratoga (CV-60) was laid down on 16 December 1952 by the New York Naval Shipyard, New York City, N.Y.; launched on 8 October 1955; sponsored by Mrs. Charles S. Thomas; and commissioned on 14 April 1956, Capt. R. J. Stroh in command.

For the next several months, Saratoga conducted various engineering, flight, steering, structural, and gunnery tests. On 18 August, she sailed for Guantanamo and her shakedown cruise. On 19 December, she reentered the New York Naval Shipyard and remained there until 28 February 1957. Upon completion of yard work, she got underway on a refresher training cruise to the Caribbean before entering her homeport, Mayport, Fla.

On 6 June, President Dwight D. Eisenhower and members of his cabinet boarded Saratoga to observe operations on board the giant carrier. For two days, she and eighteen other ships demonstrated air operations, antisubmarine warfare, guided missile operations, and the Navy's latest bombing and strafing techniques. Highlighting the President's visit was the nonstop flight of two F8U "Crusaders," spanning the nation in three hours and twenty-eight minutes, from the Bon Homme Richard (CVA-31) on the west coast to the flight deck of the Saratoga in the Atlantic.

The carrier departed Mayport on 3 September 1957 for her maiden transatlantic voyage. Saratoga sailed into the Norwegian Sea and participated in operation "Strikeback," joint naval maneuvers of the North Atlantic Treaty Organization countries. She returned briefly to Mayport before entering the Norfolk Naval Shipyard for repairs.

On 1 February 1958, Saratoga departed Mayport for the Mediterranean and her first deployment with the Sixth Fleet. From this date through 31 December 1967, she was to spend a part of each year in the Mediterranean on a total of eight cruises. The remainder of the time, she either operated off the coast of Florida or was in port undergoing restricted availability.

While deployed with the Sixth Fleet on 23 January 1961, a serious fire broke out in Saratoga's number two machinery space, which took seven lives. The fire, believed to have been caused by a ruptured fuel oil line, was brought under control by the crew, and the ship proceeded to Athens, where a survey of the damage could be made.

On 2 January 1968, Saratoga sailed for Philadelphia and an overhaul and modernization program which was to last 11 months. On 31 January 1969, she departed Philadelphia for Guantanamo, via Hampton Roads and Mayport, for extensive refresher training of the crew and air detachments.

On 17 May, Armed Forces Day, she was the host ship for President Richard M. Nixon during the firepower demonstration conducted by Carrier Air Wing Three in the Virginia Capes area. On 9 July, she departed Mayport for her ninth Mediterranean deployment. En route, a Soviet surface force and a "November" class submarine passed in close proximity, en route to Cuba. Off the Azores on 17 July, Saratoga was shadowed by Kipelovo-based Soviet aircraft. They were intercepted, photographed, and escorted while in the vicinity of the carrier. She operated with Task Group 60.2 of the Sixth Fleet in the eastern Mediterranean during September in a "show of force" in response to the large build-up of Soviet surface units there, the hijacking of a Trans World Airlines plane to Syria, and the political coup in Libya. Numerous surveillance and reconnaissance flights were conducted by Carrier Wing Three aircraft against Soviet surface units, including the carrier Moskva, operating southeast of Crete. Saratoga operated in this area again in October because of the crisis in Lebanon. She returned to Mayport and the Florida coast from 22 January until 11 June 1970, when she again sailed for duty with the Sixth Fleet.

On 28 September, President Richard M. Nixon and his party arrived on board. That night, word was received that Gamal Abdul Nasser, President of the United Arab Republic, had died; an event that might plunge the entire Middle East into a crisis. The intelligence and communications personnel of the Saratoga were required to supply the President, Joint Chiefs of Staff, and the

Secretaries of State and Defense with the essential intelligence information to keep them abreast of the deteriorating situation. The Presidential party departed the ship the next evening, and Saratoga continued on patrol in the eastern Mediterranean until she sailed for the United States on 2 November. From her arrival at Mayport until 10 March 1971, she was in a "cold iron" status. She then operated off the Florida coast until 7 June, when she departed for her eleventh deployment with the Sixth Fleet, via Scotland and the North Sea where she participated in exercise "Magic Sword II." She returned to Mayport on 31 October for a period of restricted availability and local operations.

On 11 April 1972, Saratoga sailed from Mayport en route to Subic Bay, P.I., for her first deployment to the western Pacific. She arrived in Subic Bay on 8 May and departed for Vietnam the following week, arriving at "Yankee Station" on 18 May for her first period on the line. Before the year's end, she was on station in the Tonkin Gulf a total of seven times: 18 May to 21 June; 1 to 16 July; 28 July to 22 August; 2 to 19 September; 29 September to 21 October; 5 November to 8 December; and 18 to 31 December.

During the first period, Saratoga lost four aircraft and three pilots. On the plus side, on 21 June, two of her "Phantoms" attacked three MiG 21s over North Vietnam. Dodging four surface-to-air missiles, they managed to down one of the MiG aircraft. Saratoga's planes attacked targets ranging from enemy troop concentrations in the lower panhandle to petroleum storage areas northeast of Hanoi. On her second line period, she lost an F-4 to enemy fire northeast of Hanoi with the pilot and radar intercept officer missing in action. During this period, her aircraft flew 708 missions against the enemy.

On 6 August, Lt. Jim Lloyd, flying an A-7 on a bombing mission near Vinh, had his plane shot out from under him by a SAM. He ejected into enemy territory at night. In a daring rescue by helicopters supported by CVW-3 aircraft, he was lifted from the midst of enemy soldiers and returned to Saratoga. On 10 August, one of the ship's CAP jet fighters splashed a MiG at night using "Sparrow" missiles.

During the period 2 to 19 September, Saratoga aircraft flew over 800 combat strike missions against targets in North Vietnam. On 20 October, her aircraft flew 83 close air support sorties in six hours in support of a force of 250 Territorials beleaguered by the North Vietnamese 48th Regiment. Air support saved the small force, enabled ARVN troops to advance, and killed 102 North Vietnamese soldiers. During her last period on station, Saratoga's aircraft battered targets in the heart of North Vietnam for over a week.

Saratoga departed "Yankee Station" for Subic Bay on 7 January 1973. From there she sailed for the United States, via Singapore and arrived at Mayport on 13 February 1973.

On 21 January 1975, Saratoga, on a Mediterranean deployment, was released from a response alert for possible evacuation of U.S. citizens from Cyprus during a period of strife on that island.

In March of 1980, Saratoga and embarked air wing CVW-17 departed on their 16th Mediterranean deployment. Highlights of the deployment included major exercises with the USS Forrestal (CV 59) battle group, and visits by the Chief of Naval Operations, Adm. Thomas B. Hayward, and Master Chief Petty Officer of the Navy Thomas C. Crow. Then-commanding officer, Capt. James H. Flatley III, made naval aviation history on 21 June 1980 when he completed his 1,500th carrier arrested landing. To make the event special, Midshipman James H. Flatley IV, the Captain's son, rode in the back seat.

On 28 September 1980, only one month after her return from deployment, Saratoga departed Mayport and headed north to the Philadelphia Naval Shipyard where she underwent the most extensive industrial overhaul ever performed on any Navy ship. Saratoga was the first ship to go through the Service Life Extension Program (SLEP) overhaul that would last 28 months. She conducted sea trials on 16 October 1982 and left Philadelphia with much fanfare on 2 February 1983 with her new nickname "Super Sara."

Saratoga departed the Mayport Basin yet again for her 17th Mediterranean deployment on 2 April 1984.

Saratoga's 18th deployment was anything but ordinary. After departing Mayport in August 1985, Saratoga steamed toward the Mediterranean for what was scheduled to be a routine deployment. But on 10 October, Saratoga was called into action.

Arab terrorists hijacked the Italian luxury liner Achille Lauro, which had just departed Alexandria, Egypt, on a pleasure cruise of the Mediterranean. After tense negotiations and the killing of an American tourist, the hijackers traveled in a battered tugboat to Port Said, Egypt, after Achille Lauro anchored just off the coast. Egyptian authorities made hasty arrangements for the terrorists to leave the country. They boarded an Egypt Air 737 jumbo jet at the Al Maza Air Base, northeast of Cairo.

On orders from President Ronald Reagan, seven F-14 Tomcats from the VF-74 "Bedevilers" and VF-103 "Sluggers" were launched from Saratoga. Supporting the Tomcats were VA-85 KA-6D air tankers and VAW-125 E-2C Hawkeye aircraft.

Off the coast of Crete, the F-14s, without using running lights, eased up beside and behind the airliner. On command, the Tomcats turned on their lights and dipped their wings — an international signal for a forced landing. The E-2C Hawkeye radioed the airliner to follow the F-14s. Realizing they were in a "no-win" situation, the hijackers allowed the pilot to follow the Tomcats to Naval Air Station, Sigonella, Italy.

One hour and 15 minutes later, the jumbo jet landed, and the hijackers were taken into custody. Seven hours after the fighter jets were scrambled, all Saratoga aircraft returned home without a shot fired.

On 23 March 1986, while operating off the coast of Libya, aircraft from Saratoga, USS Coral Sea (CV 43), and USS America (CV 66) crossed what Libyan strongman Muammar Gaddafi had called the "Line of Death." The next day at noon, three U.S. Navy warships crossed the same 32° 30' navigational line.

Two hours later, Libyan forces fired SA-5 surface-to-air missiles from the coastal town of Surt. The missiles missed their F-14 Tomcat targets and fell harmlessly into the water. Later that afternoon, U.S. aircraft turned back two Libyan MiG-25 fighter planes over the disputed Gulf of Sidra. Soon after, aircraft from the three supercarriers fought back in defense.

A heavily armed A-6E Intruder fired Rockeye cluster bombs and a Harpoon anti-ship cruise missile at a Libyan missile patrol boat operating on the "Line of Death." Later that night, two A-7E Corsair II jets attacked a key radar installation at Surt. By the conclusion, three Libyan patrol boats and a radar site were destroyed by Navy aircraft.

Following Saratoga's 19th Mediterranean deployment in June 1987, she was overhauled once again at the Norfolk Naval Shipyard at a cost of $280 million.

Saratoga departed Mayport for her 20th deployment on 7 August 1990, just days after Iraqi tanks invaded Kuwait. Saratoga and Carrier Air Wing 17 rapidly crossed the Atlantic, the Mediterranean, and transited the Suez Canal on 22 August to take up station in the Red Sea.

In the early morning hours of 17 January 1991, Operation Desert Shield became Operation Desert Storm. Aircraft from Saratoga flew against Iraq in the first step to knock out the Arab nation's military power and drive it from conquered Kuwait. CVW-17 aircraft dropped more than four million pounds of ordnance on enemy targets.

However, on 17 January, Lt. Cmdr. Michael Scott Speicher, flying an F/A-18C Hornet of VFA-81 aboard Saratoga, was shot down by an Iraqi surface-to-air missile, the first U.S. casualty of the Gulf War. He was placed in MIA status the next day. On May 22, 1991, following a Secretary of the Navy status review board that found "no credible evidence" to suggest he had survived the shootdown, his status was changed to Killed in Action/Body Not Recovered (KIA/BNR). On 11 January 2001, Secretary of the Navy Richard Danzig changed the status of Cmdr. Michael Scott Speicher from KIA/BNR to Missing in Action (MIA), based on new information.

On 21 January 1991, an F-14 Tomcat of VF-103 aboard Saratoga was shot down by a surface-to-air missile. Pilot Lt. Devon Jones and Radar Intercept Officer Lt. Lawrence Slade were reported missing. Lt. Jones was recovered the following day, but Lt. Slade was captured as a prisoner of war.

Flying on 30 January, all 18 F/A-18s aboard Saratoga delivered 100,000 pounds of MK-83 1,000-lb. bombs on Iraqi positions in occupied Kuwait. This was the largest amount of bomb tonnage carried on a single mission.

Saratoga departed the Gulf on 11 March 1991. After seven months and 21 days, 11,700 arrested landings, 12,700 sorties flown, 36,382 miles traveled, and a record six Suez Canal transits, Saratoga returned home on 28 March to a hero's welcome.

Saratoga's 21st Mediterranean deployment, which began on 6 May 1992, was much more than normal operations in the Med. Her six-month deployment found her in the Adriatic Sea, providing close-air support for humanitarian relief flights flying into the war-torn former Yugoslavia. Thousands of support missions were flown, but more importantly, not one single piece of ordnance was dropped, proving that U.S. military presence is a powerful deterrent. On 27 August, aircraft from Saratoga and USS Independence (CV 62), both in the Arabian Gulf, began enforcing the no-fly zone south of the 32nd parallel in Iraq under Operation Southern Watch. Any Iraqi warplanes violating that airspace would be shot down. This was to prevent the Iraqis from attacking Shiite Moslem ethnic groups in the marshes of southern Iraq. From the Gulf, Saratoga returned to the Mediterranean and was relieved on 7 October by USS John F. Kennedy (CV 67).

"Super Sara," along with Carrier Air Wing 17 (CVW-17), began her final deployment on 12 January 1994, entering the Mediterranean Sea on January 26 after a choppy Atlantic Ocean crossing. The ship and air wing spent that day, east of Gibraltar, receiving "turnover briefs" from Saratoga's predecessor in the Mediterranean, USS America (CV 66). As America ended her deployment and set sail for home, Saratoga headed east for the area the crew would come to call "Groundhog Station" in the Adriatic Sea.

Entering the Adriatic on February 1, Saratoga and CVW-17 launched the first of thousands of sorties in support of U.N. and NATO operations Deny Flight and Provide Promise over Bosnia-Herzegovina. Actress Halle Berry, star of films Boomerang, Jungle Fever, and The Last Boy Scout, spent an afternoon with Sailors on

station in the Adriatic Sea aboard Saratoga on 3 February. Then, after 44 consecutive days at sea, Saratoga visited the northern Italian city of Trieste for some eagerly awaited liberty.

The ship departed Trieste on 28 February 1994, taking up station in the southern Adriatic once again, in response to the U.S. Air Force downing of four Bosnian Serb Super Galeb attack aircraft. The four jets had been flying in defiance of the U.N.-NATO "No Fly Zone" over the former war-torn Yugoslavia.

Saratoga remained on station until 10 March. Departing for the eastern Mediterranean, the ship and air wing participated in exercises over land and sea with U.S. allies in that part of the world. Finishing up on 18 March, Saratoga returned to Trieste for another well-deserved period of recreation, then to the Greek island of Crete for bombing exercises at the Avgo-Nisi bombing range.

Completing the exercises, the Saratoga/CVW-17 team returned to the Adriatic for five more days of flying in support of Deny Flight and Provide Promise. Departing "Groundhog Station" on 7 April, Saratoga transited the Straits of Messina between Sicily and the toe of the "boot" of southern Italy, for a port call at Naples, Italy, before returning to the Adriatic for the fourth time on 17 April.

Saratoga anchored off the resort city of Palma de Mallorca, Spain, on 3 May 1994, for one week of the most eagerly-awaited port calls of the deployment. After a week of liberty on the sun-drenched Spanish island, it was back to business as Saratoga participated in the Mediterranean exercise Dynamic Impact. This exercise featured joint maneuvers with several NATO-member navies, as well as the U.S. Air Force.

After finishing six days of liberty in Valencia, Spain, Saratoga participated in Iles D'Or, or "Islands of Gold" — an exercise with the French Navy, lasting until 9 June. Saratoga, the Navy's oldest active duty carrier, then headed for the rendezvous point with her relief, the nuclear-powered aircraft carrier USS George Washington (CVN-73), the Navy's newest carrier at the time.

Saratoga arrived pierside at Naval Station, Mayport, Fla., early in the morning of 24 June 1994. With the end of the 164-day deployment, the last in the carrier's 38-year career, Saratoga's crew prepared to deactivate the ship, offloading material and closing out each of the ship's more than 3,500 spaces.

After being decommissioned on 20 August 1994 and stricken from the Navy List the same day, Saratoga was initially moved to the inactive fleet in Philadelphia. In 1998, the ship was towed to Newport, Rhode Island, where it remained for several years.

In 2014, the Saratoga was sold for scrapping and towed to Brownsville, Texas. On 21 August 2014, the ship departed Newport, Rhode Island, and was dismantled by ESCO Marine in Brownsville.

Saratoga received one battle star for service in the Vietnamese conflict.

CVA-61 • USS RANGER

(Displacement: 56,300 tons; length: 1,046 feet; beam: 130 feet 4 inches; extreme width: 249 1/2 feet; draft: 37 feet; speed: 34 knots; complement: 3,826 crew; armament: 8 5-inch guns; class: Forrestal.)

The seventh Ranger (CVA-61), a Forrestal-class aircraft carrier, was laid down on 2 August 1954 by Newport News Shipbuilding and Drydock Co., Newport News, Va.; launched on 29 September 1956; sponsored by Mrs. Arthur Radford, wife of Admiral Radford, Chairman of the Joint Chiefs of Staff; and commissioned at the Norfolk Naval Shipyard on 10 August 1957, with Capt. Charles T. Booth II in command.

Ranger joined the Atlantic Fleet on 3 October 1957. Just prior to sailing on 4 October for Guantanamo Bay, Cuba, for shakedown, she received the men and planes of Attack Squadron 85. She conducted air operations, individual ship exercises, and final acceptance trials along the eastern seaboard and in the Caribbean Sea until 20 June 1958. She then departed Norfolk, Va., with 200 Naval Reserve officer candidates for a two-month cruise that took the carrier around Cape Horn. She arrived at her new homeport, Alameda, Calif., on 20 August and joined the Pacific Fleet.

The carrier spent the remainder of 1958 in pilot qualification training for Air Group 14 and fleet exercises along the California coast. Departing on 3 January 1959 for final training in Hawaiian waters until 17 February, she next sailed as the flagship of Rear Adm. H. H. Caldwell, ComCarDiv 2, to join the Seventh Fleet. Air operations off Okinawa were followed by maneuvers with SEATO naval units out of Subic Bay. A special weapons warfare exercise and a patrol along the southern seaboard of Japan followed. During this first WestPac deployment, Ranger launched more than 7,000 sorties in support of Seventh Fleet operations. She returned to San Francisco Bay on 27 July.

During the next six months, Ranger kept herself in a high state of readiness through participation in exercises and coastal fleet operations. With Carrier Air Group 9 embarked, she departed Alameda on 6 February 1960 for a second WestPac deployment and returned to Alameda on 30 August. From 11 August 1961 through 8 March 1962, Ranger deployed to the Far East a third time.

The next seven months were filled with intensive training along the western seaboard in preparation for operations in the troubled waters of Southeast Asia. Ranger departed Alameda on 9 November 1962 for brief operations off Hawaii, and then proceeded, via Okinawa, to the Philippines. She steamed to the South China Sea on 1 May 1963 to support possible Laotian operations. When the political situation in Laos relaxed on 4 May, she resumed her operations schedule with the Seventh Fleet. Arriving at Alameda from the Far East on 14 June 1963, she underwent overhaul in the San Francisco Naval Shipyard from 7 August 1963 through 10 February 1964. Refresher training out of Alameda commenced on 25 March, interrupted by an operational cruise to Hawaii from 19 June to 10 July.

Ranger again sailed for the Far East on 6 August 1964. This deployment came on the heels of the unprovoked assault against USS Maddox (DD-731) on the night of 2 August and, two nights later, against both Maddox and USS Turner Joy (DD-951) by North Vietnamese motor torpedo boats. In retaliation for this aggression on the high seas by North Vietnam, President Lyndon B. Johnson, on 5 August, directed the Navy to strike bases used by the North Vietnamese naval craft. As Ranger steamed from the western seaboard, some 60 attack sorties rose from the decks of USS Ticonderoga (CVA-14) and USS Constellation (CVA-64).

Ranger made only an eight-hour stop in Pearl Harbor on 10 August 1964, then hurried on to Subic Bay, and then to Yokosuka, Japan. In the latter port on 17 October 1964, she became the flagship of Rear Adm. Miller, who commanded Fast Carrier Task Force 77. In the following months, she helped the Seventh Fleet continue its role of steady watchfulness to keep open the sea lanes for the Allies and stop Communist infiltration by sea. On 7 February 1965, in retaliation for a damaging Viet Cong attack on installations around Pleiku, a fighter bomber strike, launched from Ranger, USS Coral Sea (CV 43), and USS Hancock (CV 19),

blasted the military barracks and staging areas near Dong Hoi in the southern sector of North Vietnam.

Gen. William Westmoreland, commanding the Military Advisory Command in Vietnam, visited Ranger on 9 March 1965 to confer with Rear Adm. Miller. Ranger continued air strikes on enemy inland targets until 13 April, when a fuel line broke, ignited, and engulfed her No. 1 main machinery room in flames. The fire was extinguished in little over an hour. There was one fatality. Ranger put into Subic Bay on 15 April and sailed on the 20th for Alameda, arriving home on 6 May. She entered the San Francisco Naval Shipyard on 13 May 1965 and remained there under overhaul until 30 September. Following refresher training, Ranger departed Alameda on 10 December 1965 to rejoin the Seventh Fleet. She and her embarked Carrier Air Wing 14 received the Navy Unit Commendation for exceptionally meritorious service during combat operations in Southeast Asia from 10 January to 6 August 1966.

Ranger departed the Gulf of Tonkin on 6 August for Subic Bay, and steamed via Yokosuka for Alameda, arriving on the 25th. She stood out of San Francisco Bay on 28 September and entered Puget Sound Naval Shipyard two days later for overhaul. The carrier departed Puget Sound on 30 May 1967 for training out of San Diego and Alameda. On 21 July 1967, she logged her 88,000th carrier landing.

From June until November, Ranger underwent a long and intensive period of training designed to make her fully combat ready. Attack Carrier Air Wing 2 (CVW-2) embarked on 15 September 1967, with the new A-7 Corsair II jet attack plane and the UH-2C Seasprite turboprop rescue helicopter, making Ranger the first carrier to deploy with these powerful new aircraft. From carrier refresher training for CVW-2, Ranger proceeded to fleet

exercise Moon Festival. From 9 to 16 October 1967, the carrier and her air wing participated in every aspect of a major fleet combat operation. Her efficiency honed to a fine edge, Ranger departed Alameda on 4 November 1967 for WestPac. Arriving in Yokosuka on 21 November, she relieved USS Constellation and sailed for the Philippines on the 24th. After arriving at Subic Bay on 29 November, she made final preparations for combat operations in the Tonkin Gulf. Commander, Carrier Division 3, embarked on 30 November as Commander, TG 77.7; and Ranger departed Subic Bay on 1 December for Yankee Station.

Arriving on station on 3 December 1967, Ranger commenced another period of sustained combat operations against North Vietnam. During the next 5 months, her planes hit a wide variety of targets, including ferries, bridges, airfields, and military installations. Truck parks, rail facilities, antiaircraft guns, and SAM sites were also treated to doses of Air Wing 2's firepower. Bob Hope's "Christmas Show" came to Ranger in Tonkin Gulf on 21 December. Another welcome break in the intense pace of operations came with a call at Yokosuka during the first week of April 1968. Returning to Yankee Station on 11 April 1968, Ranger again struck objectives in North Vietnam.

After five months of intensive operations, Ranger called at Hong Kong on 5 May 1968 and then steamed for home. There followed a shipyard availability at Puget Sound that ended with Ranger's departure on 29 July for San Francisco. Three months of leave, upkeep, and training culminated in another WestPac deployment from 26 October 1968 through 17 May 1969. She departed Alameda on yet another WestPac deployment in December 1969 and remained so employed until 18 May 1970 at which time she returned to Alameda, arriving on 1 June 1970.

Ranger spent the rest of the summer engaged in operations off the west coast, departing for her sixth WestPac cruise on 27 September 1970. On 10 March 1971, Ranger, along with USS Kitty Hawk (CV 63), set a record of 233 strike sorties for one day in action against North Vietnam. During April, the three carriers assigned to Task Force 77 — Ranger, Kitty Hawk, and Hancock — provided a constant two-carrier posture on Yankee Station. Hours of employment remained unchanged with one carrier on daylight hours and one on the noon to midnight schedule. Strike emphasis was placed on the interdiction of major Laotian entry corridors to South Vietnam. She returned to Alameda on 7 June 1971 and remained in port for the rest of 1971 and the first five months of 1972 undergoing regular overhaul.

On 27 May 1972, she returned to west coast operations until 16 November, when she embarked on her seventh WestPac deployment. On 18 December 1972, Linebacker II operations were initiated when negotiations in the Paris peace talks stalemated. Participating carriers included Ranger, USS Enterprise (CVN 65), USS Saratoga (CV 60), USS Oriskany (CV 34), and USS America (CV 66).

The Linebacker II operations ended on 29 December when the North Vietnamese returned to the peace table. These operations involved the resumed bombing of North Vietnam above the 20th parallel and were an intensified version of Linebacker I. The reseeding of the minefields was resumed, and concentrated strikes were carried out against surface-to-air missile and antiaircraft artillery sites, enemy army barracks, petroleum storage areas, Haiphong naval and shipyard areas, and railroad and truck stations. Navy tactical air attack sorties under Linebacker II were centered in the coastal areas around Hanoi and Haiphong. There were 505 Navy sorties in this area

during Linebacker II. Between 18 and 22 December, the Navy conducted 119 Linebacker II strikes in North Vietnam. Bad weather was the main limiting factor on the number of tactical air strikes flown during Linebacker II.

On 27 January 1973, the Vietnam cease-fire, announced four days earlier, came into effect, and Oriskany, America, Enterprise, and Ranger, on Yankee Station, canceled all combat sorties into North and South Vietnam. During the U.S. involvement in the Vietnam conflict (starting in 1961 and ending on 2 January 1973), the Navy lost 726 fixed-wing aircraft and 13 helicopters to hostile action. The Marine Corps lost 193 fixed-wing aircraft and 270 helicopters to enemy action during the same period. Operation Homecoming, the repatriation of U.S. POWs between 27 January and 1 April 1973, began, and North Vietnam and the Viet Cong released 591 POWs. Of the 591 POWs released during Operation Homecoming, 145 were Navy personnel, all but one of whom were Naval Aviation personnel.

Ranger returned to Alameda in August 1973 and remained in that area through 7 May 1974, when she deployed again to the western Pacific, returning to homeport on 18 October. On 28 May 1976, while on deployment, helicopter crews from HS-4 aboard Ranger, detachments from HC-3 on USS Camden (AOE 2), USS Mars (AFS 1) and USS White Plains (AFS 4), and helicopters from NAS Cubi Point, Republic of the Philippines, assisted in Philippine disaster relief efforts in the flood-ravaged areas of central Luzon. Over 1,900 people were evacuated; more than 370,000 pounds of relief supplies and 9,340 gallons of fuel were provided by Navy and Air Force helicopters.

On 12 July 1976, Ranger and her escort ships of Task Force 77.7 entered the Indian Ocean and were assigned to operate off the coast of Kenya in response to a threat of military action in Kenya by Ugandan forces.

Ranger entered the history books on 21 March 1983, when an all-woman flight crew flying a C-1A Trader from VRC-40 "Truckin' Traders" landed aboard the carrier. The aircraft was commanded by Lt. Elizabeth M. Toedt and the crew included Lt. (j.g.) Cheryl A. Martin, Aviation Machinist's Mate 3rd Class Gina Greterman, and Aviation Machinist's Mate Airman Robin Banks.

On 24 July 1987, Tactical Electronics Warfare Squadron 131 (VAQ 131) began the first Pacific Fleet deployment of the EA-6B Prowler equipped with HARM missiles, deployed in Ranger.

On 3 August 1989, Ranger rescued 39 Vietnamese refugees, adrift for 10 days on a barge in heavy seas and monsoon rains in the South China Sea, about 80 miles from NAS Cubi Point, R.P. SH-3 Sea Kings from HS-14 assisted. An A-6 Intruder from VA-145 spotted the barge, which had apparently broken loose from its mooring near a small island off the coast of Vietnam with 10 men on board. Twenty-nine other refugees from a sinking refugee boat climbed aboard the barge when it drifted out to sea. After examination by medical personnel, all were flown to NAS Cubi Point for further processing.

President George H. W. Bush addressed the nation on 16 January 1991 at 9 p.m. EST and announced that the liberation of Kuwait from Iraq, Operation Desert Storm, had begun. The Navy launched 228 sorties from Ranger and USS Midway (CV 41) in the Persian Gulf, from USS Theodore Roosevelt (CVN 71) en route to the Gulf, and from USS John F. Kennedy (CV 67), USS Saratoga (CV 60), and USS America (CV 66) in the Red

Sea. In addition, the Navy launched more than 100 Tomahawk missiles from nine ships in the Mediterranean Sea, the Red Sea, and the Persian Gulf. On 6 February 1991, an F-14A Tomcat from VF-1, off Ranger, piloted by Lt. Stuart Broce, with Cmdr. Ron McElraft as Radar Intercept Officer, downed an Iraqi MI-8 Hip helicopter with an AIM-9M Sidewinder missile. At 9 p.m. EST on 27 February, President Bush declared Kuwait had been liberated and Operation Desert Storm would end at midnight.

On 21 April 1992, in harmony with other World War II 50th Anniversary festivities, Ranger participated in the commemorative re-enactment of the Doolittle Raid on Tokyo, Japan. Two World War II-era B-25 bombers were craned on board, and over 1,500 guests (including national, local, and military media) were embarked to witness the two vintage warbirds thunder down Ranger's flight deck and take off. In June, Ranger made a historic port visit to Vancouver, British Columbia, in conjunction with her final phase of pre-deployment workups.

Fully combat ready, Ranger began her 21st and final western Pacific and Indian Ocean deployment on 1 August 1992. On 18 August, she entered Yokosuka, Japan, for a six-day port visit and upkeep. Ranger entered the Arabian Gulf on 14 September by transiting the Straits of Hormuz. The next day, Ranger relieved USS Independence (CV-62) in an unusual close-aboard ceremony, and along with her embarked Air Wing, Carrier Air Wing TWO, immediately began flying patrol missions in support of the United Nations' declared "No Fly" zone in southern Iraq: Operation Southern Watch.

While in the Arabian Gulf, former Cold War adversaries became at-sea partners as Ranger, British, and French naval forces joined with the Russian guided missile destroyer Admiral Vinogradov for an exercise involving communication, maneuvering, and signaling drills. During joint operations, a Russian KA-27 Helix helicopter landed aboard Ranger. It was the first such landing on a U.S. Navy aircraft carrier. Ranger left the Gulf on 4 December 1992 and steamed at high speed to the coast of Somalia. Ranger played a significant role in the massive relief effort for starving Somalis in Operation Restore Hope. The Ranger/CVW-2 team provided photo and visual reconnaissance, airborne air traffic control, logistics support, and on-call close air support for Navy and Marine amphibious forces. Throughout Operations Southern Watch and Restore Hope, Ranger took 63 digital photographs, which were sent by International Marine Satellite to the Navy Office of Information within hours of being taken. This was the first time digital pictures were successfully transmitted from a ship at sea.

On 19 December 1992, Ranger was relieved on station by USS Kitty Hawk (CV 63) and began her last long journey homeward to San Diego. Ranger was decommissioned on 10 July 1993 and is at the Naval Inactive Ship Maintenance Facility, Bremerton, Wash. The decommissioned aircraft carrier arrived in Brownsville, Texas, on July 12, 2015, where it was dismantled by International Shipbreaking, a company part of the EMR Group. The dismantling project took just over two years and concluded on November 1, 2017.

Ranger earned 13 battle stars for service in Vietnam.

CVA-62 • USS INDEPENDENCE

(Displacement: 56,300 tons; length: 1,046 feet; beam: 130 feet 4 inches; extreme width: 249½ feet; draft: 37 feet; speed: 34 knots; complement: 3,950 crew; armament: 8 5-inch guns; class: Forrestal.)

The fifth Independence (CVA-62) was launched by New York Navy Yard on 6 June 1958, sponsored by Mrs. Thomas Gates, wife of the Secretary of the Navy, and commissioned on 10 January 1959, with Captain R. Y. McElroy in command.

One of the newest class of "supercarriers" at the time of her commissioning, Independence conducted shakedown training in the Caribbean and arrived at her home port of Norfolk on 30 June 1959. On 25 August, during suitability trials on board Independence, an A3D Skywarrior, piloted by Lt. Cmdr. Ed Decker, took off at a gross weight of 84,000 pounds — the heaviest aircraft ever to take off from a carrier.

Independence operated off the Virginia Capes for the next year on training maneuvers and departed on 4 August 1960 for her first cruise to the Mediterranean. There, she added her great strength to the peacekeeping power of the 6th Fleet in that troubled region, remaining in the eastern Mediterranean until her return to Norfolk on 3 March 1961. The remainder of the year was spent in training and readiness operations off the Atlantic coast.

Independence sailed on 19 April 1962 for Sixth Fleet duty in support of President John F. Kennedy's firm stand on Berlin during a reoccurrence of stress in a critical area. She returned to Norfolk on 27 August and sailed on 11 October for the Caribbean Sea. Called on by President Kennedy on 24 October during the 1962 Cuban missile crisis, Independence provided a strong, visible reminder of U.S. determination and resolve while it acted as a key participant in the U.S. naval blockade of Cuba. She arrived off Puerto Rico in response to the presence of Soviet missiles in Cuba and took part in the quarantine operations which finally forced the withdrawal of those missiles. She then returned to Norfolk on 25 November for readiness exercises along the eastern seaboard, overhaul in the Norfolk Naval Shipyard, and refresher training out of Guantanamo Bay.

Independence departed Norfolk on 6 August 1963 to take part in combined readiness exercises in the Bay of Biscay with sea-air units of the United Kingdom and France, then entered the Mediterranean on 21 August for further duty with the Sixth Fleet. Cruising throughout the Mediterranean, she gained much valuable experience during combined NATO exercises, including close air support to Turkish paratroops, reconnaissance, communications, and convoy strike support. President Makarios of Cyprus paid her a visit on 7 October 1963, after which she joined in bilateral U.S.-Italian exercises in the Adriatic with Italian patrol torpedo boats, and U.S.-French exercises which pitted her aircraft against French interceptors and a surface action with the French cruiser Colbert (C 611). She returned to Norfolk on 4 March 1964.

Following training exercises, ranging north to New York and south to Mayport, Fla., Independence departed Norfolk on 8 September 1964 for NATO Teamwork exercises in the Norwegian Sea and off the coast of France, thence to Gibraltar. She returned to Norfolk on 5 November 1964 and entered the Norfolk Naval Shipyard for overhaul.

On 10 May 1965, Independence deployed for more than seven months, including 100 days in the South China Sea off the coast of Vietnam, the first Atlantic Fleet carrier to do so. She also was the fifth U.S. carrier operating off Vietnam. Independence and her embarked Air Wing 7 received the Navy Unit Commendation for exceptionally meritorious service from 5 June to 21 November 1965. They participated in the first major series of coordinated strikes against vital enemy supply lines north of the Hanoi-Haiphong complex, successfully evading the first massive surface-to-air missile barrage in aviation history while attacking assigned targets, and executing, with daring and precision, the first successful attack on an enemy surface-to-air missile installation. The carrier launched more than 7,000 sorties in sustaining an exceptional pace of day and night strike operations against military and logistic supply facilities in North Vietnam. "The superior team spirit, courage, professional competence, and devotion to duty displayed by the officers and

men of Independence and embarked Attack Carrier Air Wing Seven reflect great credit upon themselves and the United States Naval Service."

Independence returned to her homeport, Norfolk, Va., arriving on 13 December 1965. During the first half of 1966, she operated off Norfolk, replenishing and training air groups. On 4 May 1966, she participated in Operation Strikex. The carrier departed Norfolk on 13 June for European operations with the Sixth Fleet. Independence was involved with unit and NATO exercises from July into December. She then continued her Sixth Fleet deployment into 1967.

On 25 September 1970, word was received that Gamal Abdul Nasser, President of the United Arab Republic, had died; an event that might plunge the entire Middle East into a crisis. Independence, along with USS John F. Kennedy (CV 67), USS Saratoga (CV 60), and seven other U.S. Navy ships, were put on standby in case U.S. military protection was needed for the evacuation of U.S. citizens and as a counterbalance to the Soviet Union's Mediterranean fleet.

Pilots of VMA-142, -131, and -133 began qualification landings in A-4 Skyhawks aboard Independence on 3 August 1971. For the next three days, four active duty and 20 reserve pilots operated aboard the carrier — the first time that Marine Corps Air Reserve Squadrons qualified in carrier duty.

In May 1973, President Richard M. Nixon delivered his annual Armed Forces Day address from the decks of Independence. While based in Norfolk, the ship made deployments to the Mediterranean Sea and Indian Ocean. From 8 to 13 October 1973, Task Force 60.1 with Independence, Task Force 60.2 with

USS Franklin D. Roosevelt (CV 42), and Task Force 61/62 with USS Guadalcanal (LPH 7) were alerted for possible evacuation contingencies in the Middle East. The ships were on alert as a result of the 1973 Yom Kippur war between Arab states and Israel. Independence operated off the island of Crete.

On 20 June 1979, Lt. Donna L. Spruill became the first Navy woman pilot to carrier qualify in a fixed-wing aircraft. Lt. Spruill piloted a C-1A Trader to an arrested landing aboard Independence.

In 1982, Independence provided critical support to the multinational peacekeeping force in Lebanon. On 25 June, the greatest concentration of U.S. Navy air power in the Mediterranean Sea occurred when the battle groups of USS Forrestal (CV 59) and Independence joined forces with USS Dwight D. Eisenhower (CVN 69) and USS John F. Kennedy (CV 67). After steaming together in the eastern Mediterranean Sea for several days, Forrestal and Independence relieved Dwight D. Eisenhower and John F. Kennedy, the latter sailing home to Norfolk, Va., after a long deployment.

On 25 October 1983, aircraft from Independence's embarked air wing flew missions in support of Operation Urgent Fury, the action to liberate the Caribbean nation of Grenada. Returning to Lebanon that same year, the ship's air wing conducted air strikes against Syrian positions.

On 17 February 1985, Independence arrived at the Philadelphia Naval Shipyard to undergo a modernization and overhaul program to extend her service life by 15 years. The flight deck was improved to allow the recovery of high-performance aircraft while the ship traveled at slower speeds, and the NATO Sea Sparrow launchers were upgraded. Other improvements enhanced the ship's fuel consumption. Independence completed the Service Life Extension Program (SLEP) at Philadelphia Naval

Shipyard in June 1988. Setting sail on 15 August from Norfolk, the ship transited the tip of South America and arrived at her new homeport of San Diego, Calif., on 8 October.

In August 1990, with Carrier Air Wing Fourteen embarked, Independence was sent to deter Iraqi aggression during Operation Desert Shield. Arriving on station in the Gulf of Oman on 5 August, Independence was the first carrier to enter the Arabian Gulf since 1974. The ship remained on station for more than 90 days and permanently reestablished a U.S. naval presence in the region. She returned to San Diego on 20 December 1990.

Independence changed homeports again on 11 September 1991 — this time to Yokosuka, Japan, embarking Carrier Air Wing Five and becoming the Navy's only permanently forward-deployed aircraft carrier and flagship for Commander, Carrier Group Five.

On 23 August 1992, Independence entered the Persian Gulf, prepared to enforce an Allied ban on Iraqi flights over southern Iraq below the 32nd parallel. On 26 August, President George H. W. Bush announced that the United States and its allies had informed Iraq that in 24 hours, Allied aircraft would fly surveillance missions in southern Iraq and were prepared to shoot down any Iraqi aircraft flying south of the 32nd parallel. This action was precipitated by Iraq's failure to comply with U.N. Resolution 688, which demanded that the Iraqi government stop the repression of its Shiite population in southern Iraq.

Persian Gulf allies began to enforce the ban on Iraqi planes from flying south of the 32nd parallel on 27 August in Operation Southern Watch. Any Iraqi planes that violated the ban would be shot down. Twenty Navy aircraft from CVW-5 aboard Independence in the Persian Gulf were the first coalition aircraft on station over Iraq as Operation Southern Watch

began. Southern Watch was the enforcement of a ban on Iraqi warplanes and helicopters from flying south of the 32nd parallel.

Independence became the oldest ship in the Navy's active fleet, and the first carrier in history to hold that distinction, on 30 June 1995. With this honor, Independence displayed the Revolution-era First Navy Jack, commonly called the "Don't Tread On Me" flag, from her bow until her decommissioning. The flag was presented to Independence commanding officer Capt. David P. Polatty III in a formal ceremony on 1 July. The flag was received from USS Mauna Kea (AE 22) upon her decommissioning.

In November 1995, the Independence and Carrier Air Wing Five team returned to Japan after successfully completing their third deployment to the Arabian Gulf in support of Operation Southern Watch.

In March 1996, Independence was called upon to provide a stabilizing presence during heightened tensions between Taiwan and China. Upon returning to Yokosuka in April 1996, the ship was visited by President Bill Clinton as part of an official state visit to Japan.

In 1997, Independence made a four-month deployment, covering several major exercises and seven ports of call. Included in these ports of call were two historic port visits. The first was on 28 February 1997 to the island territory of Guam. Independence was the first aircraft carrier to pull into Guam in 36 years.

The second, two months later, was to Port Klang, Malaysia. Independence became the first aircraft carrier in the world to make a port visit to Malaysia.

Before sailing back to Yokosuka, Japan, Independence made its last port call of the deployment in May 1997 to Hong Kong. Independence's port visit was the last U.S. naval port visit to the territory before its reversion to China on 1 July 1997.

Independence deployed to the Arabian Gulf in January 1998 to support negotiations between the UN and Iraq and to again participate in Operation Southern Watch.

Independence was decommissioned in ceremonies at the Puget Sound Naval Shipyard in Bremerton, Wash., on 30 September 1998. Independence's commissioning pennant was hauled down 39 years, 9 months, and 20 days after it was first proudly hoisted, and the "Don't Tread on Me" jack was transferred to the Navy's next oldest active ship, USS Kitty Hawk (CV 63).

After decommissioning, Independence was held in reserve in Bremerton, Washington. Efforts to preserve her as a museum ship did not materialize, and in 2017, she was sold for scrapping. She was towed to Brownsville, Texas, and dismantling began in 2018, marking the end of her service.

Independence earned one battle star for her service in the Vietnamese conflict.

CV-63 • USS KITTY HAWK

(Displacement: 82,538 tons (74,877 metric tons) full load; Length: 1,073 feet (327 meters); Beam: 130 feet (39.62 meters); extreme width: 282 feet (85.95 meters); Draft: 39 feet (11.89 meters); Speed: In excess of 30 knots (34.5+ miles per hour); Complement: 4,000 crew; Aircraft: 85; Armament: Three NATO Sea Sparrow launchers; four 20mm Phalanx CIWS mounts; Class: Kitty Hawk.)

The second Kitty Hawk (CVA-63) was laid down by the New York Shipbuilding Corp., Camden, N.J., on 27 December 1956; launched on 21 May 1960, sponsored by Mrs. Neil H. McElroy; and commissioned on 29 April 1961 at Philadelphia Naval Shipyard, Capt. William F. Bringle in command.

Following shakedown in the western Atlantic, Kitty Hawk departed Norfolk on 11 August 1961. After a brief stop at Rio de Janeiro, where she embarked the Secretary of the Brazilian Navy for a demonstration exercise at sea with five Brazilian destroyers, the attack carrier rounded Cape Horn on 1 October. She steamed into Valparaiso Bay on 13 October and then sailed two days later for Peru, arriving in Callao on 20 October 1961, where she entertained the President of Peru.

Back in San Diego, Adm. George W. Anderson, Chief of Naval Operations, landed on her deck on 18 November 1961 to witness antisubmarine demonstrations by USS Henry B. Wilson (DDG-7) and USS Blueback (SS-581), a Terrier missile demonstration by USS Topeka (CLG-8), and air demonstrations by Kitty Hawk.

Kitty Hawk entered San Francisco Naval Shipyard on 23 November 1961 for alterations. Following operations out of San Diego, she sailed from San Francisco on 13 September 1962. Kitty Hawk joined the 7th Fleet on 7 October 1962, relieving USS Midway (CVA-41) as flagship.

After participating in the Philippine Republic Aviation Week Air Show, Kitty Hawk steamed out of Manila Harbor on 30 November 1962 and welcomed Adm. H. D. Felt, Commander in Chief, Pacific Fleet, for a demonstration of modern naval weapons on 3 December. The ship visited Hong Kong early in December and returned to Japan, arriving at Yokosuka on 2 January 1963. During the following two months, she visited Kobe, Beppu, and Iwakuni before returning to San Diego on 2 April 1963.

On 6 June 1963, President John F. Kennedy, with top civilian and military leaders, boarded Kitty Hawk to witness a carrier task force weapons demonstration off the California coast. Addressing the men of the task group from Kitty Hawk, President Kennedy told them that, as in the past, control of the seas still means security, peace, and ultimate victory. He later wrote to President and Madam Chiang Kai-Shek, who had witnessed a similar demonstration on board USS Constellation (CVA-64): "I hope you were impressed as I was, on my visit to Kitty Hawk, with the great force for peace or war, which these mighty carriers and their accompanying escorts provide, helping to preserve the freedom of distant nations in all parts of the world."

On 30 September 1963, Kitty Hawk was at sea off the California coast for her final exercise as a unit of the First Fleet. Following a series of strike exercises and tactics along the California coast and off Hawaii, Kitty Hawk again sailed for the Far East. On 17 October 1963, she departed her homeport at San Diego for the Far East and her second tour of duty with the Seventh Fleet.

En route to the western Pacific, Kitty Hawk received her Operational Readiness Inspection in Hawaiian waters. Upon completion, Rear Adm. Duerfeldt, COMFAIRHAWAII, said the combination of Kitty Hawk/CAW-11 was "the best weapons system we have observed this year."

While approaching Japan, she learned that an assassin had shot President Kennedy. Flags were at half-mast as she entered Sasebo Harbor on 25 November 1963, the day of the President's funeral, and as the senior ship present, she had the sad honor of firing memorial salutes.

Following initial Seventh Fleet port visits to Buckner Bay, Okinawa, and Sasebo, Japan, in November, Kitty Hawk headed south to Taiwan to participate in Exercise Big Dipper. Nationalist Chinese forces combined with Seventh Fleet units for the amphibious exercise to demonstrate how American forces can answer the call from a besieged ally. Kitty Hawk aircraft supplied air support and aerial reconnaissance for the assault forces. Adm. Claude V. Ricketts, Vice Chief of Naval Operations, observed Kitty Hawk's operations on 3 December for Big Dipper. Aboard with him was Vice Adm. T. H. Moorer, Commander Seventh Fleet.

After Big Dipper, Kitty Hawk visited Kobe, Japan, for a four-day goodwill visit. On 23 December, Kitty Hawk moored at Yokosuka, Japan, for a two-week Christmas visit. On 5 January 1964, Kitty Hawk was seaward again for operations. During that at-sea period, she held joint operations with the carrier USS Oriskany (CVA-34).

Kitty Hawk returned to Yokosuka on 10 February for a two-week upkeep period. Due to inclement weather before entering port, many airplanes due to be launched to nearby NAS Atsugi for maintenance were left on board. But on 12 February, more than 20 planes were catapulted off while the ship was moored in an unusual demonstration of the flexibility of the carrier and her aircraft.

Kitty Hawk visited Hong Kong from 20-26 February 1964 and hosted many visitors aboard the ship. More than 300,000 gallons of fresh water were donated to the British government for use in the drought-stricken colony.

In late February, Kitty Hawk headed south again to Taiwan, this time to participate in the amphibious Exercise Back Pack. During the exercise, Vice Adm. J. F. D. Bush, Royal Navy, British Naval Attache to Washington, visited and was given a ride in the F-4B Phantom II jet. As in Big Dipper, Kitty Hawk aircraft provided air support and aerial reconnaissance for the Seventh Fleet Marines assaulting the beach.

Following Back Pack, Kitty Hawk spent a week at Sasebo followed by an Easter weekend visit to Buckner Bay, Okinawa.

On 6 April 1964, Rear Adm. Thomas Winfield South III was relieved by Rear Adm. William F. Bringle as Commander Carrier Division Seven aboard Kitty Hawk. Adm. Bringle was well known to the old hands aboard Kitty Hawk as he had previously served as her first commanding officer. The ship then visited Hong Kong from 10 to 17 April.

After departing Hong Kong, Capt. John "L" Butts, Jr. relieved Capt. Horace H. Epes, Jr. as Commanding Officer of Kitty Hawk on 20 April. Kitty Hawk then conducted operations in the South China Sea until she departed that area to arrive at Yokosuka, Japan, on 6 May 1964 for a three-day stay. Following this, the ship and air group engaged in joint operations with HMS Victorious on 10 and 11 May 1964.

During the period between 18 May and 10 June, Kitty Hawk was again engaged in special operations in the South

China Sea off the coast of Vietnam. While conducting photo reconnaissance flights over Communist Laotian territory, two Kitty Hawk pilots were downed by ground fire. Cmdr. D.W. Lynn, Executive Officer of VF-111, went down under Communist fire on 7 June but was rescued and returned to the ship on the 8th. Also at this time, Lt. C.F. Klusmann of the VFP-63 detachment aboard Kitty Hawk was shot down and captured by Communist forces in Laos. After almost three months in his prison camp, Lt. Klusmann managed to escape and was returned to the United States in mid-September.

Kitty Hawk arrived in Yokosuka, Japan, on 14 June 1964 after 36 continuous days at sea. On 15 June, in an impressive Change of Command ceremony aboard Kitty Hawk, Vice Adm. Thomas H. Moorer was relieved by Vice Adm. Roy L. Johnson as Commander Seventh Fleet. Almost two years earlier, in October 1962, Adm. Moorer had assumed command of the Seventh Fleet in a ceremony also held on board Kitty Hawk.

On 29 June 1964, Kitty Hawk departed Yokosuka for operations south of Japan and returned on 5 July to make final preparations for the return trip to the United States. The ship departed Yokosuka on 7 July for her return to the U.S., leaving a day early to avoid a threatening typhoon. Kitty Hawk arrived home in San Diego on 20 July 1964 after a deployment that lasted over nine months. She thus was entitled to fly her "homeward-bound" pennant, over 1,000 feet long, which she earned by being deployed for more than nine months.

On 10 August 1964, Kitty Hawk departed San Diego for a three-day trip to Bangor, Washington. While en route on 12 August, the icebreaker USS Staten Island (AGB 5) towed Kitty Hawk for several hours off Newport, Oregon, to test this capability. Kitty Hawk spent two days at Bangor offloading ammunition and departed on 15 August for a one-day trip to the Puget Sound Naval Shipyard in Bremerton, Washington. During this short inland sea cruise, local dignitaries and their families were invited aboard. On 16 August, an open house was conducted at Bremerton during which approximately 25,000 people streamed aboard. This was the largest crowd to visit Kitty Hawk in a single day.

On 16 August 1964, an eight-month overhaul and modification period began with several major modifications being installed in Kitty Hawk. These included the Naval Tactical Data System (NTDS), Integrated Operational Intelligence Center (IOIC), Automatic Handing System (AN/SPN-10), and the Airborne System Support Center (ASSC). On 4 September, she moved into Dry Dock Number 6, the world's largest dry dock, and thus provided the first full capacity load for this dry dock.

Kitty Hawk returned to San Diego in May 1965, following her extensive yard period in Bremerton. She immediately began four weeks of intensive Refresher Training during which Kitty Hawk achieved the highest rating ever given to an aircraft carrier. For five days underway in July, Walt Disney and a Hollywood crew, which included two chimpanzees, were on board to film parts of the movie "Lt. Robin Crusoe, USN," which starred Dick Van Dyke and Nancy Kwan as well as many of Kitty Hawk's crew. Along with Walt Disney were guests of the Secretary of the Navy, midshipmen, Naval Reservists, and many other observers and visitors who were able to take advantage of the hospitality for which Kitty Hawk is famous.

On 9 July 1965, she hosted 50 women flyers, all contestants in the cross-country "Powder Puff Derby" flight. That same day, some 300 members of the La Jolla, California, Regional Horseless Carriage Club of America came aboard to visit, having parked their old-time cars along the quay wall for the crew to inspect and admire. The flow of visitors continued unabated throughout the summer and fall.

From May until September, Kitty Hawk spent many long weeks at sea conducting exercises and carrier qualifications, including more than 7,000 aircraft launches and landings. On 7 and 8 August 1965, she was in San Francisco, which afforded her crew an opportunity to sight-see in that interesting city. On 8 August, she hosted over 5,000 members of the Alameda Naval Air Reserve and their families. For one period of two weeks, she acted as a testing facility for the Naval Air Test Center, Patuxent River, Maryland. During this time, a number of successful computer-controlled "no hands" landings were conducted — a "first" for the ship.

On 20 August 1965, nearly 2,500 members of Kitty Hawk families went to sea for the day. Many displays were set up, and the families watched an underway replenishment from an oiler, USS Chemung (AO-30), as well as an exciting air show from the flight deck of the ship. A month later, on 20 September, Kitty

Hawk served as host for the Change of Command ceremony for the Commander Carrier Division One in San Diego. Rear Adm. Maurice F. Weisner relieved Rear Adm. Edward C. Outlaw.

On 19 October 1965, Kitty Hawk departed San Diego, California, for her third western Pacific cruise. On board were 14 guests of the Secretary of the Navy. These guests, representatives of various press media throughout California and Utah, remained on board for six days until their arrival in Pearl Harbor, Hawaii.

On 26 October, Kitty Hawk began her Operational Readiness Inspection under the control of Fleet Training Group and Commander Fleet Air Hawaii. Results of the week of drills and tests proved Kitty Hawk the "tops." Kitty Hawk attained the highest score of any Pacific Fleet Attack Carrier undergoing this rigorous inspection during the previous two years.

In the early morning hours of 8 November, after four days of rest and relaxation in Hawaii, Kitty Hawk departed for Subic Bay in the Philippines to join the U.S. Seventh Fleet. Kitty Hawk's stay in Subic Bay was very short and the time was utilized in replenishing for her upcoming job on Dixie Station, taking part in the battle against Viet Cong insurgents in South Vietnam.

November 26th will always be a day in the history books for Kitty Hawk. It was the first time her aircraft lifted off the Flight Deck for combat operations. On that day, Kitty Hawk aircraft flew 90 attack sorties against the VC, unleashing more than 140 tons of ordnance.

While at Yankee Station on 6 December 1965, a fire swept through one of the large main engine room spaces. Despite the seriousness of the fire, Kitty Hawk was able to continue full air operations on schedule.

Kitty Hawk departed Yokosuka, en route to Yankee Station on 9 January 1966, conducting refresher flight operations and nuclear weapons loading exercises en route. On 11 January, CVW-11 aircraft, under control of CTG 70.4, conducted attacks on a USS Hornet (CVS-12) towed sled in close proximity to a USSR Task Unit near Bashi Channel. RA-5C aircraft from RVAH-13 obtained photo coverage of all surface units.

Kitty Hawk aircraft commenced Tiger Hound, Steel Tiger, Blue Tree, and in-country operations on 14 January. The Tet (Vietnamese Lunar Holiday) stand-down resulted in increased sortie requirements for CVW-11 on 21 and 23 January but provided a break in routine on 22 January. Concentration of all flight activity in the Steel Tiger/Tiger Hound areas from 20-23 January produced high-density air coverage with a resultant disappearance of targets. Intense interdiction was apparently highly effective. Post-Tet in-country operations were handicapped by frequent periods of low ceiling in the I Corps area and by the non-availability of Forward Air Controllers. Heavy sortie rates in the Steel Tiger/Tiger Hound areas apparently resulted in reduced vehicular activity as evidenced by the paucity of live targets in Laos. Planning for the possible resumption of Rolling Thunder operations was accelerated.

The pace of operations increased sharply with the resumption of Rolling Thunder operations on 31 January 1966. The weather in North Vietnam was uniformly bad until 3 February when Rolling Thunder Package III opened for a few hours. Kitty Hawk responded with a 170-sortie day, including 49 attack sorties in NVN.

On 31 January, an F-4 Phantom from VF-114 crashed near Kitty Hawk after complete hydraulic failure due to combat damage. On 1 February, an A-1 from VA-115 was shot down in the Steel Tiger area. Crew members of both aircraft were recovered uninjured. On 3 February, an RA-5C Vigilante from RVAH-13 was downed by enemy fire off the NVN coast just south of Cape Bouton. A major SAR effort, including excellent shore bombardment by USS Waddell (DDG-24) and USS Brinkley Bass (DD-887), failed to recover the crew.

Kitty Hawk departed Subic Bay on 10 February 1966 en route to Hong Kong, conducting a Surface-to-Air Missile Exercise on 10 February and an Air-to-Air Missile Exercise on 11 February. On 11 February, Rear Adm. J. F. Reedy, CTF 77, presented 61 Air Medals to pilots and crew members of Attack Carrier Air Wing Eleven, Kitty Hawk's embarked Air Wing.

Kitty Hawk arrived in Hong Kong on 12 February and departed, en route to Yankee Station, on 15 February. The exemplary conduct of Kitty Hawk crew members resulted in the following from SOPA (ADMIN) Hong Kong, "During your brief Hong Kong visit from 12 to 15 February, it was most evident to all concerned that Kitty Hawk personnel are a diplomatic force promoting an atmosphere of friendship, mutual respect, and understanding. Well done."

Kitty Hawk arrived at Yankee Station on 17 February 1966, operating there through 20 February, then moved south to Dixie Station for in-country operations from 22 February to 5 March. Extremely low ceilings and visibility throughout the area seriously limited air operations. The majority of Rolling Thunder missions after 17 February were completed by A-6A Intruder aircraft from VA-85 using radar system deliveries through the overcast. On 18 February, an Intruder was lost when it failed to complete pullout from a glide bombing attack. There were no survivors. During the period from 22 February to 5 March, Kitty Hawk aircraft averaged 100 direct air support sorties per day in support of friendly forces in South Vietnam.

Kitty Hawk returned to Yankee Station on 6 March 1966, conducting air operations while en route. On 5 March, an F4B Phantom from VF-114 was lost after being hit by enemy ground fire during in-country operations. The crew ejected due to a loss of hydraulic pressure and control effectiveness. Both the pilot and RIO were recovered safely by a SAR helicopter. All-weather A6A Intruder aircraft maintained steady pressure on North Vietnamese targets despite overcast skies and inclement weather, both day and night. Kitty Hawk aircraft provided close air support missions in defense of the beleaguered As Hau Special Forces Camp on 10 March. On 11 March, an A1H of VA-115 was lost shortly after catapult launch. The pilot was recovered on board with only minor injuries. On 14 March, Kitty Hawk aircraft and a SAR helicopter participated in the daring rescue of two USAF air crewmen after their aircraft had been shot down. Both crewmen were rescued within range of NVN shore batteries, returned to Kitty Hawk, and treated.

Kitty Hawk departed Yankee Station on 16 March 1966 and arrived in Subic Bay on 17 March for an upkeep period. Kitty Hawk departed on 29 March and arrived at Dixie Station on 31 March. On 31 March, Rear Adm. J. F. Reedy, Commander Task Force 77, presented two Distinguished Flying Crosses, 238 Air Medals, and 7 Navy Commendation Medals to pilots and crew members of Attack Carrier Air Wing ELEVEN. Aircraft from CVW-11 provided in-country and Operation Jackstay support and averaged 100 sorties per day on enemy targets.

During the period from 1 April 1966 to 23 May 1966, Kitty Hawk, with Commander Attack Carrier Striking Force, SEVENTH Fleet (CTF 77), Commander Carrier Division FIVE, and Attack Carrier Air Wing ELEVEN embarked, continued to support U.S. policy in Southeast Asia with direct combat action against insurgent Communist forces in Vietnam.

On 3 April 1966, Lt. Felix Templeton of VF-114, flying an F4B Phantom, became Kitty Hawk's first triple Centurion by making his 300th arrested landing aboard the ship. On 9 April, Lt. j.g. A. E. Johnson of VA-113, flying an A4C Skyhawk, made the 10,000th landing on Kitty Hawk since the commencement of this WESTPAC deployment on 19 October 1965.

Kitty Hawk departed Dixie Station on 11 April 1966 and arrived at Yankee Station on 12 April. Air Wing ELEVEN aircraft delivered an average of 100 tons of ordnance per day on enemy targets while conducting Rolling Thunder, Blue Tree, and Steel Tiger operations. On 12 April, a KA-3B Skywarrior (a tanker) with four crew members aboard, en route to Kitty Hawk from NAS Cubi Point, was overdue and missing. Crew member status was undetermined. On 15 April, a UH-2 helicopter from HC 1 Detachment CHARLIE was lost over the side after experiencing control difficulties soon after lift-off. One crew member was killed, and one man was killed and four injured on the Kitty Hawk flight deck by flying shrapnel from the helicopter's rotor blades. Also on 15 April, Kitty Hawk aircraft responding to a SAR effort launched for a downed USAF F4C, and silenced one 57MM and two 37MM AAA sites in the vicinity of the downed aircraft.

On 17 April 1966, an A-4C Skyhawk from VA-113 crashed into the sea immediately following launch. The pilot ejected and was recovered safely aboard with no injuries. Also on 17 April, an A-6A Intruder from VA-85 experienced hydraulic failure in flight and crashed at sea. Both the pilot and NFO ejected and were rescued at sea in good condition. An A1H aircraft from VA-115 was also downed on 17 April. Extensive SAR efforts were negative.

Also on 17 April, an attack was carried out against a primary target in North Vietnam, the Hai Doung Railroad and Highway Bridge, located approximately 20 miles east of Hanoi, and resulted in the dropping of the center span and heavy cratering of the eastern bridge abutment and approaches. On 18 April, a flight of two A6As executed a surprise midnight attack on the Uong Bi Thermal Power Plant located approximately 12 miles northeast of the seaport of Haiphong. Making radar system deliveries, the

Intruder aircraft placed 26,000 pounds of ordnance on target. On 19 April, Kitty Hawk aircraft struck the Cam Pha Port Facility. The destruction to port facilities caused by this strike was a significant economic blow to North Vietnam.

On 20 April, an A4C Skyhawk from VA-113, while orbiting a downed pilot, was also hit by ground fire. The pilot retired seaward, ejected two miles from Kitty Hawk, and was recovered safely on board after spending approximately one minute in the water. On 21 April, an A-6A Intruder from VA-85 disappeared from radar scopes at the weapons release point. His wingman observed a large flash at this time, which could have been weapons detonation. Both crew members were missing. On 22 April, an A-6A Intruder was observed to crash in the water while retiring from the target. There were no survivors.

On 26 April, an F-4B Phantom was hit in the vicinity of the starboard engine by enemy ground fire while on a bombing mission. Both the pilot and RIO ejected near Kitty Hawk and were recovered aboard in good condition by Kitty Hawk's helicopter.

On 27 April 1966, an A-6A Intruder, while on armed reconnaissance, received numerous small arms hits, one of which severely wounded the pilot. The pilot, with the NFO's assistance, flew his aircraft seaward where they both ejected and were recovered by helicopter. For this action, the NFO, Lt. j.g. B.E. Westin, USNR, received the Navy Cross. On 28 April, an F-4G Phantom was hit by enemy ground fire. Both the pilot and RIO ejected at sea and were recovered safely.

During the period from 12 through 28 April, Kitty Hawk aircraft participated in a series of strikes aimed at the North Vietnamese lines of communication (LOC). Targets hit included railroads, bridges, highways, and waterborne logistic craft. During this period, over 200 enemy waterborne logistic craft were destroyed. The strikes conducted during this period severely hampered the movement of military supplies south. Operations during this period were distinguished by aggressiveness and reliability in the face of adversity. Aircraft and crew losses were a direct reflection of the aggressiveness of CVW-11 pilots in the face of increased capabilities of enemy defenses.

Kitty Hawk departed Yankee Station on 29 April and arrived in Subic Bay on 30 April for upkeep. On 1 May 1966, Kitty Hawk, while in Subic Bay, Philippines, celebrated the fifth anniversary of her commissioning with an open house. Kitty Hawk was visited by numerous military personnel, DOD civilian personnel, and dependents from the NAS Cubi Point and Subic Bay area. The Governor of Bataan and several of his officials attended, as did the Mayor of Olongapo.

Kitty Hawk departed Subic Bay en route to Yankee Station on 6 May, conducted a Surface-to-Air Missile Exercise on 6 May,

and arrived at Yankee Station on 8 May. Air Wing aircraft averaged delivery of 110 tons of ordnance per day on enemy targets while conducting Rolling Thunder, Steel Tiger, and Blue Tree operations. On 15 May, an A6A Intruder from VA-85 was lost following fuel exhaustion due to the inability to receive fuel from tanker aircraft. Both the pilot and NFO ejected and were recovered safely. The pilot, Lt. Cmdr. John Ellison, was rescued by Kitty Hawk's embarked helicopter detachment, HC1 Detachment CHARLIE. This was the 14th rescue made by this detachment during this deployment.

On 11 May, Kitty Hawk and USS Pyro (AE-24) set a new ordnance transfer rate record by averaging 237.66 standard tons per hour. On 18 May, an F4B flying RESCAP for a downed aircraft was hit by small arms fire. The pilot and RIO ejected and were recovered uninjured by helicopter. On 19 May, an A1J suffered engine failure, suddenly and completely, following deck lift-off and crashed into the sea. The pilot was recovered uninjured by a Kitty Hawk helicopter.

On 23 May 1966, Kitty Hawk departed Yankee Station in the South China Sea and commenced the long voyage homeward after completing operations on her third WESTPAC deployment. From 27 November 1965 to 23 May 1966, the ship had conducted 9,223 combat sorties and 1,485 support sorties.

After brief stops in Subic Bay on 24 and 25 May and Yokosuka, Japan, from 29 May through 3 June 1966, Kitty Hawk sailed for the United States and arrived in San Diego on 13 June 1966. At this time, she entered a much-needed Restricted Availability (RAV) period for maintenance and repairs.

On 25 June 1966, Kitty Hawk's Hangar Bay One was transformed into a gala 1,804-seat theater for the world premiere of Walt Disney's "Lt. Robin Crusoe, U.S.N.," portions of which had been filmed onboard Kitty Hawk. The premiere was held before a host of celebrities. Simultaneously, in the South China Sea, off the coast of Vietnam, the picture was also premiered onboard Kitty Hawk's sister ship, USS Constellation (CVA-64). This marked the first time in naval history that a premiere was held aboard a ship of the line and the first time in the history of motion pictures that a double premiere was held, one at sea and the other in port.

Kitty Hawk's post-deployment RAV ended on 22 August 1966, and Kitty Hawk commenced local operations in the southern California operating area, operating in and out of San Diego. An INSERV inspection was conducted from 6 to 9 September 1966. Fleet Training Group, San Diego, conducted a Training Readiness Evaluation on 12 and 13 September 1966, and the period from 14 to 23 September was spent conducting Fleet Training Group, San Diego, Underway Training Assistance. On 26 and 27 September, COMCARDIV THREE, as Chief Inspector, conducted an Administrative Inspection. An overall ship's grade of 93.15 (Excellent) was assigned.

Kitty Hawk was awarded the Navy Unit Commendation for exceptionally meritorious service from 26 November 1965 to 14 May 1966 while participating in combat operations against the insurgent Communist guerrilla forces in the Republic of Vietnam. The valiant men of her Carrier Air Wing 11 flew over 10,000 sorties and delivered over 10,700 tons of ordnance against enemy forces. The officers and men of Kitty Hawk displayed undaunted spirit, courage, professionalism, and dedication to maintain their ship as a fighting unit under the most arduous operating conditions, enabling her pilots to destroy vital military targets in North Vietnam despite intense opposition and extremely adverse weather conditions.

On 4 November 1966, Kitty Hawk again deployed to serve the cause of freedom and national security in the waters of Southeast Asia. Kitty Hawk arrived at Yokosuka, Japan, on 19 November to relieve Constellation as the flagship for Rear Adm. David C. Richardson, Commander Task Force 77. On 26 November, Kitty Hawk departed Yokosuka for Yankee Station via Subic Bay, and on 5 December, aircraft from Kitty Hawk began their around-the-clock missions over North Vietnam. Around this time, Kitty Hawk, already accustomed to celebrities as guests, entertained a number of prominent visitors, including William Randolph Hearst, Jr., Bob Considine, Dr. Billy Graham, and John Steinbeck, among others.

Seventh Fleet carrier aircraft launched their first strikes on 24 April 1967 on MiG bases in North Vietnam with an attack on Kep Airfield, 37 miles northeast of Hanoi. The attack was delivered by A-6 Intruders and A-4 Skyhawks from Kitty Hawk and was followed up by another A-6 attack the same night.

While providing cover for the bombers during the first attack, Lt. Cmdr. Charles E. Southwick and Lt. Hugh Wisely, flying F-4B Phantom IIs of VF-114, were each credited with a probable MiG-17 kill in aerial combat.

Kitty Hawk remained in the Far East supporting the fight for freedom in Southeast Asia until departing Subic Bay on 28 May 1967. Steaming via Japan, the carrier reached San Diego on 19 June and a week later entered the naval shipyard at Long Beach for maintenance. Kitty Hawk returned to San Diego on 25 August 1967.

Kitty Hawk again deployed from San Diego for a western Pacific (WESTPAC) and Vietnam cruise on 18 November 1967, returning home on 28 June 1968. Subsequent deployments were from 30 December 1968 to 4 September 1969 and 6 November 1970 to 17 June 1971.

On this latter WESTPAC deployment, by 31 January 1971, Kitty Hawk, USS Hancock (CVA 19), and USS Ranger (CVA 61), alternating on Yankee Station, flew a total of 3,214 sorties during the month, of which 3,128 delivered ordnance in Laos. A-6 and A-7 aircraft were particularly effective in attacking truck traffic, with the enemy having put a seasonally high number of trucks on the road, averaging close to 1,000 per day.

On Yankee Station on 10 March 1971, Kitty Hawk and Ranger set a record of 233 strike sorties for one day and went on during the ensuing six-day period to mark up a strike effectiveness record that exceeded record performances by TF-77 during the previous three-year period. The carrier returned home on 17 June 1971.

Again, on 17 February 1972, Kitty Hawk deployed to the waters off Southeast Asia. By 30 March, Naval Air attack sorties in South Vietnam had dropped from 733 in February to 113 during March. On 23 March 1972, the U.S. canceled further peace negotiations in Paris, France, because of a lack of progress in the talks. This was followed by the North Vietnamese invasion of South Vietnam. This "Easter" or "Spring Offensive" was the result of the long buildup and infiltration of North Vietnamese forces during previous months and presaged some of the most intense fighting of the entire war. The North Vietnamese invasion prompted increased air operations by the carriers in support of South Vietnamese and U.S. forces. The carriers on Yankee Station when North Vietnam invaded on 30 March were Hancock and USS Coral Sea (CVA 43). During the month, four carriers rotated on Yankee Station: Kitty Hawk, USS Constellation, Coral Sea, and Hancock.

Aircraft from Kitty Hawk, as well as Hancock, Coral Sea, and Constellation, were involved in Operation Freedom Train beginning on 5 April 1972. Navy tactical air from these carriers flew sorties against military and logistic targets in the southern part of North Vietnam that were involved in the invasion of South Vietnam. The operating area in North Vietnam was initially limited to between 17° and 19°N. However, special strikes were authorized against targets above the 19th parallel on various occasions. The magnitude of the North Vietnamese offensive indicated that an extended logistics network and increased resupply routes would be required to sustain ground operations by North Vietnam in their invasion of South Vietnam. Most target and geographical restrictions that were placed in effect since October 1968 concerning the bombing in North Vietnam were gradually lifted, and the list of authorized targets expanded. Strikes in North Vietnam were against vehicles, lines of communication (roads, waterways, bridges, railroad bridges, and railroad tracks), supply targets, air defense targets, and industrial/power targets. By the end of April, operations were permitted in North Vietnam throughout the region below 20° 25' N, and many special strikes above the 20th parallel had also been authorized.

On 14 April, the Navy averaged 191 sorties per day in South Vietnam, a 97 percent increase over the previous week. Sorties concentrated west and north of Quangtri City, with interdiction and direct air support flown in the area. Carriers on Yankee Station were Kitty Hawk, Constellation, Hancock, and Coral Sea.

Two days later, on 16 April, aircraft from Kitty Hawk, Coral Sea, and Constellation flew 57 sorties in the Haiphong area in support of U.S. Air Force B-52 strikes on the Haiphong petroleum products storage area. This operation was known as Freedom Porch.

Operation Linebacker I began on 10 May 1972 and consisted of heavy strikes on targets in most of North Vietnam, evolving and lasting until restrictions on operations above 20°N were imposed on 22 October. The operation was an outgrowth of Freedom Train and President Richard M. Nixon's mining declaration, which also stated that the U.S. would make a maximum effort to interdict the flow of supplies in North Vietnam. On this first day of Linebacker I, the Navy shifted its attacks from targets in southern North Vietnam to the coastal region embracing Haiphong north to the Chinese border. In all, 173 attack sorties were flown in this region this day, although another 62 were directed into South Vietnam in continuing support of allied forces there.

It was the most intensified air-to-air combat day of the entire war. Navy flyers shot down eight MiGs. An F-4 Phantom II, from VF-96 on board Constellation, while engaged in aerial combat over Haiphong, shot down three MiGs for the first triple downing of enemy MiGs by one plane during the war. Lt. Randall H. Cunningham was the pilot, and Lt. j.g. William P. Driscoll was the RIO of the F-4. These three MiG downings, coupled with their 19 January and 8 May downing of two MiGs, made Lt. Cunningham and Lt. Driscoll the first MiG aces of the Vietnam War. Three other kills were scored by planes of VF-96, one by VF-92 off Constellation, and one by VF-51 off Coral Sea.

During the five and one-half month period of Linebacker I, the Navy contributed more than 60 percent of the total sorties in North Vietnam, with 60 percent of this effort in the "panhandle," the area between Hanoi and the DMZ. Tactical air operations were most intense during the July-September quarter, with 12,865 naval sorties flown. Most attack sorties in NVN fell into two classes: armed reconnaissance and strike. The former was usually directed against targets of opportunity within three main areas — near Hanoi, Haiphong, and the Chinese border. Strike operations were preplanned and usually directed at fixed targets. Most types of fixed targets, not associated with armed reconnaissance, required approval by the Commander-in-Chief, Pacific, or by the Joint Chiefs of Staff, prior to attack. Principal

Navy aircraft were the A-7 and A-6, which accounted for roughly 60 and 15 percent of the Navy's attack sorties, respectively. About 25 percent of the Navy's effort was at night. Carriers participating in the initial May-June operations from Yankee Station were Kitty Hawk, Constellation, Coral Sea, Hancock, Midway, and USS Saratoga (CVA 60).

On 11 May 1972, Naval aircraft flying from Kitty Hawk, Coral Sea, Midway, and Constellation laid additional minefields in the remaining ports of significance in North Vietnam — Thanh Hoa, Dong Hoi, Vinh, Hon Gai, Quang Khe, and Cam Pha, as well as the Haiphong approaches. This early mining was not confined solely to the seven principal ports. Other locations were also seeded early in the campaign, including Cua Sot, Cap Mui Ron, and the river mouths, Cua Day and Cua Lac Giang, south of Don Son and the Haiphong port complex.

Kitty Hawk, along with Constellation, Coral Sea, Hancock, Midway, Saratoga, Oriskany, and USS America (CVA 66), began night operations regularly on 24 May, and during June and July, night sorties constituted 30 percent of the total Navy attack effort in North Vietnam, relying primarily on the A-7 Corsair II and A-6 Intruder. About 45 percent of the Navy's armed reconnaissance effort was at night during June and July. The A-7 flew about as many night sorties as it did day sorties. The A-6 flew more night than day armed reconnaissance sorties during the summer months. The total number of Navy night sorties during June and July were 1,243 and 1,332, respectively. Three to four of the carriers mentioned above were maintained on Yankee Station, on a rotational basis, during the summer months.

There was a dramatic change in North Vietnam's air defense effort during the summer months of 1972. During the earlier periods of April and May, the Navy air effort in North Vietnam involved intensive air-to-air combat and a large number of surface-to-air missile (SAM) firings. In contrast, during June and July, there was an increase in Linebacker I Navy attack sorties, but a decrease in the number of air-to-air combat incidents and SAM firings. After mid-June, almost all North Vietnamese aircraft sighted or engaged were MiG-21s.

During September, the number of Navy tactical air attack sorties decreased from the level flown in August. There were 3,934 Navy tactical air attack sorties flown into North Vietnam, down by about 800 from the August total. During July and August, more than 45 percent of the Navy armed reconnaissance sorties were at night. However, in September, only 31 percent of the armed reconnaissance sorties were flown at night. In South Vietnam, the Navy flew 1,708 tactical air attack sorties, a decrease from the level flown in August. About half of the Navy's tactical air sorties were close and direct air support sorties in South Vietnam. Carriers operating on Yankee Station during the month of September were Kitty Hawk, Hancock, Midway, Saratoga, Oriskany, and America.

On 23 October 1972, the U.S. ended all tactical air sorties into North Vietnam above the 20th parallel and brought to a close Linebacker I operations. This goodwill gesture of terminating the bombing in North Vietnam above the 20th parallel was designed to help promote the peace negotiations being held in Paris, France. Air operations in South Vietnam followed the general pattern of the ground war. North Vietnam increased their small-scale attacks throughout South Vietnam in an apparent effort to gain territory before a possible cease-fire, while the main objective of Navy and Marine Corps tactical air sorties was close and direct air support in support of allied ground troops, with a view toward frustrating the enemy's desire to acquire territory before a cease-fire agreement was signed.

Kitty Hawk returned to San Diego on 28 November 1972. On 23 January 1973, a cease-fire in Vietnam went into effect. On 29 April 1973, USS Kitty Hawk was converted from an attack aircraft carrier, or CVA, to a multi-mission aircraft carrier, or CV, at Hunter's Point Shipyard in San Francisco. Most noticeably, changes to the ship's jet blast deflectors allowed Kitty Hawk to launch and recover the Navy's new F-14 Tomcat, which included moving aircraft elevator no. 1 outboard by a few feet, making it raise and lower at a slight angle. Kitty Hawk was on deployment again to the western Pacific from 23 November 1973 to 9 July 1974.

Just prior to Kitty Hawk's next WESTPAC deployment on 21 May 1975, the carriers Midway, Coral Sea, Hancock, USS Enterprise (CVAN 65), and USS Okinawa (LPH 3) responded on 19 April 1975 to the waters off South Vietnam when North Vietnam overran two-thirds of South Vietnam. Ten days later, Operation Frequent Wind was carried out by U.S. Seventh Fleet forces. Hundreds of U.S. personnel and Vietnamese were evacuated to waiting ships after the fall of Saigon to the North Vietnamese.

In March 1976, Kitty Hawk underwent a year-long, $100 million overhaul in Bremerton, Washington. Also, the ship's original Terrier missile launchers were replaced with NATO Sea Sparrow missiles.

In the late '70s, the ship teamed with CVW-15 for another WESTPAC deployment, which included search and assistance operations to aid Vietnamese refugees. Kitty Hawk also offered contingency support off the coast of Korea. The deployment was then extended to support contingency operations in the North Arabian Sea during the Iran hostage crisis. Kitty Hawk returned to San Diego in February 1980 and was awarded the Meritorious Unit Commendation and the Battle Efficiency "E" as the best carrier in the Pacific Fleet.

In October 1979, Kitty Hawk and CVW-15 departed San Diego on their last seven-month cruise to the western Pacific together. On 28 October 1979, Kitty Hawk and her escort ships were directed to operate south of the Korean peninsula in response to the assassination of South Korean President Park Chung Hee on 26 October.

On 18 November 1979, USS Midway arrived in the northern part of the Arabian Sea in connection with the continuing hostage crisis in Iran. Militant followers of the Ayatollah Khomeini, who had come to power following the overthrow of the Shah, seized the U.S. Embassy in Tehran on 4 November and held 63 U.S. citizens hostage. Spokesmen for the mob demanded that the United States return the deposed Shah to Iran. Kitty Hawk's cruise was extended two and a half months to support contingency operations in the North Arabian Sea during the Iranian hostage crisis. On 21 November, Kitty Hawk and her escort ships were directed to sail to the Indian Ocean to join Midway and her escort ships operating in the northern Arabian Sea. Kitty Hawk arrived on station on 3 December, and the two carrier forces provided the U.S. with A-6 Intruder and A-7 Corsair II attack aircraft and F-4 Phantom and the modern F-14 Tomcat fighter aircraft, which could respond to a variety of situations if called upon during the Iranian hostage crisis. This was the first time since World War II that the U.S. Navy had two carrier task forces in the Indian Ocean in response to a crisis situation.

Two weeks later, on 21 December 1979, the Defense Department announced that a three-ship nuclear-powered carrier battle group from the Sixth Fleet would deploy to the Indian Ocean to relieve the Seventh Fleet carrier battle group led by Kitty Hawk. The Sixth Fleet carrier battle group consisted of the nuclear-powered USS Nimitz (CVN 68) and her nuclear-powered escort ships. However, on Christmas Eve, 24 December 1979, a massive Soviet airlift of 5,000 Russian airborne troops and equipment into the Afghanistan capital of Kabul was conducted. The U.S. protested the large influx of Soviet troops, which the Soviet Union claimed were there at the request of the Afghanistan government. On 27 December, a Soviet-backed coup installed a new president in Afghanistan. Two carrier task forces centering around Kitty Hawk and Midway continued contingency operations in the northern Arabian Sea.

Nimitz and her escort ships joined Kitty Hawk and Midway and their escort ships on station in the Arabian Sea on 22 January 1980. The following day, Kitty Hawk departed for Subic Bay, having spent 64 days in operations connected with the Iranian crisis. For their actions in the region, Kitty Hawk and CVW-15 sailors and officers were awarded the Navy Expeditionary Medal. Kitty Hawk returned to San Diego in February 1980 and, five months later, was awarded the Meritorious Unit Commendation and the Naval Air Force Pacific Battle Efficiency 'E' as the best carrier in the Pacific Fleet.

In April 1981, Kitty Hawk left San Diego for its 13th deployment to the western Pacific. Following the cruise, the crew was awarded the Navy Expeditionary Medal and the Humanitarian Service Medal for rescuing Vietnamese refugees in the South China Sea.

In January 1982, Kitty Hawk returned to Bremerton for another year-long overhaul. The overhaul was completed on schedule in January 1983. Following the comprehensive overhaul and a vigorous training period with Carrier Air Wing NINE (CVW-9), Kitty Hawk deployed as the flagship for Battle Group Bravo. During exercise Team Spirit '84, she was struck by a submerged "Victor"-class Soviet submarine in the Sea of Japan, forcing the submarine to be towed back to its homeport. Kitty Hawk logged over 62,000 miles on this deployment and remained on station in the North Arabian Sea for more than 60 consecutive days. The ship returned to San Diego on 1

August 1984. Seven months later, in March 1985, Kitty Hawk was awarded its second Battle Efficiency 'E' award as the best carrier in the Pacific Fleet.

In July 1985, Kitty Hawk once again deployed as flagship for Battle Group Bravo, responding to tasking from the California coast to the Gulf of Aden. During the remainder of 1985, Kitty Hawk executed a hallmark cruise, completing her second consecutive fatality-free deployment while accumulating 18,000 flight hours and 7,300 arrested landings. It was also the second consecutive cruise without an accident in the launch and recovery of jet and propeller aircraft, while her catapults and arresting gear were maintained at 100 percent availability.

Kitty Hawk celebrated 25 years of proud service in 1986. She won the Admiral Flatley Award for aviation safety, the COMNAVAIRPAC Battle E for best CV AIMD in the Pacific Fleet, and the CINCPACFLT Annual Price Fighter Award. In the category of food service excellence, Kitty Hawk was the winner of the Dorrie P. Miller award as well as a semifinalist for the 1986 NEY award. Culminating an arduous work-up cycle, Kitty Hawk finished 1986 with 9,661 cat shots and 9,025 arrested landings, bringing the total traps figure to 256,586.

Kitty Hawk began 1987 with a farewell to San Diego. On 3 January, the ship departed her homeport of 25 years and set out on a six-month world cruise. During the world cruise, Kitty Hawk and CVW-9 crewmen again showed their commitment to safety by conducting a third fatality-free deployment. Kitty Hawk spent 106 consecutive days on station in the Indian Ocean and was again awarded the Navy Expeditionary Medal and the Meritorious Unit Citation for its service.

The world cruise ended at the Philadelphia Naval Shipyard on 3 July 1987. Six months later, Kitty Hawk began a Service Life Extension Program (SLEP) overhaul. Kitty Hawk emerged from the yards on 29 March 1991, her deck modified to accommodate the F/A-18 Hornet. The overhaul was estimated to have added 20 years of service to the life of the ship. Kitty Hawk commenced sea trials, the first time the 80,000-ton carrier moved under her own power since arriving in Philadelphia 3½ years earlier to begin the SLEP. She departed Philadelphia on 30 July.

With the return of CVW-15 to its decks, Kitty Hawk began its second cruise around 'the Horn' of South America to its original homeport of San Diego on 11 December 1991. On 1 August 1992, Kitty Hawk was appointed as Commander, Naval Air Forces, Pacific's 'ready carrier.' The ship embarked the Commander, Cruiser-Destroyer Group FIVE flag staff, the Commander, Destroyer Squadron SEVENTEEN staff, and Carrier Air Wing FIFTEEN for three months of work-ups before deploying to the western Pacific on 3 November 1992.

While on deployment, Kitty Hawk spent nine days off the coast of Somalia supporting U.S. Marines and coalition forces involved in Operation Restore Hope. On 16 December 1992, five air traffic controllers from Kitty Hawk were sent aboard USS Leahy (CG 16) to establish approach control services in and out of Mogadishu, Somalia, in support of Operation Restore Hope. Approaching aircraft were picked up from a VAW-114 E-2C Hawkeye, which tracked flights and issued advisories from about 200 miles out. Once the flights were within 50 miles, the Leahy team took over and led them to within visual range of the airport, about 10 miles away.

In response to increasing Iraqi violations of the United Nations sanctions, the ship was subsequently rushed to the Arabian Gulf on 27 December 1992. Just 17 days later, on 13 January 1993, Kitty Hawk, with 35 of her CVW-15 aircraft, led a joint, coalition offensive strike against missile sites in southern Iraq. The successful strike sent Saddam Hussein a clear message that continued violations of U.N. resolutions would not be tolerated.

Kitty Hawk's battle group was relieved by the USS Nimitz battle group on 18 March 1993 and headed for home after having operated in the Indian Ocean and Arabian Gulf and participated in Operations Restore Hope and Southern Watch. On 20 September 1993, while in its homeport of San Diego, Kitty Hawk turned its flight deck into a stage to host a live taping of The Nashville Network's (TNN) 10th anniversary "Southern California Spectacular" country music concert. Among the performers were the band Restless Heart and singers Martina McBride, Aaron Tippin, Shenandoah, Larry Stewart, Lari White, and Clint Black.

Lady Margaret Thatcher, former Prime Minister of Great Britain, addressed Kitty Hawk's crew during a Veterans' Day observance held on board the carrier on 11 November 1993, during an in-port period in San Diego. Lady Thatcher was in the U.S. promoting her book, "The Downing Street Years," and wanted to speak to U.S. military personnel, offering her personal admiration for the burdens and sacrifices made by the military around the world.

On 14 December 1993, while Kitty Hawk was conducting flight operations ten miles southwest of San Clemente Island, the crew of an airborne search-and-rescue (SAR) helicopter spotted three distress flares. Upon investigating the flares, the helicopter found the 40-foot Silver Eagle taking on water. Of the three passengers on board, one had a serious back injury and could not be moved. Kitty Hawk's Commanding Officer, Capt. William W. Picavance, ordered the carrier to the scene with ship's medical doctors, corpsmen, and boat crew standing by. Dispatching three helicopters to Silver Eagle's location, Kitty Hawk was able to retrieve one of the stranded mariners, but rough seas prevented the helicopters from rescuing the remaining two men.

Kitty Hawk then launched a rigid-hull inflatable boat (RIB) crew and corpsman to the sinking boat. Battling 15 to 20-foot seas, the RIB crew tied up alongside the foundering vessel. On-scene rescue swimmers boarded the Silver Eagle and transported the remaining passengers aboard the RIB. The two men were moved to Kitty Hawk, where all the survivors reported in stable condition. The men said they had left Long Beach, Calif., early in the morning of 13 December for a three-day fishing trip. This was Kitty Hawk's second rescue at sea in less than a week. On 7 December, the ship teamed up with a Coast Guard helicopter crew to save a Philippine sailor suffering from severe internal injuries on board a Philippine merchant ship.

Kitty Hawk departed her homeport of San Diego for the 17th time in June 1994 for a scheduled deployment to the western Pacific. Late on the evening of 11 July, as it approached for an arrested landing, an F-14 fighter jet struck the ramp of Kitty Hawk's flight deck and exploded, turning the deck into a sea of flames and wreckage. The jet's pilot, who, along with his radar intercept officer, had ejected seconds after impact, landed in the flames. Five Kitty Hawk flight deck personnel immediately advanced into the flames, rescued the pilot, and extinguished the fire. The five — Chief Aviation Boatswain's Mate (Aircraft Handling) Leroy Danielly, Aviation Boatswain's Mates (Aircraft Handling) First Class Larry Spradlin and Tim Goode, Aviation Boatswain's Mate (Aircraft Handling) Second Class Jose Dickson, and Aviation Electronics Technician Second Class Brandon Liesemeyer — were presented the Navy and Marine Corps Medal for heroism by Secretary of the Navy John Dalton on 15 October 1994 during the carrier's port visit to Yokosuka, Japan.

In the early summer of 1996, Kitty Hawk participated in Exercise Rim of the Pacific '96 (RIMPAC 96). She then departed San Diego on her next scheduled six-month deployment to the western Pacific on 11 October 1996. This was the 18th deployment for the 35-year-old carrier and marked the first time that a psychologist had ever been assigned to a combatant. Lt. Helen Napier, Medical Service Corps, from the Naval Hospital, Bremerton, Wash., participated in a precedent-setting pilot program developed in response to a request from the carrier's Senior Medical Officer, Capt. Homer Moore, Medical Corps, and the ship's Commanding Officer, Capt. Steven Tomaszeski. Kitty Hawk and her battle group, USS Cowpens (CG 63), USS Antietam (CG 54), USS Reid (FFG 30), and the attack submarine USS Salt Lake City (SSN 716), deployed to the western Pacific, Indian Ocean, and the Arabian Gulf. They spent three months in the Gulf in support of Operation Southern Watch and U.N. sanctions in the region. During that period, Air Wing Eleven aircraft flew 1,775 sorties, accumulating 4,065 flight hours. Battle group ships conducted Maritime Interception Operations (MIO) in which crew members boarded and searched merchant ships believed to be carrying cargo in violation of U.N. sanctions against Iraq. Kitty Hawk and her battle group returned to homeport on 11 April 1997.

On 18 July 1998, USS Independence (CV 62) turned over forward-deployed duties in Yokosuka, Japan, to Kitty Hawk while the two aircraft carriers were in Pearl Harbor, Hawaii. By this time, Kitty Hawk had added new computer technology, making it compatible with the Navy's latest advancements in information technology for the 21st century, or IT-21. Upon reaching Japan, Kitty Hawk took on a new air wing. Carrier Air Wing (CVW) FIVE had operated as a forward-deployed unit out of Atsugi Naval Air Station, Japan, since 1973.

Kitty Hawk arrived in her new homeport of Yokosuka, Japan, on 11 August 1998, amid a backdrop of banners, flags, and balloons, with hundreds of dignitaries, family members, and sailors lining the pier to welcome the crew to their new home. Welcoming remarks were made by Ryohei Olamoto, President of the Yokosuka Japan-American Society, and Vice Adm. Makoto Yamazaki, Commander-in-Chief of the Japanese Self Defense Fleet. Kitty Hawk became the third aircraft carrier to be permanently forward-deployed to Japan, after USS Midway (CV 41) and Independence.

The carrier did not remain pierside at Yokosuka for long. Kitty Hawk participated in Exercise Foal Eagle '98, the largest joint/combined exercise in the world, which began on 24 October and ran through 4 November off the coast of Korea. It was during this at-sea period that the carrier lost a petty officer first class on 17 October. He apparently was lost at sea while the ship was conducting routine operations in the Pacific Ocean approximately 345 miles east of Okinawa. Two SH-60 helicopters from HS-14 searched the surrounding waters along with USS Mobile Bay (CG 53) and USS Chancellorsville (CG 62), both forward-deployed to Yokosuka, Japan, and the U.S. Coast Guard Cutter Jarvis (WHEC 725). A memorial service was held for the missing sailor on 23 October.

On 20 November 1998, having returned from Foal Eagle '98, Kitty Hawk received the First Navy Jack during ceremonies in Yokosuka, Japan, designating the 37-year-old aircraft carrier as the oldest ship in the fleet. This distinction allowed her to display the First Navy Jack in place of the Union Jack flown aboard other Navy ships. The First Navy Jack, a flag consisting of 13 horizontal, alternating red and white stripes with a rattlesnake across the center, bears the motto, "Don't Tread On Me." Conceived in 1775 by Commodore Esek Hopkins of the Continental Navy, the flag was first used as a signal among ships to engage the enemy. In 1977, the Secretary of the Navy directed the ship with the longest total period of active service to display the First Navy Jack until decommissioned or transferred to the inactive reserve. At that time, the flag shall be passed to the next ship in line with appropriate honors. Kitty Hawk received the flag from Independence following its decommissioning on 30 September in Bremerton, Wash.

Kitty Hawk's next deployment, her 19th and first since arriving in Yokosuka, began on 2 March 1999. The ship and her embarked Carrier Air Wing (CVW) FIVE participated in Exercise Tandem Thrust with a port visit to Agana, Guam. During her visit to Agana, having just completed Exercise Tandem Thrust, Kitty Hawk was visited by Chief of Naval Operations Adm. Jay L. Johnson on 3 April. The CNO delivered the news that Kitty Hawk was being directed to the Arabian Gulf along with USS Chancellorsville (CG 62) and USS Curtis Wilbur (DDG 54) after President Clinton ordered the USS Theodore Roosevelt (CVN 71) battle group to the Adriatic Sea to support NATO forces in Yugoslavia instead of relieving the USS Enterprise (CVN 65) battle group, which was completing a regularly scheduled deployment to the region. On April 20, Kitty Hawk, Curtis Wilbur, and Chancellorsville transited the Strait of Hormuz, relieving the Enterprise's battle group in the Arabian Gulf to participate in Operation Southern Watch, enforce the "no-fly zone" over southern Iraq, and conduct Maritime Interception Operations supporting United Nations sanctions.

On 15 June 1999, two aviators were rescued from the waters of the Arabian Gulf after they ejected safely from an F-14 Tomcat. The aircraft was returning to Kitty Hawk when the crew declared a mechanical emergency. Rescue swimmers from Helicopter Anti-Submarine Squadron (HS) 14 embarked aboard the carrier, and Helicopter Anti-Submarine Squadron (Light) (HSL) 51 embarked aboard USS Chancellorsville (CG 62), recovered both crewmen from the water and transported them to Kitty Hawk. Both aviators were uninjured and released after undergoing an extensive medical examination.

Kitty Hawk, Chancellorsville, and Curtis Wilbur were relieved by the USS Theodore Roosevelt (CVN 71) battle group and departed the Arabian Gulf on 19 July 1999, having launched more than 5,400 sorties during her three months in the Gulf. On her return to Yokosuka, Japan, she made port visits to Perth, Australia, and Pattaya, Thailand, and was back at Yokosuka in late August.

After participating in October in the multi-national Exercises Foal Eagle '99 and ANNUALEX-11G off the Korea Peninsula and Japan, Kitty Hawk returned to Yokosuka on 10 November 1999. Following a period of an aggressive regimen of repairs, upgrades, and personnel training, the carrier returned to sea on the morning of 23 February 2000 for 12 days of sea trials.

During her next regularly scheduled two-month deployment to the western Pacific, Kitty Hawk participated in Exercise Cobra Gold 2000 following a port call to Pattaya, Thailand, on 17 May 2000. The exercise ran from 9 to 23 May and is a regularly scheduled joint/combined U.S.-Thai military exercise designed to ensure regional peace and strengthen the ability of the Royal Thai armed forces to defend Thailand or respond to regional contingencies. In addition to flying from Kitty Hawk, her embarked air wing, Carrier Air Wing Five, flew F/A-18 Hornets and F-14 Tomcats from Korat Royal Thai Air Force Base, approximately 165 miles northeast of Bangkok. These aircraft were assigned to act as aggressor forces, as well as to conduct local bombing exercises and provide air-to-air training with squadrons from Singapore, Thailand, the U.S. Air Force, and U.S. Marines.

Kitty Hawk and Carrier Air Wing Five again participated in the annual Exercise Foal Eagle in the Sea of Japan. She joined the exercise on 25 October 2000 as the striking arm of Battle Force 7th Fleet. From 7 November to 17 November, the carrier trained with the Japanese Self Defense Force in Exercise Keen Sword. Following the exercise, the battle group returned to Yokosuka on 20 November for the holiday leave and maintenance period, during which approximately 118,000 square feet of fresh non-skid was applied on the flight deck, the number three catapult's launch valve was replaced, and one of the ship's boilers received a five-year overhaul.

Following six days of sea trials in mid-February, Kitty Hawk remained in port until the morning of 2 March 2001, when she and her battle group — USS Chancellorsville (CG 62), USS Vincennes (CG 49), USS Gary (FFG 51), USS Vandegrift (FFG 48), and USS John S. McCain (DDG 56) — got underway for her next routine, scheduled three-month deployment. During this deployment, the battle group participated in the annual Tandem Thrust exercise beginning on 10 May.

In the wake of the terrorist attacks on the World Trade Center in New York City and on the Pentagon in Arlington, Va., on 11 September, Kitty Hawk was once again ordered to sea in support of Operation Enduring Freedom, getting underway on 1 October after an accelerated sea trials and carrier qualifications period, carried out on short notice following the events of 11 September.

The ship transited more than 6,000 miles in 12 days and reported on station in the North Arabian Sea, where it served as an afloat forward staging base for U.S. joint forces. While on station, pilots from Carrier Air Wing Five flew more than 600 missions over Afghanistan in support of the United States' war on terrorism, including more than 100 combat sorties.

The carrier's missions received attention at the highest levels. U.S. Army General Tommy Franks, commander in chief, Central Command, visited Kitty Hawk on 23 October, bringing with him a direct order from the Commander-in-Chief. "I told the President that I was coming out here and asked if there was anything I needed to do for him," Gen. Franks told the ship's crew, who gathered on the flight deck for his five-minute speech. "The President looked at me and said, 'When you get out there, give them a hug. And that's exactly what this is. I came out here to give you a hug."

"The United States of America owes you a debt," Gen. Franks continued. "You stand tall. You serve where you're told. Without [the Navy], we could not have done what has been done. And without you, we cannot do what we are going to do. Every story has a beginning, a middle, and an end. We all know who made the beginning on Sept. 11, 2001. And all of you are going to be what makes an end."

The Secretary of the Navy, Gordon R. England, visited Kitty Hawk on 30 October and expressed appreciation for the crew's service to the nation during "this critical mission."

The beginning of December brought a close to Kitty Hawk's missions in the North Arabian Sea. After 74 consecutive days at sea, the crew made a port visit to the island of Phuket, Thailand, from 13 to 15 December for rest and relaxation. They then continued on to their forward-deployed port of Yokosuka, Japan, arriving on 23 December 2001, after 83 days at sea in support of Operation Enduring Freedom.

Following an intensive 11-week Ship's Restricted Availability (SRA) period, receiving upgrades to its defensive systems and scheduled maintenance to its flight deck and engineering plant, Kitty Hawk departed Yokosuka on 12 March 2002 to begin four days of sea trials in preparation for the ship's scheduled upcoming extended sea period. After a two-day in-port period,

Kitty Hawk stood out to sea again on 18 March, this time to complete scheduled carrier qualifications (CQ) and integrated battle-group training near Guam, returning to Yokosuka on 1 April. The carrier departed Yokosuka again about three weeks later for another period of underway training.

Battle group ships participating in this underway training period included USS Chancellorsville (CG 62), USS O'Brien (DD 975), USS Vandegrift (FFG 48), USS Cowpens (CG 63), USS Curtis Wilbur (DDG 54), USS John S. McCain (DD 56), and USS Cushing (DD 985). The submarine USS Helena (SSN 725) and replenishment ships USNS Rappahannock (T-AO 204), USNS Kiska (T-AE 35), and USNS Concord (T-AFS 5) supported the training operations.

During this underway period, Kitty Hawk made port visits to Hong Kong, Singapore, and Guam, the latter on 28 May 2002. The ship celebrated its 41st birthday just prior to pulling into Hong Kong, and before pulling into Singapore, the Kitty Hawk/CVW 5 team conducted military maneuvering drills with the Singapore navy and air force. The carrier returned to its forward-deployed port of Yokosuka, Japan, on 5 June.

In a rare move, the Commander, U.S. 7th Fleet, Vice Adm. Robert F. Willard, relieved the commanding officer of Kitty Hawk, Capt. Thomas Hejl, on 3 September 2002, citing a loss of confidence in Capt. Hejl's ability to lead his crew and carry out essential missions and taskings. Capt. Hejl was temporarily assigned to the Naval Air Forces, Pacific staff in San Diego. Another Kitty Hawk officer, Lt. Cmdr. Klas Ohman, an F/A-18C Hornet pilot, was selected to take part in celebrating 100 years of controlled flight by flying a replica of an aircraft the Wright brothers flew in Kitty Hawk, N.C., on Oct. 8, 1902. The event, called "Return to Kitty Hawk," took place from 5 to 8 October at Jockey's Ridge State Park, three miles south of Kitty Hawk, N.C. Lt. Cmdr. Ohman flew a replica of Orville and Wilbur Wright's 1902 glider, the first manned aircraft with yaw, pitch, and roll control. Several aviation historians attribute this technology developed by the Wrights to be one of the most significant milestones in aviation history. It was the success of this technology that enabled the famous 17 December 1903 flight of the Wright Flyer in Kitty Hawk, the first powered flight.

Kitty Hawk left its forward operating port of Yokosuka, Japan, on 25 October 2002, for a scheduled underway period in the western Pacific. While at sea, the ship's crew, along with the embarked Carrier Air Wing (CVW) 5 and Destroyer Squadron (DESRON) 15, engaged in combined military exercises with regional allies and conducted unit-level training. During this seven-week at-sea period, Kitty Hawk and her crew joined other U.S. Navy units and units of the Japan Maritime Self Defense Force (JMSDF) from 1 to 22 November by participating in ANNUALEX 14G. ANNUALEX is a maritime exercise designed to promote teamwork and mutual defense of the waters surrounding Japan. While anchored off Hong Kong, Kitty Hawk was visited by Jackie Chan, a Hong Kong native and international film star. The carrier returned to its port of Yokosuka, Japan, on 13 December for the holiday leave and maintenance period.

The Navy's oldest active warship once again got underway for a routine deployment on 23 January 2003. At noon (EST) on 8 February 2003, during this at-sea period, the crews of Kitty Hawk and Carrier Air Wing Five reached a milestone when the ship's bow catapult team launched an aircraft from the carrier's No. 1 catapult for the 150,000th time in the ship's nearly 42 years of service.

The deployment turned out not to be routine. On 12 February, the ship was directed to the Arabian Gulf to once again deal with the regime of Saddam Hussein in Iraq. The carrier and her embarked air wing, arriving on 22 February, spent more than 100 consecutive days underway in support of Operations Southern Watch and Iraqi Freedom. The Kitty Hawk Strike group consisted of USS O'Brien (DD 975), USS Cowpens (CG 63), USS Curtis Wilbur (DDG 54), USS John S. McCain (DD 56), and USS Cushing (DD 985). The replenishment ships USNS Rappahannock (T-AO 204) and USNS Flint (T-AE 32) provided logistics support.

While operating in the Gulf, the ship lost two officers. Lt. Tom Adams, a recently transferred crew member of Carrier Airborne Early Warning Squadron (VAW) 115, died in a helicopter crash during the opening hours of Iraqi Freedom as he was participating in a military foreign exchange program, and Lt. Nathan White, of Strike Fighter Squadron (VFA) 195, was killed in combat actions during a night close-air support mission on 2 April 2003.

During Kitty Hawk's participation in Operation Iraqi Freedom, the air wing flew 5,375 sorties during 11,800 flight hours, and expended 864,860 pounds of ordnance. Approximately 39 million gallons of water were produced, and 9 million gallons of fuel were expended. With the successful conclusion of the naval portion of Operation Iraqi Freedom, the Kitty Hawk Strike Group departed the Gulf on 16 April and returned to their forward-deployed port of Yokosuka, Japan, on 6 May 2003.

2004 was an eventful year that involved a series of inspections, exercises, and port visits. On February 19, a new chapter in the book of Kitty Hawk Strike Group's history began with the first landing of an F/A-18F Super Hornet on board Kitty Hawk's 4.1-acre flight deck during the ship's 12th FDNF underway period. The VFA-102 "Diamondbacks" introduced the improved F/A-18 E/F "Super Hornet" to the 7th Fleet area of operation, replacing the F-14 Tomcat, after more than 30 years of service.

Kitty Hawk capped off the year with Annual Exercise 2005, which ran from November 9 to 18. ANNUALEX provided Kitty Hawk with the opportunity to increase its military partnership with the Japan Maritime Self-Defense Force. Kitty Hawk was one of 61 naval vessels which participated, including two U.S. submarines, 10 other Navy ships, and 49 JMSDF ships.

The ship departed Fleet Activities Yokosuka on June 8, 2006, for its 16th FDNF underway period. During the 99-day deployment, the ship took part in Exercise Valiant Shield, a multi-service war game involving three carrier strike groups, 22,000 personnel, and 280 aircraft from June 19 to 23. It was the largest military exercise conducted by the United States in Pacific waters since the Vietnam War.

The carrier then pulled into Otaru, Japan, on Hokkaido Island from July 1 to 5 after Valiant Shield. Also during the deployment, the crew made three more port visits: Singapore; Fremantle, Australia; and Laem Chabang, Thailand.

Dozens of distinguished visitors boarded the carrier during this underway period for tours. Visitors included the U.S. ambassador to Thailand, the Royal Thai army commander in chief, and various officials from Indonesia, Australia, Singapore, and Japan.

The ship returned to Yokosuka in September for a short period before departing for its summer deployment.

During this two-month deployment, Kitty Hawk and embarked Carrier Air 5 traveled more than 15,200 nautical miles and launched more than 8,000 aircraft.

After a stop in Sasebo, Japan, the strike group took part in the 18th Annual Exercise, a week-long exercise which had more than 100 American and JMSDF ships training together between November 9 and 14.

The deployment's last stop was Hong Kong, from November 23 to 27. Kitty Hawk's Morale, Welfare and Recreation (MWR) Division organized 20 tours of Hong Kong and its surrounding areas, including mainland China, for 702 Sailors.

The ship also hosted Japanese author Hiromi Nakamura who interviewed 41 Kitty Hawk Sailors for a book about Kitty Hawk's flight deck.

After returning to its homeport on December 10, the ship settled down for the holiday season and the New Year.

Kitty Hawk then went through a four-month maintenance period, during which the ship hosted Vice President Dick Cheney.

The carrier then departed on May 23 after completing sea trials and pilot refresher training, known as carrier qualifications.

Kitty Hawk kicked off the summer cruise with Talisman Saber 2007, in which the United States and Australia combined land, sea, and air forces. The exercise brought together more than 12,000 Australian and 20,000 U.S. personnel from all branches of the armed services.

The ship made port visits to Brisbane and Sydney, Australia. Then-Prime Minister John Howard, current Prime Minister Kevin Rudd, and Academy Award-winning actor Russell Crowe made visits to Kitty Hawk while it was moored in Sydney.

Kitty Hawk then participated in Exercise Valiant Shield 2007, one of the largest annual exercises in the Western Pacific. The week-long exercise involved about 30 ships, 280 aircraft, and 22,000 U.S. Airmen, Sailors, Soldiers, and Marines who worked together to build joint combat skills.

The 30 ships involved with Valiant Shield were from three carrier strike groups: Kitty Hawk's, USS Nimitz's (CVN 68), and USS John C. Stennis's (CVN 74). During the exercise, Rear Adm.

Rick Wren, commander of the Kitty Hawk strike group and Task Force 70, had command of all three strike groups.

The ship also took part in Malabar, a six-day exercise that took place in the Indian Ocean's Bay of Bengal, involving more than 20,000 personnel on 28 ships and 150 aircraft from the United States Navy, Indian navy, Royal Australian navy, Japan Maritime Self-Defense Force, and the Republic of Singapore navy.

The ship returned to Yokosuka on September 21. After a short in-port period, Kitty Hawk set out for its final fall deployment on October 21.

Kitty Hawk participated in the 19th Annual Exercise, the maritime component of Exercise Keen Sword 2008. The exercise was the largest joint exercise for the Navy and Japan Maritime Self-Defense Force. Kitty Hawk also had a port visit in Muroran, Japan. This was the first time a U.S. Navy ship made a visit to the port.

The carrier returned to its homeport on November 27 after 38 days at sea. Kitty Hawk stayed in port for a 5-month maintenance period before setting out to complete sea trials and carrier qualifications.

Before heading out to Puget Sound Naval Shipyard in Bremerton, Washington for decommissioning in 2009, the Kitty Hawk returned to San Diego, California in July 2008 to turn over with the USS George Washington (CVN-73). The USS Kitty Hawk, with a significantly reduced crew and approximately 60 former crew members, departed San Diego for her last cruise on August 28, 2008, en route to Bremerton for her temporary resting place.

After decommissioning on May 12, 2009, Kitty Hawk was placed in reserve status.

On 20 October 2017, Kitty Hawk was stricken from the Naval Vessel Register. On 25 October 2017, the Navy announced its intention to dispose of her by scrapping.

On 9 March 2021, Kitty Hawk arrived at Puget Sound Naval Shipyard to be put in dry dock and have the hull scraped of marine life before being towed to her final destination.

On 6 October 2021, Kitty Hawk and John F. Kennedy were sold for one cent each to International Shipbreaking Limited.

On 15 January 2022, Kitty Hawk left the Puget Sound Naval Shipyard, under tow, en route to Brownsville, Texas, for scrapping. As she is too big to transit the Panama Canal, she went instead by way of the Straits of Magellan.

On 31 May 2022, Kitty Hawk arrived in Brownsville, Texas, for scrapping.

CVA-64 • USS CONSTELLATION

(Displacement: 82,538 tons (74,877 metric tons) full load length: 1,073 feet (327 meters) beam: 130 feet (39.62 meters); extreme width: 282 feet (85.95 meters) draft: 39 feet (11.89 meters). Speed: In excess of 30 knots (34.5+ miles per hour) complement: 4,000 crew aircraft: 85 armament: Three NATO Sea Sparrow launchers; four 20mm Phalanx CIWS mounts class: Kitty Hawk.)

The second Constellation (CVA 64) was built by New York Naval Shipyard, Brooklyn, N.Y.; christened on 8 October 1960 by Mrs. C. A. Herter, wife of the Secretary of State; and commissioned on 27 October 1961, with Capt. T. J. Walker in command. She was named for one of the six frigates purchased by the Continental Congress in the late 1790s. The first of these frigates, which were to make American naval history, was named for the ring of 13 stars that formed a "new Constellation" on the flag of the new United States.

During the final stages of her construction at the New York Naval Shipyard, a fire broke out on Constellation's hangar deck on 19 December 1960, resulting in the deaths of fifty civilian workers.

Constellation deployed to the western Pacific from her homeport of San Diego on 5 May 1964. The first three months of that deployment involved normal operations, training, and port calls. However, on 2 August, while operating in international waters in the Gulf of Tonkin, USS Maddox (DD-731) reported being attacked by units of the North Vietnamese Navy. Within minutes, USS Ticonderoga (CVA 14) dispatched four rocket-armed F8E Crusaders to assist the destroyer. Upon arrival, the Crusaders launched Zuni rockets and strafed the North Vietnamese craft with their 20-millimeter cannons. The Ticonderoga airmen teamed up with Maddox gunners to thwart the North Vietnamese attack, leaving one boat dead in the water and damaging the other two.

Two days later, late in the evening of 4 August, Ticonderoga received urgent requests from USS Turner Joy (DD-951), by then on patrol with Maddox, for air support in resisting what the destroyer alleged to be another torpedo boat foray. The carrier again launched planes to aid the American surface ships, and Turner Joy directed them. The Navy surface and air team believed they had sunk two boats and damaged another pair. President Johnson responded with a reprisal to what he felt at the time to be two unprovoked attacks on American seapower and ordered retaliatory air strikes on selected North Vietnamese motor torpedo boat bases. On 5 August, Ticonderoga and Constellation launched 60 sorties against four bases and their supporting oil storage facilities. Those attacks reportedly resulted in the destruction of 25 PT-type boats, severe damage to the bases, and almost complete razing of the oil storage depot. The strikes lasted for four hours. Constellation lost an A-1H Skyraider, whose pilot, Lt. j.g. Richard A. Sather, became the first Navy pilot to be killed in Vietnam, and an A-4E Skyhawk, flown by Lt. j.g. Everett Alvarez, who became the first Navy POW. On 7 August 1964, Congress authorized President Johnson to "take all necessary measures to repel any armed attack against the United States... [and] to assist any member or protocol state" of the Southeast Asia Treaty Organization (SEATO). This resolution passed in the House of Representatives by a vote of 416 to 0 and in the Senate by 88 to 2.

Constellation returned on 1 February 1965. Her crew was awarded the Navy Unit Commendation and the Armed Forces Expeditionary Medal for actions in the Gulf of Tonkin. Constellation again deployed to the western Pacific from her home of San Diego on 12 May 1966 with Carrier Air Wing 15. On 1 July, three North Vietnamese torpedo boats attacked USS Coontz (DLG 9) and USS Rogers (DD 876) operating about 40 miles offshore on search and rescue missions. Aircraft from Constellation and USS Hancock (CVA 19) quickly neutralized the attackers, sinking all three with bombs, rockets, and 20mm cannon fire. After the attack, Coontz pulled 19 survivors from the water. Constellation returned from the WESTPAC deployment on 3 December 1966.

Constellation made her third deployment to the western Pacific and Vietnam in 1967. She departed San Diego with a new air wing, CVW 14, on 29 April 1967 and returned home on 4 December.

Constellation began her fourth deployment to the western Pacific and Vietnam on 29 May 1968. It was during the initial stages of this deployment that she was visited in June by President Lyndon B. Johnson. On 1 November, as directed by President Johnson, all bombing of North Vietnam was halted at 2100 Saigon time. The last Navy mission over the restricted area was flown earlier in the day from Constellation by Cmdr. Kenneth E. Enney in an A-7 Corsair II. Constellation remained on deployment, returning home to San Diego on 31 January 1969. Following maintenance and training periods, the carrier once again stood out from southern California, this time on her fifth deployment to the western Pacific and Vietnam on 11 August 1969. On 28 March 1970, Lt. Jerome E. Beaulier and

Lt. (j.g.) Stephen J. Barkley in an F-4 Phantom II of VF-142 off Constellation shot down a MiG-21 while escorting an unarmed Navy reconnaissance plane on a mission near Thanh Hoa, North Vietnam. This was the first North Vietnamese MiG kill since the 1 November 1968 bombing halt. Constellation returned home on 8 May 1970.

On 1 October 1971, Constellation sailed from San Diego to begin her sixth combat deployment to Vietnam. Seven months later, Lieutenant Randy Cunningham and Lieutenant (junior grade) Willie Driscoll became America's first aces of the war by downing three MiG fighters during vicious dog-fighting over North Vietnam, bringing their total to five enemy aircraft in four months. Alternating on Yankee Station, Constellation, USS Oriskany (CVA 34), and USS Enterprise (CVAN 65) provided 22 two-carrier days on the line by 1 November, delivering 1,766 ordnance-bearing strike sorties, twelve and nine of them into North Vietnam and South Vietnam, respectively. Two reconnaissance missions were flown during the month, with the airfield at Vinh being the mission assignment. Escort aircraft on both missions expended ordnance in a protective reaction role against firing antiaircraft artillery sites near the field. Other protective reaction strikes were executed.

Constellation and Enterprise operated on Yankee Station together during the month until 10 December, when Enterprise was unexpectedly directed to transit to the Indian Ocean where she operated as flagship for the newly formed Task Force 74 for the possible evacuation of U.S. citizens from East Pakistan in connection with the Indo-Pakistani war. Constellation's tour was extended to the end of the month due to these new contingency operations. USS Coral Sea (CVA 43) joined Constellation on the line on 15 December.

A total of 2,462 ordnance delivery strike sorties were flown during December 1971. The number of surface-to-air missile firing incidents increased, and bold excursions by MiG aircraft into Laos prompted both the U.S. Air Force and U.S. Navy to develop new tactics, combining efforts, to suppress the MiG threat. A major protective reaction strike effort by both USAF and USN commenced on 26 December and terminated on 30 December. During this period, TF-77 flew 423 strike sorties employing all-weather A-6A systems backed up by A-7Es as pathfinders, with Dong Hoi, Quang Khe, and Vinh being the major targets assigned to the Navy. During the month, Laser Guided Bombs (LGBs) were introduced by squadrons aboard Constellation. Initially, 16 trial LGB drops were road cuts, with subsequent targets being antiaircraft artillery sites. In the coming year, LGBs were to be used effectively against heretofore seemingly indestructible targets in NVN, such as heavy steel bridge structures built into solid rock.

USS Enterprise rejoined Constellation on Yankee Station on 18 January 1972, following her tour in the Indian Ocean in December 1971. The next day, 19 January, Lieutenants Randall H. Cunningham and William P. Driscoll, flying an F-4 Phantom of VF-96 off Constellation, shot down a MiG-21. This was the first enemy aircraft downed since the 28 March 1970 shoot-down of a MiG-21 by Lt. Jerome E. Beaulier and Lt. (j.g.) Stephen J. Barkley in an F-4 Phantom II of VF-142 off Constellation. The 19 January action occurred during a protective reaction strike in response to earlier antiaircraft artillery and surface-to-air missile firings from the area that had menaced an RA-5C reconnaissance plane and its escorts. This accounted for the Navy's 33rd MiG shot down in the Vietnam War since the first shoot-down on 17 June 1965 by Commanders Louis C. Page and John J. Smith in an F-4 of VF-21 off USS Midway (CVA 41).

Throughout January 1972, Constellation, Coral Sea, and Enterprise served intermittently on Yankee Station. With only light ground action, limited troop contacts, and the continued withdrawal of U.S. ground troops, the level of air operations remained low, a situation that persisted generally throughout the first three months of the year. During January, only eight Navy tactical air attack sorties were flown in South Vietnam. In North Vietnam, there was very little attack effort except for some protective reaction strikes.

During the following month, naval air attack sorties in South Vietnam rose to 733 compared to the eight during January. The increase was due to preemptive operations by allied forces in preparation for an expected large-scale enemy offensive during Tet, which did not materialize. Constellation, Coral Sea, and Hancock served overlapping tours on Yankee Station, ensuring two to three carriers were on station at a time during most of February 1972.

Naval air attack sorties in South Vietnam in March 1972 dropped again to 113. On 23 March, the U.S. canceled further peace negotiations in Paris, France, due to a lack of progress in the talks. This was followed by the North Vietnamese invasion of South Vietnam. This "Easter" or "Spring Offensive" was the result of the long buildup and infiltration of North Vietnamese forces during previous months and led to some of the most intense fighting of the entire war. The North Vietnamese invasion prompted increased air operations by the carriers in support of South Vietnamese and U.S. forces. The carriers on Yankee Station when North Vietnam invaded on 30 March were Hancock and Coral Sea, which had rotated with Constellation and USS Kitty Hawk (CVA 63).

Beginning on 5 April 1972, aircraft from Constellation, along with those from Hancock, Coral Sea, and Kitty Hawk, took part in Operation Freedom Train, which involved Navy tactical air sorties against military and logistic targets in the southern part of North Vietnam that were involved in the invasion of the south. The operating area in North Vietnam was initially limited to between 17° and 19°N. However, special strikes were authorized against targets above the 19th parallel on various occasions. The magnitude of the North Vietnamese offensive indicated that an extended logistics network and increased resupply routes would be required to sustain ground operations by North Vietnam in their invasion of South Vietnam. Most

target and geographical restrictions that had been in effect since October 1968 concerning the bombing in North Vietnam were gradually lifted, and the list of authorized targets expanded. Strikes in North Vietnam were directed against vehicles, lines of communication (roads, waterways, bridges, railroad bridges, and railroad tracks), supply targets, air defense targets, and industrial/power targets. By the end of April, operations were permitted in North Vietnam throughout the region below 20° 25'N, and many special strikes above the 20th parallel had also been authorized.

By mid-April, the Navy was averaging 191 sorties per day in South Vietnam, a 97 percent increase over the previous week. Sorties concentrated west and north of Quang Tri City, with interdiction and direct air support flown in the area. Carriers on Yankee Station were Constellation, Hancock, Coral Sea, and Kitty Hawk. On 16 April, in Operation Freedom Porch, aircraft from Constellation and the other three carriers on Yankee Station flew 57 sorties in the Haiphong area in support of U.S. Air Force B-52 strikes on the Haiphong petroleum products storage area. From 25 through 30 April, Constellation's VA-146, VA-147, and VA-165 hit areas around the besieged city of An Loc in support of South Vietnamese troops, some only 40 miles outside the capital of Saigon. Targets attacked included artillery fire bases, enemy tanks, bunkers, troop positions, ammunition caches, and gun emplacements.

On 9 May 1972, Operation Pocket Money, the mining campaign against principal North Vietnamese ports, was launched. Early that morning, an EC-121 aircraft took off from Da Nang airfield to provide support for the mining operation. A short time later, Kitty Hawk launched 17 ordnance-delivering sorties against the Nam Dinh railroad siding as a diversionary air tactic. Poor weather, however, forced the planes to divert to secondary targets at Thanh Hoa and Phu Qui, which were struck at 090840H and 090845H, Vietnam time, respectively. Coral Sea launched three A-6A and six A-7E aircraft loaded with mines and one EKA-3B in support of the mining operation directed against the outer approaches to Haiphong Harbor. The mining aircraft departed the vicinity of Coral Sea at 090840H to execute the mining at precisely 090900H to coincide with President Richard Nixon's public announcement in Washington that mines had been seeded. The A-6 flight, led by CAG Cmdr. Roger E. Sheets, comprised Marine Corps aircraft from VMA-224 and headed for the inner channel. The A-7Es, led by Cmdr. Leonard E. Giuliani and made up of aircraft from VA-94 and VA-22, were designated to mine the outer segment of the channel. Each aircraft carried four MK 52-2 mines. Capt. William R. Carr, USMC, the bombardier/navigator in the lead plane, established the critical attack azimuth and timed the mine releases.

The first mine was dropped at 090859H, and the last of the field of 36 mines at 090901H. Twelve mines were placed in the inner segment and the remaining 24 in the outer segment. All MK 52-2 mines were set with 72-hour arming delays, thus permitting merchant ships time for departure or a change in destination consistent with the President's public warning. It was the beginning of a mining campaign that planted over 11,000 MK 36 type destructor and 108 special MK 52-2 mines over the next eight months. It is considered to have played a significant role in bringing about an eventual peace arrangement, particularly since it hampered the enemy's ability to continue receiving war supplies.

On 10 May 1972, Operation Linebacker I, the heavy strike of targets in most of North Vietnam, evolved and lasted until restrictions on operations above 20°N were imposed on 22 October. The operation was an outgrowth of Freedom Train and the President's mining declaration, which also stated that the U.S. would make a maximum effort to interdict the flow of supplies in North Vietnam. On this first day of Linebacker I, the Navy shifted its attacks from targets in southern North Vietnam to the coastal region from Haiphong north to the Chinese border. In all, 173 attack sorties were flown in this region this day, although another 62 were directed into South Vietnam in continuing support of allied forces there.

It was the most intensified air-to-air combat day of the entire war. Navy flyers shot down eight MiGs. An F-4 Phantom II from VF-96 on board Constellation — Lt. Randall H. Cunningham, the pilot, and Lt. (j.g.) William P. Driscoll — while engaged in aerial combat over Haiphong, shot down three MiGs for the first triple downing of enemy MiGs by one plane during the war. These three MiG downings, coupled with their 19 January and 8

May downing of two MiGs, made Lt. Cunningham and Lt. (j.g.) Driscoll the first MiG aces of the Vietnam War. Three other kills were scored by planes of VF-96, one by VF-92 off Constellation, and one by VF-51 off Coral Sea.

During the five and one-half month period of Linebacker I, the Navy contributed more than 60 percent of the total sorties in North Vietnam, with 60 percent of this effort focused on the "panhandle," the area between Hanoi and the DMZ. Tactical air operations were most intense during the July-September quarter, with 12,805 naval sorties flown. Most attack sorties in North Vietnam fell into two classes: armed reconnaissance and strike. Armed reconnaissance usually targeted opportunities near Hanoi, Haiphong, and the Chinese border, while strike operations were preplanned and directed at fixed targets. These fixed targets required approval by the Commander-in-Chief, Pacific, or by the Joint Chiefs of Staff before attack. Principal Navy aircraft included the A-7 and A-6, accounting for roughly 60 and 15 percent of the Navy's attack sorties, respectively. About 25 percent of the Navy's effort occurred at night. Carriers participating in the initial May to June operations from Yankee Station included Constellation, Coral Sea, Hancock, Kitty Hawk, Midway, and USS Saratoga (CVA 60).

On 11 May 1972, naval aircraft from Constellation, Coral Sea, Midway, and Kitty Hawk laid additional minefields in North Vietnam's significant ports: Thanh Hoa, Dong Hoi, Vinh, Hon Gai, Quang Khe, and Cam Pha, as well as the Haiphong approaches. This mining was not confined solely to the seven principal ports. Other locations, including the Cua Sot, Cap Mui Ron, and the river mouths, Cua Day and Cua Lac Giang, south of Don Son and the Haiphong port complex, were also seeded early in the campaign. On 12 May, the 72-hour delay arming time on the initial mines laid at Haiphong expired at 120900H Vietnam time. Nine ships at Haiphong took advantage of the grace period

to depart the port, while twenty-seven ships remained. Both Soviet and Soviet-bloc ships diverted to different destinations, avoiding a direct confrontation with the minefields. Constellation was relieved at Yankee Station and returned home to San Diego, arriving on 30 June 1972. She received a Presidential Unit Citation from President Nixon in 1973.

On 5 January 1973, Constellation, along with Carrier Air Wing 9, began her seventh deployment to the western Pacific and Vietnam. The Vietnam cease-fire, announced on 23 January, went into effect on the 27th. Aircraft from Constellation and Oriskany operating on Yankee Station, now located off the northern part of South Vietnam, flew strikes against targets in southern Laos. Combat sorties from carriers on Yankee Station against targets in Laos continued since the Vietnam cease-fire. These combat support sorties were flown in support of the Laotian government, which had requested this assistance, and had no relationship with the Vietnam cease-fire. Constellation returned from this deployment on 11 October 1973.

Constellation again departed for WESTPAC on 21 June 1974, her eighth such deployment. On 19 November, Constellation participated in the Central Treaty Organization Exercise Midlink 74, the largest naval exercise ever held in the Arabian Sea, involving forces from the United States, United Kingdom, Iran, Pakistan, and Turkey. Constellation returned from this deployment on 22 December 1974. On 31 January 1975, Constellation departed San Diego for the Puget Sound Naval Shipyard in Bremerton, Washington, to undergo one of the most extensive carrier overhauls ever undertaken, enabling her to carry the Navy's newest air supremacy fighter, the F-14A Tomcat, and the S-3A Viking, a submarine hunter. On 1 July 1975, she, along with all U.S. carriers, was redesignated from "CVA" to "CV" to improve the accuracy of designations in modern warfare. This change allowed the new designation CV to serve a multipurpose air, surface, and ASW role, depending on the type of aircraft carried. After 14 months at Puget Sound, Constellation departed the shipyard on 26 April 1976 to rejoin the Pacific Fleet. On 3 July 1976, Constellation celebrated America's Bicentennial by hosting a nationally-telecast TV special from her flight deck featuring many major celebrities and guests.

A newly refurbished Constellation began her 10th deployment in April 1977, which included the first port call by a U.S. carrier to Pattaya, Thailand.

On 26 September 1978, "Connie" sailed west again, beginning her 11th deployment. On 27 December, Constellation and her escort ships were directed to the vicinity of Singapore in response to the internal crisis in Iran and because of vital U.S. interests in the Arabian Gulf area. However, on 2 January 1979, the President directed Constellation and her escort ships to remain on station in the South China Sea and not enter the Indian Ocean. Constellation and her escorts were released from contingency operations in the South China Sea on 28 January. The crisis in Iran abated when the Shah of Iran departed for exile on 16 January. Due to the uneasy situation in Iran, all U.S. government dependents and nonessential American citizens were ordered to evacuate the country on 30 January.

Constellation and her escorts were ordered to the Gulf of Aden on 7 March 1979 in response to the conflict between North and South Yemen. The Gulf of Aden and the Arabian Gulf were considered vital waterways for the passage of petroleum products to the U.S. and her allies. On 16 April, Constellation was relieved by Midway as the Indian Ocean contingency carrier. Constellation returned to San Diego on 17 May 1979. The deployment had twice been extended, and her service earned her the Navy and Marine Corps Expeditionary Medals.

Constellation began her 12th deployment to the western Pacific on 26 February 1980. On 18 April, the carrier and her battle group departed Subic Bay in the Republic of the Philippines to relieve Coral Sea in the Indian Ocean, doing so on 30 April. Coral Sea had been on station for 89 days in connection with the Iranian crisis. Constellation herself set a new endurance record in 1980, remaining on station in the Indian Ocean for an exhausting 110 consecutive days. Midway relieved Constellation on 17 August, and the ship began her voyage home, arriving on 15 October.

Constellation began her 13th deployment in October 1981, returning to San Diego in May 1982. Before the deployment, in late summer 1981, the carrier hosted President Ronald Reagan. During this visit, Mr. Reagan presented a Presidential Flag to the ship and proclaimed her "America's Flagship." While operating in the Gulf of Oman, Secretary of Defense Caspar Weinberger and the Chief of Naval Operations visited Constellation.

In December 1982, Constellation sailed north to Puget Sound Naval Shipyard in Bremerton, Washington, to begin a 14-month complex overhaul. During the overhaul, Constellation was modified to carry the Navy's newest strike fighter, the FA-18 Hornet, becoming the first carrier to receive the new aircraft. She was also fitted with the new Phalanx radar-guided Gatling-gun, two new flush deck catapults, and the NATO Sea Sparrow Missile System. Constellation completed the $235 million overhaul two weeks early and under budget, a first since the 1940s, according to the shipyard.

Constellation set sail on her 14th deployment to the western Pacific and Indian Ocean on 21 February 1985. This deployment included the first operational deployment of the F/A-18 Hornet strike fighter and the LAMPS, using the SH-60B Seahawk ASW helicopter. The Hornets replaced the A-7E Corsair IIs operated by two squadrons assigned to CVW-14, making Constellation the Navy's first carrier to have F/A-18s assigned to her air wing. The cruise included port visits to Singapore, Kenya, and Western Australia. Constellation returned to her homeport of San Diego in late August. For her performance during this cruise, the carrier's crew earned the Meritorious Unit Commendation. The ship also received the Secretary of the Navy's Environmental Protection Award.

With the cruise and two major awards, 1985 was a good year for Constellation, but 1986 would be even better. During the first part of the year, Constellation earned the Golden Anchor Award for retention excellence and a second consecutive Environmental Protection Award. The most important award came when the ship earned the U.S. Pacific Fleet Battle Efficiency Award covering her outstanding performance from January 1985 to June 1986.

Constellation began a two-month Northern Pacific Cruise (NOPPAC) in September 1986. In early September, the ship spent five days in Vancouver, British Columbia, where many members of the crew visited EXPO '86. In late September, the ship spent four days in Anchorage, Alaska, becoming the first carrier to visit that port. During these port visits, the ship hosted over 15,000 visitors.

Constellation's final port visit was in Seattle, Washington, where an estimated 45,000 visitors walked her decks while America's Flagship celebrated the Navy's 211th birthday. Connie returned home to Naval Air Station North Island in San Diego in time for her 25th birthday. On 27 October 1986, Constellation celebrated her Silver Anniversary on board with a concert, ceremony, and a giant cake.

On 11 April 1987, Constellation once again made her way west to the western Pacific and Indian Oceans. But this cruise, her 15th major deployment, took her up into the North Arabian Sea and the Gulf of Oman.

On 2 August 1988, Constellation successfully fought a severe fire in the main engineering space using the installed HALON firefighting equipment; this was the first carrier use of the system in fighting a fire.

In February 1990, Constellation left San Diego, returning to the East Coast for a three-year overhaul. The $800-million Service Life Extension Program (SLEP), completed in Philadelphia Naval Shipyard in March 1993, added an estimated 15 years to the carrier's operational life. The overhaul saw upgrades to virtually every system on the ship. Constellation departed the Philadelphia Naval Shipyard on 4 March 1993, the fifth and last carrier to complete SLEP. Back in San Diego, the carrier was visited on 18 August by Secretary of the Navy John H. Dalton on his first trip to Southern California. Following a similar trip two days earlier to East Coast commands, the Secretary also visited Naval Station San Diego, Camp Pendleton, the Marine Corps Recruit Depot, Naval Air Station Miramar, Naval Submarine Base San Diego, and other area commands. On 31 July 1994, Lt. Kara Hultgreen made her first qualifying landing in an F-14A on board Constellation, 110 miles southwest of San Diego. She thus became the first fully qualified female Tomcat pilot. Lt. Hultgreen was assigned to VF-213 at NAS Miramar, California. Lt. (j.g.) Carey Dunai, also in an F-14, became the second woman to reach the milestone with her qualifying trap moments later. Lt. Hultgreen was killed at sea on 25 October 1994 while approaching USS Abraham Lincoln (CVN 72) for a landing 50 miles off San Diego.

Constellation and her battle group deployed again on 10 November 1994 and spent most of December in the western Pacific. On 11 January 1995, the carrier and her seven-ship battle group entered the Arabian Gulf. With Constellation were the guided missile cruisers USS Chosin (CG 65) and USS Lake Erie (CG 70), the destroyer USS Kinkaid (DD 965), the ammunition ship USS Kiska (AE 35), the replenishment oiler USS Cimarron (AO 177), and the attack submarine USS Topeka (SSN 754). Aircraft from Constellation's air wing, Carrier Air Wing Two (CVW 2), patrolled the U.N. no-fly zone over southern Iraq in support of Operation Southern Watch.

During this period in the Gulf, when Constellation's crew learned of its hometown San Diego Chargers' American Football Conference championship victory over the Pittsburgh Steelers, it decided to send the Chargers and the fans of San Diego a personal message. On 18 January, about 850 sailors from Connie and her air wing assembled on the flight deck in their white uniforms to form the words "Go Chargers" spelled out on the ship's flight deck. Photographers in a helicopter flying above the ship took photos and shot video and sent them back to San Diego. Constellation returned to San Diego in May 1995.

Actor and comedian Tim Allen visited Constellation during an at-sea period in early 1996. Mr. Allen, star of ABC's then-top-rated sitcom Home Improvement, along with the show's producers and writers, came aboard on 19 January and scouted Constellation as a possible setting for a future show. That show, number A553, episode 127/6.01, aired on the ABC television network on 17 September 1996. In February 1997, Constellation was selected by the Navy as one of the test ships for a new working uniform. The test lasted six months. Also, from 10 through 21 February 1997, Constellation, along with her Carrier Battle Group and the USS Boxer (LHD 4) Amphibious Ready Group with the 15th Marine Expeditionary Unit embarked, took part in Pacific Joint Task Force Exercise (PAC JTFEX) 97-1 off the Southern California coast. More than 15,000 military personnel tested their ability to project combat power ashore during the 3rd Fleet exercise. The exercise included various air strike and support missions, maritime interdiction operations, humanitarian operations, operational testing of weapons systems, logistics support, search and rescue, and command and control. The amphibious operations supporting the exercise culminated with an amphibious landing at Camp Pendleton, California, involving surface and helicopter assault forces.

On 19 April 1997, while Constellation and her nine-ship battle group were on a routine six-month deployment, she was

conducting a vertical replenishment with the replenishment ship USNS Niagara Falls (T-AFS 3) when one of the replenishment ship's UH-46E Sea Knight helicopters crashed into the sea. The crash occurred approximately 2,000 miles southwest of Hawaii, and all four crew members were safely rescued. The helicopter crew members from Helicopter Support Squadron Five, homeported in Guam, were rescued by the quick reaction of other CH-46 helicopters on scene at the time of the incident. Assisting in the rescue was a motor-whale boat from Niagara Falls, also homeported in Guam. The four crew members, who escaped injury, were transported to Constellation for medical observation and were released to Niagara Falls the same day.

The Constellation Battle Group entered the Arabian Sea on 16 May 1997 and conducted high-tempo operations that included more than 4,400 sorties during more than 10 weeks in the Arabian Gulf. Operations included exercises with friendly forces in the region. Carrier Air Wing (CVW) 2, embarked in Constellation, participated in Operation Southern Watch, flying 1,460 sorties enforcing the no-fly zone over southern Iraq. On 14 July 1997, Constellation rescued two Iranian merchant seamen found adrift in international waters of the Arabian Gulf. Lookouts aboard the carrier spotted the two men clinging to floating debris about 70 nautical miles northeast of Bahrain. Constellation crewmen immediately launched a rescue boat to retrieve the stranded seamen from the water. Once on board, the two stated they had been adrift for approximately five days after their Iranian cargo dhow, Ramazan, with eight other crewmen aboard, broke up in rough seas. The two Iranian merchant seamen were later transferred ashore. An unsuccessful search for any other survivors continued using helicopters from Helicopter Antisubmarine Squadron Light (HSL) 37, Detachment 9, embarked aboard USS Lake Erie (CG 70), and Helicopter Antisubmarine Squadron Light (HSL) 42, Detachment 5, embarked aboard USS Thorn (DD 988).

Constellation and her battle group departed the 5th Fleet's Area of Responsibility on 17 August 1997 and began her journey back toward San Diego. The Battle Group included USS Merrill (DD 976), USS John Paul Jones (DDG 53), USS Chosin (CG 65), USS Lake Erie (CG 70), USS Cimarron (AO 177), and USS Mount Hood (AE 290). Beginning on 28 August 1997, Constellation participated in Fleet Battle Experiment Bravo (FBE-B) with deployed naval forces in the eastern and western Pacific to test war-fighting concepts and capabilities for the 21st century. The experiment had two main parts. The first experiment ran from 4 through 7 September and involved the Ring of Fire concept, a naval fire support "Battle LAN" that was tested during Third Fleet's Battle Experiment Alfa (FBE-A) in March 1997. The second part of the experiment, called Silent Fury, began on 2 September and lasted until the 23rd. During this phase, Third Fleet, in conjunction with Naval Strike Air Warfare Center at Naval Air Station Fallon, Nevada, examined the targeting process and the challenges inherent in the employment of new types of guided munitions soon to be in the Navy's weapons inventory. Other units participating were the USS Peleliu (LHA 5) Amphibious Ready Group with the 13th Marine Expeditionary Unit (MEU) embarked, USS Russell (DDG 59), and USS Nimitz (CVN 68).

After completing one of the most successful work-up schedules in Navy history, Constellation departed San Diego on 18 June 1999, beginning her 19th overseas deployment. She immediately put her war-fighting skills to the test by conducting a Joint Task Force Exercise (JTFEX). This marked the first time a carrier conducted JTFEX at the beginning of a deployment. With increased tensions between North and South Korea, Constellation then headed for the Korean theater to closely monitor the situation and provide a calming influence. After port calls in Pusan, Republic of Korea; Yokosuka, Japan; Singapore; and Kuala Lumpur, Malaysia, Constellation entered the Arabian Gulf on 28 August, relieving USS Theodore Roosevelt (CVN 71). She then began flying combat air patrols over the Iraqi no-fly zones in support of Operation Southern Watch. During her 10-week stay in the 5th Fleet theater, Constellation and CVW-2 flew 1,256 sorties and expended nearly 44 tons of ordnance during nine combat engagements against Iraqi ground and air targets. She left the Arabian Gulf for some well-earned rest with port visits in Fremantle/Perth and Sydney, Australia. Constellation recorded a rare deployment milestone of 10,000 aircraft "traps" just before returning to San Diego for a memorable holiday homecoming on 17 December 1999. For their extraordinary efforts during the Navy's last deployment of the 20th century, the crew of Constellation was awarded the 1999 Battle "E" as the Pacific Fleet's best carrier.

On 19 May 2000, more than 1,700 educators, school administrators, students, and community leaders from California, Nevada, and Arizona joined the crew of Constellation at Naval Air Station North Island for a day of life at sea. The ship set sail at 7 a.m. and cruised the waters off San Diego, returning at about 4 p.m.

On 16 March 2001, Constellation stood out from San Diego for its 20th deployment in its 40-year history. Nearly 12,000 Sailors and Marines began the scheduled six-month deployment as part of the Constellation Carrier Battle Group and the USS Boxer (LHD 4) Amphibious Ready Group (ARG). In addition to Constellation and the embarked air wing, Carrier Air Wing (CVW) 2, other deploying units included USS Chosin (CG 65), USS Benfold (DDG 65), USS Kinkaid (DD 965), USS Thach (FFG 43), the Canadian frigate HMCS Winnipeg (FFH 338), USS Rainier (AOE 7), and the attack submarines USS Santa Fe (SSN 763) and USS Columbia (SSN 771). Constellation returned from this deployment on 15 September 2001 to a different United States. The events of four days earlier on 11 September 2001, a date, like that of the attack on Pearl Harbor, that will "live in infamy," had changed the country. Terrorists had crashed three commercial airliners into New York's World Trade Center and the Pentagon in Arlington, Va. Constellation hosted more than 270 former "Connie" Sailors at a reception in the ship's hangar bay on 27 October 2001 to celebrate the ship's 40th anniversary. "My thoughts today do not so much concern Constellation herself, mighty as she is, as the men who in the future will write her story; whose actions aboard her will tell all," said Cmdr. C. K. Ruiz, Constellation's first executive officer on the day of the ship's commissioning at the New York Naval Shipyard, 27 October 1961. Retired Capt. Stanley W. Vejtasa, Constellation's second commanding officer, was the guest speaker for the event. More than 150 "plankowners," or members of the original commissioning crew, were in attendance.

The Constellation battle group was underway again off the coast of southern California on 19 October 2002 for Joint Task Force Exercise (JTFEX), a final battle group exercise prior to their upcoming deployment, when a Sailor from Strike Fighter Squadron (VFA) 151, was blown overboard by jet blast at five minutes past midnight as he walked behind an EA-6B Prowler on the carrier's flight deck. He had been performing routine duties during normal night flight operations at the time of the incident. Because his shipmates saw him go over the side, rescue efforts began immediately. Constellation executed a 180-degree turn to position itself in the vicinity of the Sailor. Helicopters and Rigid Hull Inflatable Boats (RHIBs) from Constellation, USS Valley Forge (CG 50), USS Bunker Hill (CG 52), and USS Kinkaid (DD 965) assisted. At approximately 7:20 a.m., Bunker Hill's helicopter detachment spotted the Sailor in the water. Constellation's helicopter squadron recovered the Sailor and returned him to the ship where he was in stable condition following medical treatment. Constellation returned to port on 29 October.

Constellation's 21st deployment began on 2 November 2002 when the carrier and her battle group deployed in support of Operation Enduring Freedom. The battle group consisted of guided-missile cruisers USS Bunker Hill (CG 52) and USS Valley Forge (CG 50); guided-missile destroyers USS Higgins (DDG 76) and USS Milius (DDG 69); and guided-missile frigate USS Thach (FFG 43). Other units included the fast attack submarine USS Columbia (SSN 771), based in Pearl Harbor, Hawaii, and the fast combat support ship USS Rainier (AOE 7), based in Bremerton, Washington.

The crew of Constellation quickly responded to a main machinery room fire on 8 November, preventing any serious injuries or degradation to the ship's ability to continue its deployment. The fire, which broke out in one of Constellation's four main machinery rooms, was caused by a fuel oil leak. The affected space was immediately evacuated, and installed firefighting equipment contained the fire until the ship's fire parties and damage control teams extinguished it. Constellation arrived in the Arabian Gulf on 14 December, its second deployment to the area in 15 months.

While on station in the Arabian Gulf, Constellation was visited by Rep. Ellen Tauscher (D-Calif.), serving her third term representing California's 10th Congressional District, San

Francisco's East Bay suburbs. The carrier was also visited by Chief of Naval Operations Adm. Vern Clark, Master Chief Petty Officer of the Navy (MCPON) Terry Scott, and Chief of Chaplains Rear Adm. Barry Black. On the hangar deck, the CNO told the crew, "Remember, the American people are watching you. Be ready to go write some history. Hit fast, hit hard, hit with precision."

On 19 January 2003, Constellation added another accomplishment to the list by recording its 390,000th trap. Lt. Frank Wittwer, flying Falcon 407, an F/A-18C Hornet from Strike Fighter Squadron (VFA) 137, caught the number two wire, at 8:53 p.m., marking the milestone trap for the ship. On 5 March, crew members joined thousands of Americans across the United States in the simultaneous playing of the National Anthem. The Star-Spangled Banner echoed throughout Constellation, approximately 1,060 feet from bow to stern, at precisely 7 p.m. local time. Additionally, several Sailors participated in interviews with hometown media. Minutes before the National Anthem was played, Commanding Officer Capt. John Miller provided comments to Constellation's crew about the meaning and significance of the project. "The National Anthem project is a wonderful example of the support we are receiving from the folks back home. We are fortunate to have the opportunity to participate in this landmark demonstration of pride and patriotism," he remarked.

Operation Iraqi Freedom, the multinational coalition effort to liberate the Iraqi people, eliminate Iraq's weapons of mass destruction, and end the regime of Saddam Hussein, began on 20 March 2003 with the firing of Tomahawk missiles from U.S. ships in the Arabian Gulf and Red Sea. On 25 March, aircraft from two squadrons of Carrier Air Wing 2 aboard Constellation successfully destroyed three significant Iraqi naval targets. The aircraft involved received short-order tasking for a time-sensitive strike to engage several naval targets in the vicinity of Basra.

Two F/A-18C Hornets from Strike Fighter Squadron (VFA) 151, "The Vigilantes," one of three F/A-18 squadrons aboard, participated in the strikes with one aircraft from Sea Control Squadron (VS) 38, "The Red Griffins," the S-3B Viking squadron, aboard Constellation. One Hornet independently identified and destroyed two naval targets. The S-3B, working in conjunction with the second F/A-18C, as the laser designator, destroyed its target by firing a laser-guided missile, the AGM-65E Maverick. The F/A-18 provided laser illumination throughout target engagements. This event was the first time the Navy's S-3 has received tasking for overland strikes in its 30-year history. In addition, it is also the first time the S-3 has fired a laser-guided

missile in combat. On 1 April 2003, an S-3B Viking from Sea Control Squadron 38 of Carrier Air Wing (CVW) 2 aboard Constellation veered off the flight deck after making an arrested landing at approximately 5:10 a.m. local time. Shortly after touching down on deck, the S-3B malfunctioned while taxiing on the carrier's flight deck and slid to the port side of the deck. The plane went over the side and hit the flight deck safety netting, with the two pilots aboard ejecting into the water. The plane then followed into the water.

A helicopter from Helicopter Anti-Submarine Squadron (HS) 2 was performing search-and-rescue (SAR) operations for the flight cycle and arrived on the scene immediately. A SAR swimmer recovered both pilots and transported them to Constellation, where they were evaluated for minor injuries. After completing SAR operations and recovering the pilots, scheduled flight operations aboard Constellation continued.

Constellation departed the Arabian Gulf on 16 April 2003 and began her journey back to her homeport of San Diego. While transiting home on her final deployment in the Arabian Gulf, Constellation received a distress call from an Indonesian inter-island ferry boat carrying civilian passengers in the Banda Sea. USS Milius (DDG 65), also transiting home with the Constellation Strike Group, was in a position to assist and was tasked by Destroyer Squadron (DESRON) 7, embarked on Constellation, to investigate and render assistance to the vessel.

Milius responded immediately and quickly launched a Rigid Hull Inflatable Boat (RHIB) into the water to assess the situation. The ferry, which had 27 civilians aboard, including children, suffered an electrical generator and diesel propulsion engine failure, which had been inoperable for three days. The vessel had attempted to hail several passing ships for two days before Milius responded. Once on the scene, several crew members from Milius embarked the ferry to make repairs. They were able to restore electrical power but found the diesel engine was not repairable. Adding further complications to the initial situation was the fact that no one on the vessel spoke English. Constellation, however, had two linguists aboard and was able to dispatch a helicopter carrying Cryptologic Technician Interpretive 3rd Class Russell Crandall, an Indonesian interpreter, to assist the Milius crew in communicating with the crew of the ferry.

After a full assessment of the situation and discussion with the vessel's captain, it was determined that the best option for ensuring the vessel's safety would be to tow it to the nearby Indonesian island of Pulau-Sanana, the closest island to the vessel. As Milius began slowly towing the disabled vessel towards the port of Sanaa, the civilians aboard the vessel showed their gratitude to the Milius crew by chanting, "USA, USA!" Following a port visit in Pearl Harbor on 22 May 2003, Constellation and other ships in its strike group got underway for San Diego on 27 May. Aircraft of Carrier Air Wing 2 flew from Constellation on 31 May and June 1. Constellation arrived at Naval Air Station North Island, California, on 2 June 2003.

During Operation Iraqi Freedom, the Constellation Carrier Strike Group flew more than 1,500 sorties (missions) and expended more than 1 million pounds of ordnance, including 408 Tomahawk cruise missiles. USS Bunker Hill was one of the first warships to conduct Tomahawk strikes against leadership targets in Iraq. Its embarked LAMPS (Light Airborne Multi-Purpose System) helicopter detachment, Helicopter Anti-Submarine Squadron Light 45 "the Wolfpack," supported the rescue of United Nations workers being forcibly removed from oil platforms in the Northern Arabian Gulf and provided medical evacuations from Umm Qasr. On 13 July, Constellation held an "Alumni Day" event, inviting shipbuilders, Sailors, and Marines who served aboard the ship during its history to self-guided tours through some of the ship's spaces. After 41 years of commissioned service and completing 21 deployments, Constellation was decommissioned in San Diego at the Naval Air Station North Island on 7 August 2003.

Beginning on 12 September 2003, the decommissioned carrier was towed from the Naval Air Station North Island to the Inactive Ships Maintenance Facility at Bremerton, Washington, by a contracted ocean-going tug operated by Foss Maritime of Seattle, Washington.

Following her decommissioning, Constellation remained in the inactive reserve fleet until 2021. On 9 March 2021, Constellation arrived at Puget Sound Naval Shipyard to be put in dry dock and have the hull scraped of marine life before being towed to her final destination. On 6 October 2021, Constellation and John F. Kennedy were sold for one cent each to International Shipbreaking Limited. On 15 January 2022, Constellation left Puget Sound Naval Shipyard, under tow, en route to Brownsville, Texas, for scrapping. As she was too big to transit the Panama Canal, she went instead by way of the Straits of Magellan. On 31 May 2022, Constellation arrived in Brownsville, Texas, for scrapping.

CVAN-65 • USS ENTERPRISE

(Displacement. 86,600 (f.) l. 1101', b. 133'; ew. 252'; dr. 36'; s. 30+ k.; cpl. 4,600; cl. Enterprise.)

A bottle of champagne shattered on the bow of the awesome new warship, USS Enterprise (CVN 65), breaking the silence of an anxious crowd at Newport News Shipbuilding and Drydock Company, Newport News, Va. It was Saturday morning, Sept. 24, 1960, when Mrs. William B. Franke, wife of former Secretary of the Navy, christened the eighth USS Enterprise, bringing to life the longest, tallest, and mightiest warship to ever sail the seas.

At the commissioning of Enterprise, the world's first nuclear-powered aircraft carrier, Secretary of the Navy John B. Connally, Jr., called it a worthy successor to the highly decorated seventh USS Enterprise of World War II. "The Fighting Gray Lady," as it was called, served in such well-known battles as the raid on Tokyo and the Battle of Midway." Secretary Connally went on to say, "The new Enterprise will reign a long, long time as queen of the seas."

Enterprise made its maiden voyage under the command of Capt. Vincent P. DePoix on Jan. 12, 1962, conducting a three-month shakedown cruise.

In August, Enterprise joined the Sixth Fleet in the Mediterranean. Soon after its return to Norfolk, Va., in October, Enterprise was dispatched to its first international crisis.

Enterprise and other ships in the Second Fleet set up a "strict quarantine of all offensive military equipment under shipment to Cuba." The blockade was put in place on Oct. 24, and the first Soviet ship was stopped the next day. On Oct. 28, Khrushchev agreed to remove the missiles and dismantle the missile bases in Cuba.

Enterprise made its second and third deployments to the Mediterranean in 1963 and 1964. During the latter deployment, on May 13, the world's first nuclear-powered task force was formed when USS Long Beach and USS Bainbridge joined Enterprise. On July 31, the three ships were designated Task Force One and sent on "Operation Sea Orbit," a historic 30,565-mile voyage around the world, accomplished without a single refueling or replenishment.

In October, Enterprise returned to Newport News Shipbuilding and Dry Dock Company for its first refueling and overhaul.

The Big E transferred to the Pacific's Seventh Fleet in November 1965 and became the first nuclear-powered ship to engage in combat when it launched bomb-laden aircraft in a projection of power against the Viet Cong on Dec. 2, 1965. Its hot decks launched 125 sorties on the first day, unleashing 167 tons of bombs and rockets on the enemy's supply lines. The next day it set a record of 165 strike sorties in a single day.

In all, Enterprise made six combat deployments to Southeast Asia from 1965 to 1972. Between combat tours, Enterprise returned to Newport News Shipbuilding and Drydock Company in 1970 for an overhaul and second refueling.

Following the 1973 cease-fire in Vietnam, Enterprise proceeded to Puget Sound Naval Shipyard, Bremerton, Wash., where "Big E" was altered and refitted to support the Navy's newest fighter aircraft -- the F-14A "Tomcat."

When Enterprise made its seventh Western Pacific (WESTPAC) deployment in September 1974, it became the first carrier to deploy with the new F-14 Tomcat fighter plane. During the deployment, in February 1975, Enterprise was called on to help in the evacuation of Saigon. During Operation Frequent Wind, "Big E" aircraft flew 95 sorties.

The ship made its eighth and ninth WESTPACs in July 1976 and April 1978, respectively. It sailed to Puget Sound Naval Shipyard in January 1979 for a 30-month comprehensive overhaul.

Enterprise made its 10th, 11th, and 12th WESTPACs in 1982, 1984, and 1986, respectively. In April of the 1986 deployment, Enterprise became the first nuclear-powered aircraft carrier to transit the Suez Canal. It then entered the Mediterranean for the first time in 22 years.

In April 1988, Enterprise, on its 13th deployment, was assigned to escort reflagged Kuwaiti oil tankers in the Persian Gulf while stationed in the North Arabian Sea. In a measured response, Enterprise and Carrier Air Wing 11 struck a decisive blow to the Iranian navy in the most intensely fought naval battle since the Korean campaign.

Enterprise began its 14th overseas deployment in September 1989. In early December, Enterprise participated in "Operation Classic Resolve," President Bush's response to Philippine President Corazon Aquino's request for air support during the rebel coup attempt. Enterprise remained on station conducting flight operations in the waters outside Manila Bay.

In March 1990, Enterprise completed its highly successful around-the-world deployment by arriving in Norfolk, Va. Enterprise had successfully and safely steamed more than 43,000 miles from its long-time homeport of Alameda, Calif. In October, Enterprise moved to Newport News Shipbuilding and Drydock Company for refueling and the Navy's largest complex overhaul ever attempted.

Work began in earnest in January 1991. While specially trained shipyard workers refueled the ship's eight nuclear reactors, crewmen and more shipyard workers began overhauling the carrier's 3,500 compartments. Enterprise completed its overhaul, the most extensive in U.S. Naval history, on Sept. 27, 1994. Sea trials began, during which Enterprise performed an extended full power run as fast as when it was new, remaining the fastest combatant in the world. The ship then returned to its homeport in Norfolk, Va.

In mid-January 1995, "Big E" returned to Newport News Shipbuilding and Drydock Company for a five-month Selected Restricted Availability. The yard period involved upgrades to all combat and communication systems, intelligence suites, command and control capabilities, ventilation systems, berthing and dining areas, and underway replenishment equipment. "Big E" departed Newport News in July and returned to Norfolk Naval Base, from which the ship conducted operations in the Caribbean Sea and participated in the highly successful Joint Task Force '96 Exercise with a British task group. Enterprise completed its

15th overseas deployment in December 1996, with operations conducted throughout the Mediterranean Sea, Arabian Gulf, and Red Sea. During this deployment, Enterprise made the fastest transit ever conducted from off Bosnia to the Persian Gulf in support of strikes against Iraq. This transit took less than six days and resulted in peace being restored to the region. "Big E" returned to its homeport of Norfolk on December 20, 1996, to a joyous holiday reunion with family and friends.

In January 1997, Big "E" returned to Newport News Shipbuilding and Drydock Company for a six-month Selected Restricted Availability. The focus of the yard period was habitability upgrades and various combat systems. The ship completed carrier qualifications and work-ups in 1998 before deploying to the Mediterranean Sea and Arabian Gulf in November. On December 16, 1998, the Enterprise Battle Group with Carrier Air Wing THREE (CVW 3) initiated Operation Desert Fox, launching more than 300 Tomahawk cruise missiles and 691,000 pounds of ordnance against military targets in Iraq. Shortly after Desert Fox, Big E sailed to the Adriatic Sea for possible military intervention in the Yugoslavian province of Kosovo. Before returning to Norfolk on May 6, 1999, Enterprise made another appearance in the Arabian Gulf, striking several Iraqi military targets in response to hostile action.

Enterprise began its 17th overseas deployment on April 25, 2001, with Carrier Air Wing Eight (CVW-8). In addition to Enterprise and its embarked air wing, the ships of the battle group included guided missile cruisers USS Philippine Sea (CG 58) and USS Gettysburg (CG 64); guided missile destroyers USS Stout (DDG 55), USS McFaul (DDG 74), and USS Gonzalez (DDG 66); destroyers USS Nicholson (DD 982) and USS Thorn (DD 988); guided missile frigate USS Nicholas (FFG 47); logistics ship USS Arctic (AOE 8); and attack submarines USS Providence (SSN 719) and USS Jacksonville (SSN 699).

During the early part of the 2001 deployment, Enterprise and four Enterprise Battle Group ships participated in the Joint Maritime Course 01-2 (JMC 01-2), a British Royal Navy joint and combined warfare training exercise, from June 18 through June 28, 2001, in the North Sea near the Hebrides Islands, as well as land and airspace around Scotland.

On September 11, 2001, Enterprise was just beginning her voyage home from the Arabian Gulf. Watching a U.S. morning news show live, although locally in the early evening, the crew saw the terrorist attacks by the al Qaeda terrorist network against New York's World Trade Center and on the Pentagon. Immediately, Enterprise turned around and headed back to the waters off Southwest Asia. On October 7, 2001, the U.S. launched air attacks against al Qaeda terrorist training camps and Taliban military installations in Afghanistan. The carefully targeted actions were designed to disrupt the use of Afghanistan as a base for terrorist operations and to attack the military capability of the Taliban regime.

For approximately the next three weeks, aircraft from Enterprise flew nearly 700 missions in Afghanistan and dropped hundreds of thousands of pounds of ordnance. Departing from Southwest Asia in late October, Enterprise arrived back at its homeport of Norfolk, Va., on November 10, 2001, about two weeks later than originally planned. During its last day at sea, the ship played host to the ABC Television Network's Good Morning America for a live two-hour broadcast of the program from the ship.

On January 7, 2002, Enterprise entered the Norfolk Naval Shipyard for a scheduled one-year Extended Docking Selected Restricted Availability (EDSRA).

In 2003-2004, the carrier provided air support for Operation Iraqi Freedom. In 2004, the ship participated in Summer Surge 2004 and several multinational exercises. In May 2006, Enterprise departed for a six-month deployment, operating in the 6th, 5th, and 7th Fleet areas, and supported both Operations Iraqi and Enduring Freedom. She returned to Norfolk on November 18, 2006.

On December 19, 2007, the carrier returned home after a six-month deployment in the Persian Gulf.

From September 2003 to February 2004, the ship was deployed to support Operation Iraqi Freedom. In June-July 2004, the ship participated in Summer Surge 2004.

In 2005, the Enterprise entered Newport News Shipyards for a routine overhaul. Upon leaving the yard, she ran onto a sandbar and was disabled for several hours. The Enterprise deployed in 2005 and 2006, including to the Persian Gulf.

In 2008, the Enterprise underwent an 18-month docking during which $655 million was spent on her refitting, making her the oldest active ship in the U.S. Navy. In April 2010, the Enterprise returned to service. She undertook two last deployments to the Persian Gulf.

On December 1, 2012, the Enterprise was decommissioned.

USS Enterprise (CVN-65) Post-2012: A Legacy Continues

The decommissioning of the USS Enterprise (CVN-65) on December 1, 2012, marked the end of an era for the world's first nuclear-powered aircraft carrier. However, the story of this iconic vessel continued as it transitioned through inactivation and dismantling, leaving a lasting legacy that endures through preservation efforts and the forthcoming USS Enterprise (CVN-80).

Following its decommissioning, the USS Enterprise was moved to Newport News Shipbuilding for inactivation. This phase involved the delicate and complex process of defueling the ship's eight nuclear reactors and removing equipment and hazardous materials. The meticulous work ensured that the ship was safe for its final journey and subsequent dismantling.

In 2017, the Enterprise was formally stricken from the Naval Vessel Register, marking its official removal from the list of active naval vessels. The ship was then prepared for towing to the Puget Sound Naval Shipyard, where further dismantling and recycling operations would take place.

In early 2018, the USS Enterprise began its final voyage,

being towed from Newport News to Brownsville, Texas. This significant move was managed by International Shipbreaking Limited, a company experienced in the safe and environmentally responsible dismantling of decommissioned naval vessels.

The dismantling process in Brownsville was extensive, involving cutting the ship into sections and recycling the steel and other materials. Given the Enterprise's size and the complexity of its nuclear-powered systems, this process spanned several years. Special attention was given to the safe handling and disposal of nuclear materials, ensuring that environmental and safety standards were strictly adhered to.

One significant aspect of the preservation efforts was the saving of the ship's island (superstructure), which is intended to be part of a future museum display.

CV-66 • USS AMERICA

(Displacement: 83,573 tons; length: 1,047 ½ feet; beam: 130 feet; extreme width: 249 feet; draft: 38 feet; speed: 20+ knots; complement: 4,582 crew; aircraft: 70; armament: Terrier surface-to-air missiles; class: Kitty Hawk.)

The third USS America (CV-66) was laid down on January 1, 1961, at Newport News, Virginia, by the Newport News Shipbuilding and Dry Dock Corporation; launched on February 1, 1964; sponsored by Mrs. David L. McDonald, wife of Adm. David L. McDonald, the Chief of Naval Operations, and commissioned at the Norfolk Naval Shipyard on January 23, 1965, with Capt. Lawrence Heyworth, Jr., in command.

After fitting out at the Norfolk Naval Shipyard until March 15, 1965, America remained in Hampton Roads for operations off the Virginia Capes until getting underway on March 25. She conducted her first catapult launch on April 5, 1965, with Commander Kenneth B. Austin, the carrier's executive officer, piloting a Douglas A-4C Skyhawk. Proceeding thence to the Caribbean, the carrier conducted shakedown training and concluded it at Guantanamo Bay on June 23.

Entering the Norfolk Naval Shipyard for post-shakedown availability on July 10, she remained there until August 21. She next operated locally through late August and then proceeded to the operating areas off the Virginia Capes and to Bermuda, arriving back at Norfolk on September 9. On September 25, Rear Adm. J. O. Cobb broke his flag as Commander, Carrier Division (CarDiv) 2.

America sailed for her first Mediterranean deployment late in 1965. New Year's Day, 1966, found her at Livorno, Italy. Over the ensuing weeks, the ship visited Cannes, France; Genoa, Italy; Toulon, France; Athens, Greece; Istanbul, Turkey; Beirut, Lebanon; Valletta, Malta; Taranto, Italy; Palma, Majorca, Spain; and Pollensa Bay, Spain. She sailed on July 1, 1966, for the United States. Early in the deployment, from February 28 to March 10, America participated in a joint Franco-American Exercise Fairgame IV, which simulated conventional warfare against a country attempting to invade a NATO (North Atlantic Treaty Organization) ally. She arrived at the Naval Operating Base, Norfolk, on July 10, remaining there for only a short time before shifting to the Norfolk Naval Shipyard on July 15 for availability.

America operated locally in the Norfolk area from August 29 to September 19, 1966, after which time she proceeded to Guantanamo Bay to carry out training. After Hurricane "Inez" swirled through the region, her sailors spent an estimated 1,700 man-hours helping the naval base at Guantanamo recover and return to normal operations.

The following month, America initiated into carrier service the Ling-Temco-Vought A-7A Corsair II, conducting its flight qualifications off the Virginia Capes, while she also conducted automatic carrier landing system trials that demonstrated the feasibility of "no hands" landings of McDonnell-Douglas F-4 Phantom and Vought F-8 Crusader aircraft.

From November 28 to December 15, 1966, America took part in "LANTFLEX 66," gaining experience in the areas of anti-air, anti-submarine, and carrier strike operations. The ship also participated in a mine drop, missile shoots, and provided air support for amphibious operations. She returned to NOB Norfolk on December 15, remaining there through the end of the year 1966.

On January 10, 1967, America departed Norfolk for her second Mediterranean cruise and relieved USS Independence (CV-62) at Pollensa Bay on January 22. While crossing the Atlantic, America conducted carrier qualifications for her SH-3A crews, missile shoots in the mid-Atlantic, day and night air operations, and various other exercises. Upon nearing Gibraltar, she received a visit from Soviet long-range reconnaissance aircraft, Tupolev TU-95 Bears on January 18. Two F-4B Phantom jets met the Bears as they approached and escorted them past the ship.

Before anchoring at Athens, on February 4, America participated with Italian control and reporting centers in an intercept-controller exercise. Shortly afterwards, America again met with Italian forces in an exercise involving raids upon an attack carrier by fast patrol boats.

In March 1967, America and her consorts, operating as Task Group (TG) 60.1, participated in the United States/United Kingdom Exercise Poker Hand IV with the British aircraft carrier HMS Hermes. America and Hermes provided raid aircraft to test each other's anti-aircraft defenses.

On April 1, Dawn Clear, a two-day NATO exercise, commenced with TG 60.1 units participating. During the first day, America provided raid aircraft against Greek and Turkish "targets." The following day, the exercise continued as Greek aircraft flew raids against TG 60.1 surface units. Following Dawn Clear, the ship conducted routine training operations in the Ionian Sea.

America anchored at Valletta at 1000 on April 5, 1967, for a five-day visit. Weighing anchor on April 10, the carrier departed Malta to sail for task group operations in the Ionian Sea. She conducted an open sea missile exercise with the guided missile destroyers USS Josephus Daniels (DLG-27) and USS Harry E. Yarnell (DLG-17). Other operational aspects of the at-sea period consisted of routine day/night flight operations and a major underway replenishment with other units of TG 6 0.1. The following days saw the threat of civil war in Greece commencing with the military coup that ended parliamentary rule in that country. Although King Constantine II held his throne, the possibility of violence in the streets of Athens loomed as a potential threat to the American citizens suddenly caught up in the turmoil. It seemed that evacuation by ship might be necessary and Commander, 6th Fleet, ordered the formation of a special operations task force.

Under the command of Rear Adm. Dick H. Guinn, TF 65, with America as flagship, sailed eastward to stand by for evacuation, should that step be necessary. Fortunately, violence never materialized in Greece, and the task force was not called upon to act. On April 29, 1967, Rear Adm. Lawrence R. Geis relieved Rear Adm. Guinn as Commander, CarDiv 4, Commander, TF 60, Commander, TF 65, and Commander, TF 502 (NATO). With a new admiral on board, and the Greek political crisis behind her, America sailed into Taranto Harbor, Italy, on the first day of May for eight days of relaxation. During three days of general visiting in Taranto, America hosted 1,675 visitors who came aboard to tour the hangar and flight decks. America departed Taranto on May 8 for routine task group operations in the Ionian and Tyrrhenian Seas, she followed these with a port visit to Livorno.

By May 25, there was evidence that a crisis was brewing in the Middle East. America's crew, from reading the ship's paper, The Daily Eagle, could see that tensions between Israel and the

Arab States had been rising fast. As soon as the ship was slated to finish with the last of her "Poop Deck" exercises, she would be heading back to the Sea of Crete.

For the next 48 hours, America steamed east and south from the coast of Spain, through the Malta Channel, and on to the Sea of Crete to join up with the ships of TG 60.2, the carrier USS Saratoga (CVA-60) and her destroyers. The carrier task force, under the command of Rear Adm. Geis, prepared for any contingency.

For the next week, the officers and men of America listened to the nightly news report over WAMR-TV, the carrier's closed-circuit television station, and read every bit of news in The Daily Eagle. Headlines told of a worsening situation. First, Egypt moved troops into the Gaza Strip, demanding that the United Nations (UN) Peacekeeping Force be withdrawn. Then, Israel beefed up her forces and, in turn, each of the other Arab countries put their armed forces on alert. As war clouds darkened, the United Arab Republic closed the Gulf of Aqaba to Israeli shipping.

During this time, the carrier conducted normal training operations off the island of Crete and held two major underway replenishment operations. On June 5, 1967, seven American newsmen from the wire services, the three major American television networks, and several individual newspapers across the country flew on board. These seven were soon joined by others, 29 in all including media representatives from England, Greece, and West Germany.

Their presence was evident everywhere on board the carrier. They lined the signal bridge and the flight deck, their cameras recording the cycle of flight operations, refuelings, and the tempo of shipboard routine. At night, Robert Goralski of NBC News and Bill Gill of ABC News teamed up to present the WAMR "Gill-Goralski Report," a half-hour on the latest developments in the Middle East and around the world. America's presence was soon noted, and the carrier soon attracted other, less welcome, visitors. A Soviet destroyer had joined up on the morning of June 2, 1967. Armed with surface-to-air guided missiles, the Russian ship constantly cut in and out of the carrier's formation.

Shortly after noon on June 7, Vice Adm. William I. Martin, Commander of the 6th Fleet, sent the Soviet ship a message in both Russian and English: "Your actions for the past five days have interfered with our operations. By positioning your ship in the midst of our formation and shadowing our every move, you are denying us the freedom of maneuver on the high seas that has been traditionally recognized by seafaring nations for centuries."

"In a few minutes," the message continued, "the task force will commence maneuvering at high speeds and various courses. Your present position will be dangerous to your ship as well as to the ships of this force. I request that you clear our formation without delay and discontinue your interference and unsafe practices." Although that particular Soviet guided missile destroyer left America alone, her sister ships soon arrived to dog them for days, harassing the carrier and her escorting destroyers.

On the morning of June 5, while America was refueling from the oiler USS Truckee (AO-147), with the CarDiv 4 band and the "rock 'n' roll" combo of Truckee playing against one another, the word came that the Israelis and the Arabs were at war. That afternoon, the bosun's pipe called the crew to a General Quarters drill, and the excitement of the moment was evident as all hands rushed to their battle stations. When General Quarters was secured, the word was passed over the 1-MC, the ship-wide general announcement system, to set Condition Three, an advanced state of defensive readiness.

On June 7, 1967, the destroyer USS Lloyd Thomas (DD-764), in company with America, obtained a sonar contact, which was classified as a "possible" submarine. Rear Adm. Geis immediately dispatched Lloyd Thomas and the guided missile destroyer USS Sampson (DDG-10) to investigate the contact. Sampson obtained contact quickly and coordinated with Lloyd Thomas in tracking the possible submarine. America launched

one of her antisubmarine helicopters, a Sikorsky SH-3A Sea King of Helicopter Antisubmarine Squadron Nine (HS-9), and gained sonar contact. At midnight, the contact was reclassified as a "probable" submarine. At that time, no known or friendly submarines were reported to be in the area of the contact. The destroyers maintained good sonar contact through the night.

At 0530 on June 8, a Lockheed SP-2H Neptune antisubmarine patrol plane of Patrol Squadron Seven (VP-7), coordinating with the destroyers and helicopters, obtained a magnetic anomaly detector (MAD) confirmation over the contact. The MAD equipment allows an ASW aircraft to confirm that a contact detected in the sea by other means is actually a very large metal object.

Rear Adm. Geis announced the "probable" submarine's presence at noon. The newsmen, still embarked, dashed off stories to their home offices. Other events, however, would soon overshadow the story about a 'probable' sub lurking near an American carrier task force.

At about 1400 local time, on June 8, 1967, the technical research ship USS Liberty (AGTR-5) was attacked by Israeli torpedo boats and jet fighters, approximately 15 miles north of the Sinai port of El Arish, in international waters. She had been in position to assist in communications between United States diplomatic posts in the Mideast and to aid in the evacuation of American dependents from the area if necessary.

However, the first word that reached America and the Department of Defense in Washington gave no indication as to the identity of the attackers. America's flight deck came alive. In a matter of minutes, F-4B Phantom interceptors were in the air to ward off any possible attack against task force units. At the same time, bombs and rockets moved from the magazines deep within the ship to the flight deck. Four Douglas A-4 Skyhawk attack bombers were loaded and launched together with fighter cover. As the planes sped towards Liberty's position, word was received from Tel Aviv that the attackers had been Israeli and that the attack had been made in error. The planes outbound from America were recalled with their ordnance still in the racks. The attack on Liberty had cost the lives of 34 men, with 75 wounded, 15 seriously. Adm. Martin dispatched two destroyers, USS Davis (DD-937) and USS Massey (DD-778), with Lt. Comdr. Peter A. Flynn, MC, USN, one of America's junior medical officers, and two corpsmen from the carrier on board. The destroyers rendezvoused with Liberty at 0600 on June 9, 1967, and the medical personnel, including a second doctor from one of the destroyers, were transferred immediately to the damaged research ship.

At 1030, two helicopters from America rendezvoused with Liberty and began transferring the more seriously wounded to the carrier. An hour later, about 350 miles east of Souda Bay, Crete, America rendezvoused with Liberty. The carrier's crew lined every topside vantage point, silently watching the helicopters bring 50 wounded and nine dead from Liberty to America. As Liberty drew alongside, listing, her sides perforated with rockets and cannon shells, nearly 2,000 of the carrier's crew were on the flight deck and, spontaneously moved by the sight, gave the battered Liberty and her brave crew a tremendous cheer.

America's medical team worked around the clock, removing shrapnel and treating various wounds and burns. Doctors Gordon, Flynn, and Lt. Donald P. Griffith, MC, worked for more than 12 hours in the operating room, while other doctors, Lt. George A. Lucier and Lt. Frank N. Federico, made continuous rounds in the wards to aid and comfort the wounded. Their jobs were not finished that day; for the next week and more, Liberty's wounded required constant attention. Since the fighting had started between the Israelis and the Arabs, a weary quiet had settled over the carrier's flight deck. Ready, the ship waited for any possible situation, but the planes never left the decks.

However, as the Israeli forces moved to a speedy victory in the "Six-Day War," the Arabs charged that 6th Fleet aircraft were providing air cover for Israeli ground forces. As witnessed and reported by the newsmen on board, these charges were completely false. The 6th Fleet, as with all other American forces, had remained neutral. On Wednesday morning, June 7, 1967, Adm. Martin issued a statement to the press: "It would have been impossible for any aircraft from the 6th Fleet to have flown the support missions alleged by various Middle Eastern spokesmen. No aircraft of the 6th Fleet have been within a hundred miles of the eastern coast of the Mediterranean, specifically Israel and the UAR. Furthermore, no 6th Fleet aircraft has entered the territorial airspace of any Middle Eastern or North African nation during the current period of tension."

Adm. Martin gave members of the press copies of both America's and Saratoga's flight plans for the days in question and a rundown of the task force's position at all times during the conflict. He pointed out that a check of the carriers' ordnance inventory would refute the charges, that both the number of pilots and aircraft embarked had changed only with the return of personnel and planes from the Paris Air Show. America conducted a memorial service on June 10 on the carrier's flight deck. The oft-repeated words of the Navy Hymn, "those in peril on the sea," echoed across the wind-swept deck, possessing poignant meaning for those who were aware of Liberty's travail.

As Israeli forces advanced towards the Suez Canal and the Jordan River, and appeals for a cease-fire came, the tension relaxed aboard ship. The crew took time out for an 11-bout boxing smoker in the hangar bay. With a running commentary by the Gill-Goralski team, nearly 2,000 crew members crowded around the ring while others watched the action over closed-circuit television. America continued on station for several more days, but the tension seemed to have gone. The newsmen left, the uninvited Soviet guests called no more, and regular flight operations resumed. During the crisis, the presence of America and the 6th Fleet had demonstrated once again the power, mobility, and flexibility of sea power.

On a lighter note, during the same period, other activities were happening aboard ship and in Paris, France. Two squadrons of CVW-6 participated in the 27th Paris Air Show held at the French capital's Le Bourget Airport from May 25 to June 5, 1967. A Fighter Squadron Thirty-Three (VF-33) F-4B Phantom and an Early Warning Squadron One Two Two (VAW-122) Grumman E-2A Hawkeye were on display at the airfield throughout the show.

America next hosted, commencing on June 14, 49 midshipmen from the United States Naval Academy and Naval Reserve Officer Training Corps (NROTC) units across the country. For six weeks, the "middies," under the watchful eyes of the ship's officers, filled junior officer billets in all of the departments in the ship. In late July, the second group of 41 "middies" arrived for their six-week cruise.

America transited the Dardanelles on June 21, 1967, and arrived at Istanbul, where Rear Adm. Geis laid a wreath at the foot of the grave of the Unknown Soldier as a tribute to the Turkish war dead. Three days later, however, a group of angry demonstrators burned the wreath.

Then, approximately 600 students, with 1,500 spectators and sympathizers, participated in an anti-American/6th Fleet protest march, culminating in speeches in the area of the fleet landing. Liberty for the crew was canceled for most of the afternoon; however, by early evening, the situation had quieted down enough for liberty to be resumed. All was peaceful for the remainder of the visit. America departed Istanbul on June 26 for five days of operations in the Aegean Sea. On July 1, the carrier steamed into the port of Thessaloniki, Greece, for her first visit to that port. For Independence Day celebrations aboard ship, Rear Adm. Geis and America's commanding officer, Capt. Donald D. Enen, hosted the Prefect of Thessaloniki, the Mayor of Thessaloniki, the American Consul, and approximately 75 Greek Army officers and civilians. On July 8, Rear Adm. Daniel V. Gallery, USN (Ret.), arrived on board via "COD" (Carrier Onboard Delivery) aircraft. Adm. Gallery was visiting as many 6th Fleet ships as possible during his month-long stay in the Mediterranean to gather material for articles and books. He also departed by COD on July 9. On July 16, America anchored at Athens for her second visit to that port of the 1967 cruise, before she proceeded to Valletta on July 29. On August 7, America anchored in the Bay of Naples. After visits to Genoa and Valencia, the carrier sailed into Pollensa Bay and commenced the turnover of her 6th Fleet materials to her relief, the attack carrier USS Franklin D. Roosevelt (CVA-42).

America moored at Pier 12 Naval Station, Norfolk, on September 20, 1967, and entered the Norfolk Naval Shipyard on October 6. She remained there, undergoing a restricted availability, into early January 1968. From January 6 to 8, the ship steamed for three days of sea trials in the Virginia Capes operating area. After a four-day ammunition onload at Anchorage X-ray in Hampton Bay and a brief stay at Pier 12, NOB, Norfolk, America departed for a month-long cruise to the Caribbean for the Naval Technical Proficiency Inspection (NTPI), refresher training with the Fleet Training Group, Guantanamo Bay, and type training in the Atlantic Fleet Weapons Range (AFWR) before she could proceed to the Jacksonville Operating Area for carrier qualifications. America departed Norfolk on January 16, 1968. Upon arrival at Guantanamo Bay soon thereafter, the ship conducted extensive drills and exercises, and inspections were conducted in almost all shipboard activities. General Quarters was a daily routine as the ship strove to reach the peak of proficiency required for its upcoming combat deployment to the western Pacific (WestPac).

On February 1, America departed the Guantanamo area, bound for the AFWR. The next day, February 2, representatives from the AFWR came on board to brief America representatives and Carrier Air Wing (CVW) 6 pilots on forthcoming operations. The training consisted of invaluable and highly successful exercises in environmental tracking, antimissile defense, airborne jamming against radars, emergency aircraft recovery, and simulated PT boat attacks.

With this phase of her combat training completed, America departed the AFWR on February 9 for carrier qualifications in the Jacksonville Operating Area, and held them from the 12th through the 15th.

On February 17, 1968, America moored at berths 23 and 24 at Norfolk Naval Shipyard to prepare for final type training, prior to her upcoming WestPac deployment. On March 7, America again put to sea, back to the AFWR for further type training and Exercise Rugby Match. En route to the Caribbean, the ship held various exercises in weapons loading, electronic countermeasures (ECM), and General Quarters. On March 10, America flew off the first of eight simulated air strikes. America's CVW flew "attack" sorties against "enemy" positions on Vieques Island, near Puerto Rico. A search and rescue exercise (SAREX) was conducted to test the ship and air wing response to the distress call of a downed aviator. She also held several missile defense exercises to test the ship's reflexes against a surface threat.

America's planes flew photographic reconnaissance sorties over Vieques and "found" simulated targets on film. Communications exercises simulated conditions in Tonkin Gulf, as a high volume of message traffic similar to that to be experienced in Southeast Asia was generated by Commander,

CarDiv 2, who was embarked on the ship. On March 13 and 14, the weapons department also flexed their muscles by firing two Terrier missiles.

Exercise Rugby Match, a major Atlantic Fleet exercise involving approximately eighty ships, was held in the AFWR from March 7 to 29. America and Commander, CarDiv 2 (as commander, Task Group (TG)) 26.1, participated from the 18th to the 20th. As the "Blue" Force attack carrier, America and her air wing pilots provided close air support (CAS), photo reconnaissance, and combat air patrol (CAP) sorties for Task Force (TF) 22, the "Blue" amphibious landing force, during a landing on the island of Vieques. Prior to America's main participation during this period, CVW-6 flew an aerial mining mission in the amphibious operating area on the 15th. D-day was March 19. On return from their missions as CAS and CAP, several aircraft tested the antiaircraft defenses of the task force by flying raids against America.

America moored at Pier 12 NOB, Norfolk, at 1315 on March 23, 1968. Two days later, on the 25th, she put to sea again for a dependents' cruise. Then, on the dark, rainy afternoon of April 10, America stood out of Hampton Roads, bound for "Yankee Station," half a world away. The next day, the ship's complement of men and machines was brought up to full strength as America recovered the remainder of CVW-6's aircraft off the coast of the Carolinas. En route, she conducted one last major training exercise. Rio de Janeiro, Brazil, was the next stop en route to Southeast Asia, America's first to that city and continent. Now, with her course set almost due east, America sailed through waters she had never traveled before. Across the southern Atlantic, around the Cape of Good Hope, past Madagascar, and out into the broad expanse of the Indian Ocean towards the Sunda Strait and Subic Bay, Philippine Islands. From Subic, the ship sailed northwest through the South China Sea towards "Yankee Station."

En route, on May 26, 1968, the ship participated in Exercise New Boy and the next day held carrier qualifications. At 1000 on May 30, she arrived at "Yankee Station," and at 0630 the next morning, the first aircraft since commissioning to leave her deck in anger was launched against the enemy.

During four line periods, consisting of 112 days on "Yankee Station," America's aircraft pounded at roads and waterways, trucks and waterborne logistics craft (WBLCS), hammered at petroleum storage areas and truck parks, and destroyed bridges and cave storage areas in the attempt to impede the flow of men and war materials to the south. On July 10, 1968, Lt. Roy Cash, Jr. (pilot) and Lt. (j. g.) Joseph E. Kain, Jr. (radar intercept officer), in an F-4 Phantom from VF-33, downed a MiG-21, 17 miles northwest of Vinh, North Vietnam, for the ship's first MiG "kill" in the Vietnam War. America and her embarked air wing, CVW-6, would later be awarded the Navy Unit Commendation for their work during that time.

Between line periods, America visited Hong Kong, Yokosuka, and Subic Bay. With America's mission on "Yankee Station" nearing completion, she launched the last of her attack aircraft at 1030 on October 29. The next day, she set sail for Subic Bay and the offload of various "Yankee Station" assets. In addition, a heavy attack squadron, VAH-10, and an electronic countermeasures squadron, VA-130, departed the ship on November 3 as they began a trans-Pacific movement of their entire detachments to Alameda, and 144 aviators along with several members of the ship's company departed for the United States on the "Magic Carpet" flight.

The days the ship spent en route to Australia, New Zealand, Brazil, and Norfolk were, of necessity, more relaxed than those of her six months of combat. Nine hundred ninety-three "Pollywogs" were initiated into the realm of Neptunus Rex on the morning of November 7th as the ship again crossed the Equator. On November 9, a flight deck "cookout" was sponsored by the supply department as the entire crew enjoyed char-broiled steaks and basked in the equatorial sun. After mooring at 1330 on December 16, 1968, at Pier 12, Norfolk, her "round-the-world" cruise completed, post-deployment and holiday leave began, continuing through the first day of the year 1969. Shortly thereafter, on January 8, 1969, she headed for the Jacksonville Operating Area where she served as the platform for carrier qualifications.

On January 24, America arrived at the Norfolk Naval Shipyard to begin a nine-month overhaul. Upon completion of the overhaul, the carrier conducted post-repair trials and operated locally off the Virginia Capes. During one period of local operations, between November 21 and 23, 1969, America took part in carrier suitability tests for the Lockheed U-2R reconnaissance plane.

On January 5, 1970, the carrier departed the Norfolk area to commence a nine-week cruise in the Guantanamo Bay operating area. From February 15 to 21, America participated in Operation Springboard 70, the annual series of training exercises conducted in the Caribbean. The program was established to take advantage of good weather and the extensive modern training facilities, including targets of all kinds, which are available to achieve maximum training during the period. This exercise included submarine operations, air operations, and participation by the Marine Corps. At the completion of this testing and training, America departed the Guantanamo area to arrive at the Jacksonville area on March 1 in order to conduct carrier qualification landings with the various squadrons stationed in and around the Jacksonville/Cecil Field area.

America arrived at NOB, Norfolk, on March 8, 1970, and remained there for approximately one month making last-minute preparations for an eight-month deployment.

On April 10, 1970, with CVW-9 on board, America left Norfolk and paused briefly in the Caribbean Sea for an operational readiness inspection (ORI) before proceeding on a voyage that took her across the equator to Rio de Janeiro, around the Cape of Good Hope, across the Indian Ocean, into the Pacific Ocean, and finally to Subic Bay in the Philippines.

On May 26, America began its first day of special operations in the Gulf of Tonkin, when Comdr. Fred M. Backman, commanding officer of VA-165, and his bombardier/navigator, Lt. Comdr. Jack Hawley, in a Grumman A-6C Intruder, flew the ship's first combat sortie of the 1970 WestPac cruise. On the same

day, the Navy's newest light attack aircraft, the A-7E Corsair II, received its first taste of combat. At 1201, Lt. (j. g.) Dave Lichterman of VA-146 was catapulted from the deck in the first A-7E ever to be launched in combat. He and his flight leader, Comdr. Wayne L. Stephens, the squadron's commanding officer, subsequently delivered their ordnance with devastating accuracy using the A-7E's digital weapons computer. Shortly after 1300, Comdr. R. N. Livingston, skipper of the "Argonauts" of VA-147, and Lt. Comdr. Tom Gravely rolled in on an enemy supply route to deliver the first bombs in combat in an A-7E, reportedly "all on target."

For five line periods, consisting of 100 days on "Yankee Station," America's aircraft pounded at roads and waterways, trucks and waterborne logistics craft (WBLC), hammered at petroleum storage areas and truck parks, in an attempt to impede the flow of men and war materials to the south.

On August 20, 1970, at Manila, Vice Adm. Frederic A. Bardshar, Commander, Attack Carrier Striking Force, 7th Fleet, hosted the President of the Philippines, Ferdinand E. Marcos, on board America. President Marcos was given a 21-gun salute as he and Mrs. Marcos arrived on board from their Presidential yacht to visit the ship. Accompanied by American Ambassador and Mrs. Henry A. Byiade, they were greeted by Vice Adm. Bardshar and America's commanding officer, Capt. Thomas B. Hayward, and were subsequently escorted to the ship's hangar deck where the carrier division band and the ship's Marine detachment rendered honors. Following their arrival, the visiting party dined with Vice Adm. Bardshar and Capt. Hayward, and were later given a brief tour of the ship.

On September 17, America completed her fourth line period and headed for special operations off the coast of Korea

and subsequently, the Sea of Japan. On September 23, the carrier entered the Tsushima Straits, remained in the Sea of Japan for approximately five days, and exited on September 27 through the Tsugaru Strait. During this period, America and CVW-9 engaged in three exercises: Blue Sky, with elements of the Chinese Air Force from Taiwan; Commando Tiger, conducted in the Sea of Japan, involving air units of the Republic of Korea (ROK) Air Force (ROKAF); and, after exiting the Tsugaru Straits, Autumn Flower, air defense exercises with the Japanese Air Self Defense Force (JASDF) and the United States Fifth Air Force. On November 7, 1970, America completed her fifth line period and departed for her last visit to Subic Bay. Through five line periods, the carrier had flown off 10,600 sorties (7,615 combat plus combat support), 2,626 actual combat sorties, completed 10,804 carrier landings, expended 11,190 tons of ordnance, moved 425,996 pounds of cargo, handled 6,890 packages, and transferred 469,027 pounds of mail. She had accomplished this without a single combat loss and only one major landing accident, fortunately with no fatalities. Considering sustained combat operations in prevailing immoderate weather and highly successful 7th Fleet exercises without one day's loss in operations due to any material casualty, America left the Pacific Ocean justifiably proud of her accomplishments.

On the long trip home, America welcomed approximately 500 more "Pollywogs" into the realm of "Neptunis Rex." The day before the carrier arrived at Sydney, Australia, for a three-day rest and recreation visit, United States Ambassador to Australia and his wife, the Honorable and Mrs. Walter L. Rice, flew on board to accompany the ship into Sydney.

With so much to be thankful for, America celebrated two Thanksgivings. At exactly 2329 on November 26, America crossed the International Date Line. Moments later it became Thanksgiving Day again. On both days, crew members feasted on turkey, beef, lobster tails, Virginia ham, and roast duckling.

After rounding Cape Horn on December 5, 1970, America headed north, stopped briefly at Rio de Janeiro for fuel, and arrived at Pier 12, NOB Norfolk, on December 21. She remained there until January 22, 1971, when the ship entered the Norfolk Naval Shipyard for a three-month restricted availability. She departed the yard, on schedule, on March 22. Over the ensuing weeks, the ship operated locally in the Virginia Capes operating areas. She then carried out exercises in Puerto Rican waters, with United States Navy as well as Royal Navy warships—including HMS Ark Royal (R-09), HMS Cleopatra (F-28), and HMS Bacchante (F-69).

After a return to Norfolk, America stood out of Hampton Roads on July 6, 1971, for the Mediterranean. On July 16, 1971, America dropped anchor at Rota, Spain, in order to receive her turnover information from the ship she was relieving on station, USS Franklin D. Roosevelt. America then entered the Mediterranean for the third time since her commissioning. Between the time the ship left Rota and reached Naples, she participated in three major exercises.

Following a port call at Naples, America proceeded on a course toward Palma, Mallorca. While en route, she participated in "PHIBLEX 2-71," in which she covered a mock amphibious landing at Capoteulada, Sicily. After a port visit at Palma, Mallorca, America participated from August 16 to 27 in National Week X, one of the largest exercises conducted in the Mediterranean. During the exercise, on August 26, an E-2B Hawkeye was flown by VAW-124 nonstop across the Atlantic. The Hawkeye left NAS Norfolk, Va., flew over Newfoundland, Canada, and Lajes, Azores, to reach America in the Med. At the termination of the exercise, America proceeded to Corfu, Greece, her next liberty port. She then visited Athens shortly thereafter.

After conducting routine operations in the eastern Mediterranean and making a port call at Rhodes, Greece, the ship proceeded to the Aegean Sea to participate in Operation Deep Furrow 71, with America and CVW-8 providing close air support for almost the entire exercise.

Proceeding thence to Thessaloniki, Greece, for a port visit, America then participated in National Week XI, in the central Mediterranean. The carrier subsequently visited Naples before she steamed into the western Mediterranean to participate in exercises with British, Dutch, Italian, and French forces in Exercise Ile D'Or, completing her part in the evolutions by November 19. America then conducted port visits to Cannes and Barcelona before proceeding to Rota. There, on December 9, 1971, she was relieved on station by USS John F. Kennedy (CVA-67).

Arriving back at Norfolk on December 16, America moored at Pier 12, NOB, Norfolk, for post-deployment stand-down before unloading ammunition in preparation for availability at the Norfolk Naval Shipyard. After the two-month overhaul, the carrier conducted sea trials. Soon thereafter, America embarked on a program of training, accelerated due to the fact that the date of her deployment had been advanced one month, and participated in Exercise Exotic Dancer V. She returned to Pier 12, Naval Station Norfolk, upon conclusion of the exercises.

On June 2, 1972, three days before America was to sail, Adm. Elmo R. Zumwalt, the Chief of Naval Operations, visited the ship and explained the reason why her orders had been changed, sending her to the Gulf of Tonkin instead of the Mediterranean. Sailing on June 5, America crossed the equator on June 12 and held the usual initiation of "Pollywogs" into the realm of Neptune.

Escorted by USS Davis (DD-937) and USS Dewey (DD-937), and accompanied by the fleet oiler USS Waccamaw (AO-109), America proceeded toward Southeast Asia and rounded Cape Horn on June 21. Joining the 7th Fleet later in June, America relieved the attack carrier USS Coral Sea (CVA-43) on station and commenced combat operations on July 12. A ruptured main feed pump, however, prompted an early return to Subic Bay on July 25 for repairs, the ship arriving in the Philippines during a time of natural devastation—floods and landslides.

The repair work was delayed for two weeks while needed parts were rushed to Subic Bay. America stood out on August 9, 1972, to return to the line and soon resumed carrying out strike operations against Communist targets in North Vietnam. On September 11, VMFA-333 flying off America downed a MiG-21 near Phuc Yen airfield in North Vietnam. This was the only MiG kill for the Navy/Marine Corps during September and brought the grand total of MiGs downed by Navy/Marine Corps pilots to 55 since the war began. On October 6, bombs from America's planes dropped the Thanh Hoa Bridge, a major objective since the bombing of the North had begun years before. Completing her line period and stopping over briefly at Subic Bay, America steamed to Singapore, departing that port on October 20 to resume operations on "Yankee Station." On October 23, the U.S. ended all tactical air sorties into NVN above the 20th parallel and brought to a close Linebacker I operations. This goodwill gesture of terminating the bombing in NVN above the 20th parallel was designed to help promote the peace negotiations being held in Paris, France. During May through October, the Navy flew a total of 23,652 tactical air attack sorties into NVN. U.S.

tactical air sorties during Linebacker I operations helped stem the flow of supplies into NVN, thereby limiting the operating capabilities of North Vietnam's invading army.

Carriers involved in Linebacker I operations were America, USS Enterprise (CVAN 65), USS Constellation (CVA 64), USS Coral Sea (CVA 43), USS Hancock (CVA 19), USS Kitty Hawk (CVA 63), USS Midway (CVA 41), USS Saratoga (CVA 60), and USS Oriskany (CVA 34).

On October 28, 1972, America lost a pilot. Cmdr. James W. Hall took off in his A-7C Corsair on a surface-to-air missile suppression mission. Over the target area in Nghe An province, North Vietnam, Cmdr. Hall was heard to radio to his wingman, "Two SAMs (surface-to-air missiles) lifting at 12 o'clock." No other radio messages were heard. The first missile missed his wingman, but the second struck Hall's aircraft. No parachute was observed, and no emergency radio beepers were heard. Less than a month later, a fire broke out on board America at 1410 on November 19, 1972, in the number two catapult spaces. The ship went to General Quarters as smoke began to fill the 03 level, and damage control parties soon had the blaze extinguished. Clean-up and repair work ensued, and despite not having the services of one of her catapults, America remained on the line and continued to meet her commitments. After an extended line period of 43 days, America reached Subic Bay on December 2, where the number two catapult was repaired, and departed the Philippines on December 8 to return to "Yankee Station." A week before Christmas, America learned that the breakdown of peace talks in Paris had led to a resumption of bombing of targets in North Vietnam. America swung into action, and the pace proved hectic until the Christmas cease-fire.

"Christmas away from home is never good," America's historian wrote, "but the men of America made the best of it with homemade decorations." There were services to celebrate the season, "and carolers were noted strolling through the passageways ...." On December 28, 1972, the carrier anchored in Hong Kong harbor and remained there until January 4, 1973, when she stood out for the Philippines and the period of rest and repairs at Subic Bay that would precede the ship's return to the line. All hands avidly followed the progress of the peace talks as America returned to "Yankee Station," and resumed operations. After two weeks on the line, the ship learned that peace had been secured and that an agreement was to be signed in Paris. At 0800 on January 28, 1973, the Vietnam War—at least that stage of it—was at an end. Rumors swept the ship that her deployment would be shortened because of the cessation of

hostilities, and hope ran high as the ship moored at Subic Bay on February 3.

America did return to "Yankee Station" one last time, but her time on station proved short. America was ordered to depart the Far East for the U.S. This was the initial move in reducing the number of carriers serving in Southeast Asia from six to three by mid-June 1973. She returned to Subic Bay on February 17, 1973, and sailed thence for the United States three days later, on February 20. The carrier arrived at Mayport, Fla., disembarking men from CVW-8 and embarking the teenage sons of some of the ship's company officers and men, thus allowing them to ride the ship back to Norfolk with their fathers, something thoroughly enjoyed by all who took part.

On March 24, 1973, America arrived back at NOB, Norfolk, mooring at Pier 12 and bringing to a close her sixth major deployment since commissioning. She immediately began preparations for a 30-day stand-down and the restricted availability to follow at the Norfolk Naval Shipyard. She entered the yard on May 11 and emerged after that period of repairs and alterations on August 10. America conducted local operations out of Norfolk into October, and during this period the ship celebrated a significant milestone in the life of a carrier: she logged her 100,000th landing on August 29, 1973, when her COD aircraft (nicknamed "Miss America"), piloted by Lt. Comdr. Lewis R. Newby and Lt. Comdr. Ronnie B. Baker, landed on board. Cake cuttings on the hangar deck and in the wardroom celebrated the occasion.

On October 29, America cleared Hampton Roads for Jacksonville and a period of carrier qualifications. She was conducting routine training operations on November 1, 1973, when she went to the assistance of the crippled sailing schooner Harry W. Adams of Nova Scotia. The 147-foot schooner, her engine disabled and without power for her pumps, was taking on water. Helicopters from America sped to the scene, and the ship provided rescue specialists and underwater demolition experts to assist in the effort. The ship's captain and his crew of nine all escaped serious injury, although the carrier's helicopters brought three of the crew on board for medical examinations and a warm meal. America stood by until the late afternoon, when the Coast Guard cutter USCG Port Roberts arrived to assist Harry W. Adams into port at Jacksonville. After concluding her operations in the Jacksonville area, America paid a port call at Ft. Lauderdale, Fla., from November 4 to 8, 1973. She proceeded thence to sea for exercises of various kinds to hone the skills of the ship-air wing team and, following her operational readiness inspection off Mayport, proceeded back to Norfolk, mooring at Pier 12, NOB, on November 21. America then steamed south after the Thanksgiving holiday for Atlantic Fleet readiness exercises, returned via Mayport to Norfolk on December 13, and remained in her home port until sailing for the Mediterranean on January 3, 1974.

Relieving Independence at Rota, Spain, on January 11, she became the flagship for Rear Adm. Frederick C. Turner, Commander, TF 60. America commenced operations in the western Mediterranean that day and, over the next few weeks, divided her time between at-sea periods and port visits to Toulon, Barcelona, and Valencia. From February 15 to 19, the carrier participated in Exercise National Week XVI, and upon the conclusion of that evolution, anchored in Souda Bay, Crete. She proceeded thence for a port call at Athens.

Standing out of the waters of that Greek port on March 1, 1974, America participated in "PHIBLEX 9-74," in which the ship's air wing, CVW, practiced supporting an amphibious landing. The carrier then operated north of Crete on exercises in early April, after which time she put into Athens on April 9.

America then participated in the NATO exercise, Dawn Patrol, in which units of the navies of the United States, United Kingdom, Portugal, Holland, France, Italy, and West Germany took part. During one phase of this exercise, the carrier's Marine detachment embarked on USS El Paso (LKA-117) and stormed ashore from that amphibious ship while America's planes provided close air support. Upon the conclusion of Dawn Patrol, the carrier paid another visit to Athens, proceeding on May 19 for a four-day period of exercises, after which she steamed to Istanbul, arriving there on May 23.

Immediately following this port call, the ship returned to Athens and sailed for Exercise Shahbaz to test the air defense capability of NATO ally Turkey early in June. America then anchored off the island of Rhodes, Greece, on June 6 for a four-day port visit, after which she returned to Athens to embark Naval Academy midshipmen for their summer training cruise. America then participated in Exercise Flaming Lance, off the coast of Sardinia, during which USS Leahy (DLG-16) controlled over 1,000 intercepts by America's aircraft.

Making her last port call at Athens for the deployment, the carrier steamed to Souda Bay on July 1, loading minesweeping equipment that had been used in Operation Nimbus Star, the clearance of the Suez Canal. America then proceeded to Corfu and began the transit out of the eastern Mediterranean on July 6, 1974, arriving at Palma, Mallorca, three days later. America anchored off Rota on July 15, for what was scheduled to have been an off-load of the equipment of Commander, TF 60, staff. Clashes between Greek and Turkish forces on Cyprus, however, prompted the Joint Chiefs of Staff to order America to remain at Rota until the arrival of her relief, USS Independence (CVA-62), on July 28. As soon as that attack carrier entered the 6th Fleet operating area, America commenced her homeward voyage, ultimately reaching Pier 12, Naval Station Norfolk, on August 3.

A little over a month later, America sailed for the North Sea to participate in a NATO exercise, Northern Merger, departing Norfolk on September 6, 1974. America joined with HMS Ark Royal in providing air support for a NATO task force and for an amphibious landing. Throughout the exercise, Soviet surface units, as well as Bear and Badger aircraft, conducted surveillance missions over and near the NATO force. Upon the conclusion of Northern Merger, America steamed to Portsmouth, England, arriving there on September 29 to commence a five-day port visit. The carrier then proceeded back to the United States, reaching Pier 12, NOB, Norfolk on October 12, to commence preparations for a major overhaul at the Norfolk Naval Shipyard. Entering the yard on November 27, 1974, America remained there until September 27, 1975, when the ship got underway to conduct post-overhaul sea trials. America departed Norfolk Naval

Shipyard on October 16, 1975, for local operations off the Virginia capes and, after a few weeks alongside her familiar berth, Pier 12, NOB, Norfolk, departed Hampton Roads for Cuban waters and refresher training. While steaming north of Cuba and preparing for the operational readiness inspection that concluded refresher training, America picked up distress calls and immediately deployed helicopters and fixed-wing aircraft to search for a disabled motorized sailboat, Ruggentino. One of the carrier's helicopters located a boat in distress and guided a tug to the scene, which soon took the disabled craft in tow.

That boat, however, proved to be named Content, so America and her aircraft resumed the search for Ruggentino. One of America's planes located the craft in question soon thereafter, and the ship dispatched a motor whaleboat to assist. America's sailors soon had the boat pumped out and headed for port. This effort, two successful search-and-rescue missions in one night under adverse weather conditions, earned the ship a "well done."

America completed her schedule of training in Cuban waters and then returned north, arriving back at Norfolk on December 16, 1975. Following the year-end stand-down, the carrier resumed local operations out of Norfolk in January 1976 and, in March, participated in Exercise Safe Pass '76 with ships of the Canadian, West German, Dutch, and British navies. She ultimately sailed for the Mediterranean on April 15, 1976, with CVW-6 and Commander, Carrier Group (CarGru) 4, Rear Adm. James B. Linder, embarked.

Soon after her arrival in the turnover port of Rota, America participated in the NATO exercise "Open Gate," before entering the Mediterranean. Passing the Pillars of Hercules on May 3, the ship entered into the eastern Mediterranean in support of Operation Fluid Drive, a contingency operation for the evacuation of non-combatants from war-torn Lebanon. For the next three months, the carrier maintained a high state of readiness. In conjunction with Fluid Drive, the ship and her air wing maintained continuous surveillance of the Soviet Mediterranean fleet, which at that point was at its largest since the Yom Kippur War of 1973.

On May 24, 1976, America anchored in Rhodes, Greece, to commence her first liberty of the deployment, but violent anti-American demonstrations prevented the carrier's crew from going ashore, and the ship stood out two days later. America conducted a port visit to Taranto, Italy, instead, but the deteriorating situation in the eastern Mediterranean required the ship to sail sooner than scheduled.

The assassination of the United States ambassador to Lebanon Francis E. Meloy, and Economic Counselor Robert O. Waring as they were on their way to visit Lebanese President Elias Sarkis on June 13, 1976, prompted the evacuation of Americans from that nation a week later, on the 20th. America remained on alert while landing craft from the dock landing ship USS Spiegel Grove (LSD-32) transferred the evacuees from the beach to safety. Following the successful evacuation, the carrier proceeded westward for a few days of liberty in Italian ports, celebrating our nation's bicentennial Independence Day, July 4, 1976, at Taranto.

Proceeding back into the eastern Mediterranean on July 11, 1976, to conduct a missile exercise north of Crete, the ship continued to maintain responsibility for Fluid Drive. On July 27, as more Americans were evacuated from Lebanon on board USS Portland (LSD-37), the carrier provided support. Relieved of her responsibilities in the eastern Mediterranean on August 2, America reached Naples soon thereafter, and remained in port for two weeks. The carrier returned to sea on August 18 and participated in Exercise National Week XXI with other 6th Fleet units.

Upon the termination of National Week XXI, America proceeded to Palma de Mallorca, whence she proceeded to participate in Poop Deck 76 with Spanish Air Force units and United States Air Force units based in Spain. Then, following visits to the Spanish ports of Barcelona and Malaga, America took part in the final exercise of her Mediterranean cruise, Exercise Display Determination. HMS Ark Royal teamed with America, and ships from the navies of Italy, Greece, Portugal, and Turkey participated as well. The American carrier conducted convoy escort duties, simulated close air support for amphibious operations, and simulated strikes against military targets. Upon conclusion of Display Determination, the carrier proceeded to Rota, where she was relieved by USS Franklin D. Roosevelt.

America ultimately reached Norfolk on October 25, 1976. On November 6, the carrier proceeded up the Elizabeth River to the Norfolk Naval Shipyard, where she remained into February 1977. America then operated locally out of Norfolk into the spring of 1977 until sailing for the Mayport, Fla., Operating Area on May 3. Following her participation in Exercise Solid Shield 77, a joint service amphibious training exercise, the carrier returned to Norfolk on May 4.

America sailed from Hampton Roads on June 10, 1977, for a five-week South Atlantic deployment as a unit of TG 20.4. Other ships in company included USS South Carolina (CGN-37), USS Claude V. Ricketts (DLG-5), USS Dupont (DD-941), and USS Neosho (AO-143). Following her return to Norfolk, America operated locally before she sailed to conduct operations in the Caribbean.

Thence returning to Norfolk on August 27, 1977, America sailed for the Mediterranean on September 29, with CVW-6 embarked, and reached Rota on October 9. Departing that port on October 14, the carrier proceeded to the Tyrrhenian Sea, where she operated until October 26. Following a port call at Brindisi, Italy, America began operations in the Ionian Sea on November 7, and anchored at Souda Bay, Crete, two days later. She operated locally in these waters until November 12, when she sailed for Kithira Island, Greece, anchoring there on the 19th.

Weighing anchor the following morning, America sailed for the Adriatic Sea, bound for Dubrovnik, Yugoslavia. Visiting this seaport from November 22 to 26, the carrier transited the Adriatic for a port call at Trieste, staying there from November 28 to December 3, 1977. Returning to operate in the waters of Souda Bay for more exercises, America subsequently departed Crete on December 12 for Palma de Mallorca, where she spent Christmas.

Departing Palma two days later, America proceeded through the Ligurian Sea to her next port of call, Genoa, which she reached on December 30. She remained there until January 8, 1978, when she sailed to carry out antisubmarine exercises in the Tyrrhenian Sea. Upon the conclusion of these exercises, she anchored in Golfo di Palma, Sicily. Operations in the western Mediterranean and again in the Tyrrhenian Sea rounded out most of January 1978, and the ship rested briefly at Catania, Italy, before getting underway for Exercise National Week on February 5.

She returned to the Tyrrhenian Sea and western Mediterranean for further exercises during March, and then visited Barcelona before concluding the deployment with more exercises in the western Mediterranean. At Rota, she was relieved by USS Forrestal (CV-59) and sailed for Norfolk, arriving home on April 25, 1978. Following post-deployment stand-down, America conducted carrier qualifications off the Virginia Capes and then entered Norfolk Naval Shipyard for an availability. Upon the conclusion of that period of repairs and alterations, the carrier conducted post-availability sea trials on September 19 and 20, 1978, and conducted carrier qualifications with CVW between October 12 and 20.

Tragedy marred the last day of operations when a Grumman S-3A Viking antisubmarine aircraft went over the side upon landing, hung by the safety nets momentarily, then plunged into the sea soon thereafter. Although the pilots, Lt. Comdr. Ziolowski and Lt. (j.g.) Renshaw, ejected clear of the plane, they were not recovered.

America subsequently conducted refresher training out of Guantanamo Bay early in November before calling at Ft. Lauderdale on November 10, 1978, to commence a four-day stay. Returning to Norfolk soon thereafter, the carrier remained in the Norfolk area, alternating periods of time in port alongside Pier 12 with type training and exercises off the Virginia Capes.

The carrier cleared Norfolk on January 5, 1979, for the Caribbean operating areas and conducted type training there from January 5 to 23. The ship visited St. Thomas in the Virgin Islands from January 24 to 29. America then resumed type training in the waters of the Caribbean and West Indies, concluding those evolutions on February 12 to return to Norfolk.

After bringing CVW-11 on board off the Virginia Capes on March 8 and 9, 1979, America spent the next two days moored at Pier 12, making final preparations for her departure for the Mediterranean. The carrier sailed on March 13. Two days later, on the 15th, America conducted a "BearEx" with a Lockheed P-3 Orion from Bermuda simulating a Russian Bear reconnaissance aircraft. This practice proved timely, for the following day, A-7 Corsair II and Grumman F-14A Tomcat aircraft from America intercepted a pair of the long-range Tupelov TU-95 Bear-D planes that were en route

to Cuba from their bases in the Soviet Union. The Bears never came within visual range of the carrier's battle group.

Reaching Rota on March 24, America relieved USS Saratoga (CV-60) and commenced operations in the western Mediterranean on March 29. During this deployment, the ship visited a variety of ports, starting with Naples, Taranto, and Catania. Moving into the Adriatic, the carrier stopped at Split, Yugoslavia, before moving north to Venice and Trieste. In the eastern Mediterranean, America called at Alexandria, Egypt, and Souda Bay, Crete. Returning west, she visited Palma de Mallorca and Barcelona in Spain, Marseilles on the coast of France, Genoa in northern Italy, and Valencia in Spain before heading for Rota. She completed turnover proceedings at Rota on September 10 and 11, 1979, and got underway immediately to commence the homeward voyage.

Highlighting this period were numerous multilateral and unilateral exercises, as in previous Mediterranean deployments. During one phase of National Week XXVII, America and her consorts took part in an open sea exercise that took them into the waters of the Gulf of Sidra (Sirte) — claimed by Libya as territorial waters since October 11, 1973. With the Libyan government serving notice that any ship or aircraft operating south of the 32° 30' north latitude would be violating its territory, America's battle group maintained an alert, in view of the proximity of Libyan airfields and Soviet-made aircraft operating therefrom. Departing Augusta Bay, Sicily, on July 26, 1979, the task group arrived in its exercise area on the 28th. As planes from CVW-11 maintained nearly continuous fighter cover, the ships conducted their exercise unhindered.

Ultimately departing Rota on September 12, 1979, to conduct a "blue water" turnover with USS Nimitz (CVN-68), America encountered her second pair of Bears. F-14A Tomcats of VF-213 intercepted the two, however, and caused them to turn away to the north, having never sighted a single ship in the carrier's battle group. Reaching Norfolk on September 22, America stood down after her 6th Fleet deployment. The carrier departed Norfolk again on October 15, 1979, for Mayport and conducted local operations off the coast of Florida before moving into the Gulf of Mexico to conduct carrier qualifications. Returning north upon completion of those evolutions, America put to sea on October 30 for more carrier qualifications. These involved the first arrested carrier landings of the new McDonnell-Douglas F/A-18 Hornet on that date. This aircraft underwent rigorous testing over the days which followed, with 32 catapult shots and arrested landings being completed before America returned to Norfolk on November 3.

Entering the Norfolk Naval Shipyard on November 6, 1979, America underwent repairs and alterations for much of 1980, commencing her post-repair trials on September 23, 1980. Among the work performed during the availability was the installation of the NATO Sea Sparrow missile and close-in weapon systems such as the multi-barreled Phalanx machine gun.

The ship carried out a second period of post-repair trials from October 16 to 21, after which she returned to NOB, Norfolk, where she conducted sea trials from October 27 to 29, 1979. Subsequently conducting refresher training out of Guantanamo Bay, America returned to the Virginia Capes Operating Area to conduct carrier qualifications in early December. She spent the remainder of the year 1980 undergoing upkeep at NOB, Norfolk. America operated locally in the Virginia Capes area into January 1981 and, during these operations on January 14, 1981, brought on board a Grumman C-1A Trader COD aircraft piloted by Ens. Brenda Robinson, USNR. Ens. Robinson became the first black female naval aviator to be carrier qualified. The ship later conducted carrier qualifications for CVW-11. On January 29, 1981, as America was returning to NOB, Norfolk, she received a message from a Greek motor vessel, Aikaterini, in distress. America, diverted to the scene to render assistance until the Coast Guard could arrive, sent helicopters from her embarked HS-12 with damage control equipment, members of the ship's fire department, and damage control assistance to the stricken ship. Returning to Pier 12, NOB, Norfolk on February 2, America proceeded thence for carrier qualifications off the Virginia Capes, and thence to the Caribbean for type training. Returning to Norfolk on March 19, America, in the company of her consorts USS California (CGN-36) and USS Preble (DLG-46), subsequently sailed for the Mediterranean on April 14, 1981, destined ultimately for the Indian Ocean. Reaching Palma de Mallorca on April 23, America then participated in NATO Exercise Daily Double, with the amphibious assault ship USS Nassau (LHA-4), as well as with Greek and Italian navy units on the 28th before she steamed to Port Said, Egypt. Originally scheduled to have commenced her transit of the Suez Canal on May 5, 1981, the tense situation in Lebanon prompted a 24-hour "hold" on the evolution. Given the go-ahead soon thereafter, America made the 104½-mile transit on May 6, in ten hours — the first United States Navy carrier to steam through the Suez Canal since USS Intrepid (CVA-11) had made the passage shortly before the Arab-Israeli "Six-Day War" of 1967. It was also the first "supercarrier" to transit the canal since it had been modified to permit the passage of supertankers.

America operated in the Indian Ocean on "Gonzo" Station for the first time between May 12 and June 3, after which she visited Singapore. On June 18, the carrier departed that port for her second stint on "Gonzo Station," a 35-day deployment.

On July 15, 1981, America was requested to provide search and rescue (SAR) aircraft to assist in locating a merchant ship in distress in the northern Arabian Sea. The Greek merchantman Irene Sincerity was reportedly afire. America's planes located the ship, and USS California rescued the 39 crewmen, disembarking them in good condition in Karachi, Pakistan, soon thereafter.

Upon completion of her second northern Arabian Sea line period on August 4, America set a course for Australian waters, conducting a "Weapons Week" exercise in the vicinity of Diego Garcia. During "Weapons Week," a Lockheed P-3 Orion of Patrol Squadron Fifty (VP-50) requested assistance from two F-14 Tomcats from America, flying near Pierre Island, Diego Garcia. California sped to the island and located an individual stranded on Pierre Island. He had been on a treasure-hunting expedition bound from Sri Lanka to Mauritius. The cruiser took the man to Diego Garcia.

Departing the Diego Garcia Operating Area on August 15, America conducted a unique burial-at-sea on the 18th, when the remains of the late Lt. Stephen O. Musselman were consigned to the ocean. Lt. Musselman had been shot down on September 10, 1972, in an A-7 Corsair from America over North Vietnam. His remains had been returned by the Vietnamese government on July 8, 1981. Lt. Musselman's widow requested that these remains be consigned to the last ship he had served on and buried there.

America anchored at Fremantle on August 25, 1981, and remained there for six days, sailing for "Gonzo Station" on the 31st. During her third line period, the ship spent 34 days on station. On September 23, a fire broke out in a steam trunk line that carries steam from the main engineering spaces to the flight deck catapult system at about 1745. Soon after America's fire party arrived on the scene to isolate the fire, smoke began filling the areas adjacent to the crew berthing areas, so Capt. James F. Dorsey Jr. ordered General Quarters sounded.

America's firefighters soon managed to quell the blaze, and the ship secured from battle stations at 2316. The carrier resumed normal flight operations the next morning at sunrise and remained on station until relieved by USS Coral Sea (CV-43) on October 16. Two days later, while America steamed toward the Bab el Mandeb Strait, the ship went to General Quarters due to threats issued by the People's Democratic Republic of Yemen. The ship passed without incident and continued her journey through the Red Sea unhindered.

On October 21, 1981, America commenced the northbound transit of the Suez Canal. This transit, unlike the comparatively light-hearted one of May 6, proved more tense. As a result of the unsettled conditions in Egypt following the October 6, 1981, assassination of President Anwar Sadat, the Egyptian government accorded America's passage through the Suez Canal the utmost security considerations. The Egyptian Navy provided a patrol vessel to escort the carrier, while an Egyptian Air Force helicopter conducted reconnaissance flights over both banks of the waterway. Egyptian Army units patrolled the adjacent canal roads. Additionally, liaison officers on board the carrier maintained constant touch with the security forces by radio. Making the passage of the canal without incident, America continued across the Mediterranean, reaching Palma de Mallorca on October 25. After a three-day port call, the carrier conducted exercises with Spanish forces, and subsequently sailed for home on November 1, departing the Mediterranean the following day. She arrived at Norfolk on November 12, 1981.

Following a short stand-down, America conducted carrier qualifications in the Virginia Capes operating area, before

mooring at the Norfolk Naval Shipyard on December 14. Emerging from the naval shipyard on April 20, 1982, America operated locally off the Virginia Capes. Departing Norfolk on May 10, the ship steamed to the Guantanamo Bay Operating Area, returning to her home port on May 28. Following further carrier qualifications off the Virginia Capes, the carrier then steamed south to conduct type training in the West Indies, interspersing these evolutions with a port visit to St. Thomas. Returning to Norfolk on July 8, America operated locally between July 22 and 24, before sailing on August 22, with CVW-1 embarked, to participate in joint NATO exercises United Effort and Northern Wedding 82. America visited Edinburgh, Scotland, from September 15 to 21 and proceeded thence to Portsmouth, England, arriving there on the 23rd. Sailing for the Mediterranean on the 26th, the carrier operated briefly with the 6th Fleet, participating in Exercise Display Determination between September 30 and October 8. She then sailed for the United States, and, following her operational readiness evaluation in the Caribbean operating areas, reached Mayport to disembark CVW-1. America returned to Norfolk on November 4, 1982.

America departed Norfolk on December 8, 1982, proceeded to the Virginia Capes Operating Area, embarked CVW-1, and set out across the Atlantic. Visiting Palma de Mallorca on December 22, America remained there through the Christmas holiday, weighing anchor on December 28 to sail for the Lebanese coast, where she was to take up duty in support of the Multinational Peacekeeping Force in strife-torn Lebanon. Relieving USS Nimitz on station on January 2, 1983, America spent the next 18 days off Lebanon before Nimitz took over on January 20. Steaming thence to Pireaus, Greece, America, along with USS Dale (DLG-19) and USS Savannah (AOR-4), anchored there on January 23 for a five-day port visit to Athens.

Underway on January 29, 1983, the carrier transited the Sea of Crete en route to an overnight anchorage at Port Said. Transiting the Suez Canal on January 31, America reached the Red Sea the same day and reported for duty with the 7th Fleet on February 4. On February 9, the carrier and her accompanying battle group conducted Exercise Beacon Flash 83. Subsequently, on February 28, America and her consorts conducted a "Weapons Week" exercise in the vicinity of Diego Garcia. Following those evolutions, the carrier visited Colombo, Sri Lanka, anchoring on March 7. Weighing anchor on March 12, America resumed operations in the Indian Ocean soon thereafter, culminating in "Beacon Flash 83-4," and a subsequent port visit to Masirah Island, Oman.

Steaming thence to Mombasa, Kenya, and a five-day port visit, America departed that port for a week of intense flight operations, followed by participation in Beacon Flash 85 on April 19, 1983. Returning to anchor at Masirah Island again three days later, the carrier and her battle group operated in the northern Arabian Sea, en route to the Suez Canal. Transiting that waterway on May 4, America headed for Souda Bay, reaching an anchorage there on May 7. Five days later, the carrier got underway for Malaga, Spain, reaching her destination on May 14 for a nine-day port visit. The ship subsequently departed Malaga on May 23 and reached Norfolk on June 2, 1983.

America then entered the Norfolk Naval Shipyard on July 8. For four months, the ship underwent a period of repairs and alterations, emerging from the yard on October 28, 1983. She then operated locally off the Virginia Capes with CVW-1 embarked before she proceeded to Mayport and ultimately to Puerto Rican waters for refresher training. Subsequently visiting Nassau in the Bahamas for a five-day port visit, America returned to the east coast of the United States, reaching Mayport on December 8. She then conducted carrier qualifications for both east and west coast squadrons en route to her home port, reaching Norfolk on December 14, 1983. The carrier operated locally from Norfolk into February 1984, alternating periods of upkeep in port with carrier qualifications and exercises. She then conducted two periods of type training (February 6 to 20 and March 25 to April 8), interspersing these with an in-port period at Ft. Lauderdale from February 21 to 24 and then calling at St. Thomas upon conclusion of the second period of training. Returning to Norfolk on March 22, America spent the next month preparing for her next deployment and got underway to participate in Exercise Ocean Venture on April 24. Visiting Caracas, Venezuela, upon conclusion of that evolution, America departed on May 9 for the Mediterranean.

Reaching Malaga, Spain, on 21 May 1984, the carrier

commenced her transit of the Mediterranean on 29 May and reached Port Said on 3 June. Transiting the Suez Canal the following day, she passed through the Red Sea and joined the 7th Fleet on 8 June, relieving USS Kitty Hawk (CV-63). Returning to the 6th Fleet on 29 August, America transited the Suez Canal on 2 September bound for Naples. The carrier visited Monaco from 13 to 22 September 1984 before participating in one phase of NATO Exercise Display Determination. After stopping briefly in Naples, America returned to sea soon thereafter and took part in the second phase of Display Determination before visiting Catania. Ultimately reaching Augusta Bay on 27 October, she was relieved by USS Dwight D. Eisenhower (CVN-69) on that date, sailing soon thereafter for the United States.

Arriving at Norfolk on 14 November 1984, America conducted carrier qualifications in the Virginia Capes operating areas from 29 November to 17 December before returning to port on the 18th. The ship remained in an upkeep status until 18 January 1985, when she shifted to the Norfolk Naval Shipyard for overhaul. Emerging from the yard on 13 May 1985 for sea trials off the Virginia Capes, America remained at Norfolk until 28 May, when she sailed to conduct refresher training. Then, following a port call at Port Everglades, Fla. (13 to 17 June), America conducted carrier qualifications before returning to Norfolk on 25 June. The ship operated locally out of Norfolk through mid-August.

America sailed on 24 August 1985 to participate in Ocean Safari, a six-week NATO exercise which ultimately took her to Norwegian waters. After visiting Portsmouth, England, upon conclusion of her training, America returned to Norfolk on 9 October. She spent the remainder of the year 1985 alternating periods of upkeep at the Naval Station, Norfolk, with local operations in the Virginia Capes operating area. As the new year, 1986, began, tensions in the Mediterranean basin would result in America's sailing to deploy with the 6th Fleet one month earlier than planned. On 7 January 1986, President Ronald Reagan ordered all American citizens out of Libya and broke off all remaining ties between the two nations. At the same time, the President directed the dispatch of a second carrier battle group to the Mediterranean and directed the Joint Chiefs of Staff to look into military operations against Libya, a country strongly suspected of fomenting terrorist activity.

Operations near Libya began at the end of January 1986. These evolutions, collectively named Attain Document, were carried out between 24 and 31 January 1986 and between 10 and 15 February by surface ships and aircraft. America, with CVW-1 embarked, and her accompanying battle group departed Norfolk on 10 March 1986, a month earlier than previously scheduled, and arrived in the Mediterranean in time to participate in the third phase of Attain Document, a freedom of navigation (FON) exercise in the Gulf of Sidra.

Late on 23 March, American planes flew south of latitude 30° 30' N — the "Line of Death" proclaimed by Libyan leader Muammar Qaddafi. On 24 March, USS Ticonderoga (CG-47), accompanied by two destroyers, USS Scott (DDG-995) and USS Caron (DD-970), moved south of the "Line," covered by fighter aircraft, at 0600. A Libyan missile installation near Surt (Sirte) launched two Soviet-made SA-5 Gammon surface-to-air missiles (SAMs) at 0752 toward F-14A Tomcats of America's VF-102. Later that afternoon, the installation at Surt (Sirte) fired additional SAMs at American planes, but, like the first pair, they went wide of their mark. At about 1430, a Libyan missile-equipped Combattante II G-type patrol craft sortied from Misratah, Libya, and approached Ticonderoga and her consorts. Two Grumman A-6E Intruders from America's Attack Squadron Thirty-Four (VA-34) fired Harpoon missiles at the craft and sank her in the first use of the Harpoon in combat. Shortly thereafter, when American radars detected the Libyan installation at Sirte activating its target acquisition radars, two A-7E Corsairs from USS Saratoga's VA-81 put the site out of action with HARMs (high-speed anti-radiation missiles).

One hour after the first patrol boat had sortied, a Soviet-built Nanuchka-type patrol craft began heading out into the Gulf of Sidra. Intruders from VA-34 and Saratoga's VA-85 attacked with Rockeye cluster bombs, but the craft sought refuge alongside a neutral merchant ship and avoided destruction. Damaged, she

returned to the port of Benghazi after nightfall. The following day, 25 March 1986, at 0200, another Nanuchka II-type patrol boat entered international waters and came under attack from Intruders from VA-85 and Coral Sea's VA-55; the latter utilized Rockeyes in the attack, the former then sank the craft with a Harpoon. The same squadrons then attacked and damaged a second Nanuchka II, forcing her to put into Benghazi.

Attain Document III came to a close at 0900 on 27 March 1986, three days ahead of schedule and after 48 hours of largely unchallenged use of the Gulf of Sidra by the United States Navy. Thence steaming to Augusta Bay, Sicily, America relieved USS Saratoga on station, and subsequently visited Livorno, Italy, from 4 to 8 April 1986. In the meantime, intelligence information in the wake of the strikes designed to let Col. Qaddafi know that the United States had not only the desire but the capability to respond effectively to terrorism indicated that Qaddafi intended to retaliate. Such retaliation occurred soon thereafter.

On 5 April 1986, two days after a bomb exploded on board a Trans World Airways (TWA) flight en route to Athens from Rome, killing four American citizens, a bomb exploded in the La Belle Discotheque in West Berlin, killing two American servicemen and a Turkish civilian. Another 222 people were wounded in the bombing, 78 Americans among them. Col. Qaddafi threatened to escalate the violence against Americans, civilian and military, throughout the world.

Repeated efforts by the United States to persuade the Libyan leader to forsake terrorism as an instrument of policy, including an attempt to persuade other western nations to isolate Libya peacefully, failed. Rumors of retaliation by the United States were soon followed by Qaddafi's threat to take all foreigners in Libya hostage, to use them as a shield to protect his military installations. In light of that threat, and the failure of means to gain peaceful sanctions against Libya, and citing "incontrovertible evidence" of Libyan complicity in the recent terrorist acts, President Reagan directed that attacks on terrorist-related targets in Libya be carried out.

Operation Eldorado Canyon commenced early on the afternoon of 14 April 1986, as tanker aircraft took off from bases in England to support the Air Force North American F-111F and EF-111 planes that soon followed them into the air and began the long 3,000-mile trip to the target. Later that afternoon, between 1745 and 1820, America launched six Intruders (strike aircraft) from VA-34 and six A-7E Corsair IIs (strike support); Coral Sea launched her strike/strike support aircraft, eight A-6Es from VA-55 and six F/A-18 Hornets between 1750 and 1820. Both carriers

launched additional aircraft to support the strike and to provide a combat air patrol (CAP) and other functions.

"In a spectacular feat of mission planning and execution," the Navy and Air Force planes, based 3,000 miles apart, reached their targets on time at 1900. The Hornets from Coral Sea and Corsair IIs from America launched air-to-surface Shrike missiles and HARMs against Libyan SAM sites at Benghazi and Tripoli. Moments later, VA-34's Intruders, roaring in at low-level in the blackness, dropped their MK. 82 bombs with near surgical precision on the Benghazi military barracks, reckoned to be an alternate command and control facility for terrorist activities and a billeting area for Qaddafi's elite Jamahiriyah Guard as well as a warehouse for components for MiG aircraft. VA-34's attack heavily damaged the warehouse, destroying four crated MiGs and damaging a fifth.

Following that counter-terrorist strike, America visited Naples between 28 April and 4 May and then participated in NATO Exercise Distant Hammer with units of the Italian and Turkish Air Forces and visited Cannes upon conclusion of the evolution. During June, the carrier operated with USS Coral Sea and the newly arrived USS Enterprise (CVN-65), and took part in a "Poop Deck" exercise with Spanish and United States Air Force units off the coast of Spain, arriving at Palma de Mallorca soon thereafter. Participating in a NATO Exercise Tridente, in late June

1986, America visited Naples before participating in a National Week exercise. Subsequently visiting Catania and operating in the central and western Mediterranean, the carrier wound up the month of July at Benidorn, Spain, before returning to sea for further operations in that region. Visiting Naples between 11 and 17 August, America spent the rest of her deployment in operations in the western and central Mediterranean before USS John F. Kennedy relieved her at Rota between 28 and 31 August.

When America returned from its Mediterranean deployment on 10 September, it marked the first battle group to spend no more than six months overseas as part of the Navy's efforts to reduce deployments. Having deployed to the Sixth Fleet on 10 March 1986, the carrier was relieved by USS John F. Kennedy (CV 67) with Carrier Air Wing 3 (CVW-3) embarked. America then went to the Norfolk Naval Shipyard on 20 November 1986 for an overhaul which lasted until 11 February 1988.

America left the shipyard on 15 February 1988 for sea trials and workups in preparation for the next round of intensive operations. In April, after completing a shakedown cruise, America participated in "Fleet Week '88." Sailors and ships were sent to New York City to demonstrate the Navy to the citizens in preparation for the USS Iowa (BB 61) battle group's move to Staten Island in 1989.

In February 1989, America departed for exercises in the Caribbean and the North Atlantic. America again operated in the Vestfjord before making a port visit to Le Havre, France, and returning to Norfolk on 3 April 1989. On 16 April, the VS-30 "Diamondcutters," embarked in America, became the first fleet S-3 squadron to fire a Harpoon anti-ship missile. The launch resulted in a direct hit on the target by a detachment assigned to VS-30 as it participated in Exercise North Star '89.

America departed Norfolk on 11 May 1989 for her sixteenth major deployment, to the Mediterranean and Indian Ocean. On 11 August, USS Coral Sea and America departed early from separate port visits when they were diverted to the eastern Mediterranean as a show of force in the wake of the suspected hanging of Marine Corps Lt. Col. William R. Higgins by Middle East terrorists, and threats to other hostages. Lt. Col. Higgins had been kidnapped in February 1988 while a member of the United Nations peacekeeping forces in Lebanon. After operations in the Mediterranean, Arabian Gulf, and Indian Ocean, America returned home on 10 November 1989.

America spent the first part of 1990 conducting local operations. After a three-and-a-half-month shipyard availability, America conducted Refresher Training, Advanced Phase Training, and FLEETEX prior to deploying five months early in support of Operation Desert Shield on 28 December 1990. America transited the Straits of Gibraltar on 9 January 1991, and on 15 January, the America battle group transited the Suez Canal and arrived on station in the Red Sea to participate in Operation Desert Storm. While deployed, the America/CVW-1 team distinguished themselves, on 15 February, as the only carrier and air wing to fight in both the Red Sea and Arabian Gulf. On 20 February, America's VS-32 became the first S-3 squadron to engage, bomb, and destroy a hostile vessel — an Iraqi gunboat.

On 23 February, aircraft from America destroyed a Silkworm (anti-ship) missile battery after Iraq unsuccessfully fired a missile at USS Missouri (BB 63). America departed the Arabian Gulf on 4 March, having launched over 3,000 combat sorties, significantly contributing to the liberation of Kuwait.

From 16 through 22 March 1991, America conducted a port visit to Hurghada, Egypt, making the first port call of the deployment after 78 consecutive days at sea. The ship transited the Suez northward on 3 April and five days later went through the Straits of Gibraltar to enter the Atlantic Ocean on her way home, reaching Norfolk on 18 April. Returning amid a heroes' welcome, the carrier participated in Operation Welcome Home/Fleet Week '91 in New York City from 6 through 11 June 1991, taking part in the largest victory parade since World War II. After an abbreviated in-port period and compressed work-ups, America deployed to the North Atlantic for two months in support of North Star '91, then departed on 2 December 1991 for the Mediterranean and Arabian Gulf once again, her eighteenth major deployment. She also became the first carrier to earn an unprecedented third campaign star on the Southwest Asia Service Medal when she returned to the Arabian Gulf in early 1992 in support of United Nations sanctions against Iraq, returning home from the extended deployment on 6 June 1992.

America entered the Norfolk Naval Shipyard in July 1992 for the start of a six-month shipyard availability.

America returned to sea in December 1992 for sea trials. In January and February 1993, the carrier conducted training carrier operations off the coast of Florida for new pilots, and continued with refresher training in February and March. After COMPUTEX, the exercise Ocean Venture, and a port visit to St. Thomas, USVI, in April and May, America and her air wing continued work-ups for deployment. The USS America Joint Task Group departed Norfolk and other East Coast ports on 11 August 1993 for another major deployment.

The America Battle Group displayed its versatility in October 1993 when, after several weeks supporting United Nations peacekeeping efforts over Bosnia, orders came on four hours' notice to transit the Suez Canal and relieve USS Abraham Lincoln (CVN 72) on Groundhog Station 90 miles north of the equator in the Indian Ocean in support of UN efforts in the beleaguered African nation of Somalia. America transited the Suez on 29 October 1993. She was followed, on 1 November, by members of her battle group, USS Simpson (FFG 56) and the replenished oiler USS Savannah (AOR 4). The transit took America over 2,500 miles in a week. The turnover with Abraham Lincoln permitted the west coast carrier to return to Alameda, Calif., thereby ending a scheduled six-month deployment on time.

Supporting the U.N. humanitarian efforts in Somalia was the Naval Battle Force Somalia, commanded by Rear Adm. Arthur Cebrowski, Commander, Carrier Group Six in America. Other elements of Naval Battle Force Somalia included Simpson, USS New Orleans (LPH 11), USS Denver (LPD 9), USS Comstock (LSD 45), USS Cayuga (LST 1186), and the Thirteenth Marine Expeditionary Unit.

Before leaving the Adriatic, the eight squadrons of America's air wing, CVW-1, flew 863 sorties in support of humanitarian efforts in Bosnia. Before returning to the Mediterranean, CVW-1 aircraft flew missions into southern Iraq from the Red Sea in support of Operation Southern Watch. On 12 December 1993, America passed through the Suez Canal, once again arriving in the Sixth Fleet area of operations and conducted various exercises with allied forces before returning to Norfolk in February 1994. In April 1994, America commenced a four-month availability at Norfolk Naval Shipyard in Portsmouth, Virginia.

America departed Norfolk on 28 August 1995 for a routine six-month deployment to the Mediterranean and the Indian Ocean. This was her 20th and final deployment in her 30-year history. The carrier participated in Operations Deny Flight and Deliberate Force, under the control of the U.N. and NATO from 9 through 30 September, and visited various ports, including Trieste, Italy, for a five-day port visit from 30 September through 5 October.

America visited the capital city of Valletta, Malta, from 23 through 28 January 1996 — the first U.S. Navy carrier to visit this historical port in over 24 years. America and elements of her battle group were operating in the Adriatic Sea in support of the NATO Implementation Force (IFOR) in Bosnia and Herzegovina for Operation Joint Endeavor. Secretary of the Navy John H. Dalton visited the carrier on 6 February. Speaking of the carrier's role in Joint Endeavor, the Secretary said, "I would like you to know I appreciate the sacrifices you've made. You're actually saving lives. You've brought the warring parties to the peace table. Without you, it wouldn't have happened." America returned to the pier in Norfolk, Va., ending her Mediterranean Sea deployment on 24 February 1996. After more than three decades of proud and historic naval service, America was decommissioned in a ceremony at Norfolk Naval Shipyard in Portsmouth, Va., on 9 August 1996.

During the ceremony, and as dictated by Navy tradition, America's final commanding officer, Capt. Robert E. Besal, was presented with the ship's commissioning pennant, marking the end of active service. The guest speaker for the ceremony was Adm. Leighton W. Smith, a former American commanding officer and the recent Commander-in-Chief of Allied Forces Southern Europe, where he was in charge of all NATO military operations in Bosnia. Following the decommissioning, America was transferred to the Ready Reserve Fleet in Philadelphia.

Stricken from the Navy List on the day of her decommissioning, America was planned to be sold for scrapping. However, the carrier was sunk in the Atlantic Ocean, approximately 300 miles off the Virginia coast, on 14 May 2005, following a series of tests consisting of underwater and surface simulated attacks on the ship. In a letter to a coalition of veterans and former crewmembers who offered to make the carrier a museum, Vice Chief of Naval Operations Adm. John B. Nathman explained: "America will make one final and vital contribution to our national defense, this time as a live-fire test and evaluation platform. America's legacy will serve as a footprint in the design of future carriers — ships that will protect the sons, daughters, grandchildren, and great-grandchildren of America veterans. We will conduct a variety of comprehensive tests above and below the waterline, collecting data for use by naval architects and engineers in creating the nation's future carrier fleet. I

CVA-67 • USS JOHN F. KENNEDY

(Displacement 75,000; length 1046'; beam 129'4"; extreme width 249'; draft 35'7"; speed 30 knots; complement 3,297; class Kitty Hawk.)

USS John F. Kennedy (CV 67) is named for the 35th President of the United States. The ship's keel was laid on October 22, 1964, at the Newport News Shipbuilding and Drydock Company in Virginia. President Kennedy's 9-year-old daughter, Caroline, christened the ship in May 1967 in ceremonies held at Newport News, Virginia. The ship subsequently entered naval service on September 7, 1968.

Kennedy was originally designated as CVA 67, an attack aircraft carrier. In the early 1970s, the classification was changed to CV 67, indicating the ship was capable of supporting anti-submarine warfare aircraft, making it an all-purpose, multi-mission aircraft carrier.

Kennedy's maiden voyage was to the Mediterranean Sea. She subsequently made another seven deployments to this area of the world through the '70s in response to a deteriorating situation in the Middle East. The ship's fourth Mediterranean cruise included her first visit to a North Atlantic port, Edinburgh, Scotland.

By the mid-'70s, Kennedy had been upgraded to handle both the F-14 "Tomcat" and the S-3 "Viking." Kennedy underwent her first year-long major overhaul in 1979. The ship's ninth deployment, in 1981, was her first to the Indian Ocean. Kennedy transited the Suez Canal, hosted the first visit aboard a United States ship by a Somali head of state, and achieved its 150,000th arrested landing.

In 1982, Kennedy won an eighth Battle "E" efficiency award and a fourth Golden Anchor retention award. In 1983, as a result of the growing crisis in Beirut, Lebanon, Kennedy was called upon once again to support efforts that would define the ship's operations into the next year. Awards received during that period included a ninth Battle "E," the Silver Anchor Award for Retention, the RADM Flatley Award for Safety, and the Battenburg Cup for being the overall best ship in the Atlantic Fleet.

Kennedy spent the winter of 1984 in drydock for a complex overhaul at the Norfolk Naval Shipyard. In 1985, the ship received a fifth Golden Anchor Retention Award and several departmental efficiency awards. While in the shipyard, the ship was also awarded the inaugural Department of Defense Phoenix Award, signifying a level of maintenance excellence above all other Department of Defense components worldwide. In July 1986, KENNEDY served as the centerpiece for a vast

international naval armada during the international Naval Review in honor of the 100th Anniversary and Rededication of the Statue of Liberty. Kennedy departed for the Mediterranean in August 1986 and returned in March 1987.

Kennedy departed Norfolk, Virginia, for her 12th major deployment to the Mediterranean in August 1988. On January 4, 1989, while conducting routine operations in international waters, F-14 Tomcats from the embarked air wing shot down two Libyan MIG-23s that were approaching the battle group in a hostile manner.

After spending the first half of 1990 participating in a variety of exercises, Kennedy paid visits to New York for Fleet Week '90 and Boston for the Fourth of July. In August, with just four days' notice, Kennedy deployed in support of Operation Desert Shield.

Kennedy entered the Red Sea in September 1990 and became the flagship of the Commander, Red Sea Battle Force. On January 16, 1991, aircraft from the ship's Carrier Air Wing THREE began Operation Desert Storm with attacks on Iraqi forces. The ship launched 114 strikes and 2,895 sorties, with aircrews of CVW-3 flying 11,263 combat hours and delivering more than 3.5 million pounds of ordnance in the conflict.

After the ceasefire, Kennedy transited the Suez Canal for the fourth time in seven months and began its journey home. Kennedy arrived in its homeport of Norfolk on March 28, 1991, to the greatest homecoming celebration and outpouring of public support since World War II.

Kennedy then entered a four-month shipyard restricted availability at Norfolk Naval Shipyard. The ship departed the shipyard at the end of September with extensive repairs and maintenance accomplished on engineering systems, flight deck systems, and equipment. Additionally, the ship was ready to handle F/A-18 Hornet aircraft to replace the A-7E Corsair IIs that had flown on their last deployment from the deck of Kennedy.

The 1992-93 deployment, from Oct. 7, 1992, until April 7, 1993, marked Kennedy's 14th deployment to the Mediterranean area. The tone of the deployment was set by turmoil in the former Republic of Yugoslavia. The ship conducted multiple exercises with the armed forces of Mediterranean littoral nations, hosted a great number of visitors in port and at sea, and spent substantial operating time in the Adriatic Sea. On December 8, 1992, Kennedy passed a milestone by making its 250,000th trap of an aircraft.

USS John F. Kennedy completed a two-year comprehensive overhaul in the Philadelphia Naval Shipyard on September 13, 1995. Following the overhaul, Kennedy moved to its new homeport at the Mayport Naval Station in Mayport, Florida.

Kennedy departed Mayport in April 1997 for its 15th deployment to the Mediterranean Sea and returned to Mayport in late October 1997.

After their return, Kennedy entered a three-month ship's restricted availability at Naval Station Mayport in February 1998 for many upgrades. In April 1998, after extensive repairs and maintenance, Kennedy got underway for several at-sea periods for carrier qualifications, weapons onload and offload training, student pilot and general shipboard damage control training. Kennedy also participated in the week-long Fleet Week '98 activities, strengthening the Navy's relationship with New York City residents.

During 1999, continuing at-sea periods prepared Kennedy for its 16th deployment to the Mediterranean Sea/Arabian Gulf. After a heroic rescue of the crew from the foundered tug Gulf Majesty, during Hurricane Floyd in mid-September, Kennedy carried the banner of freedom to our friends and allies overseas, making history once again. The ship made the first carrier port call to Jordan and hosted the King of Jordan, allowing him to experience life at sea. JFK then participated in Operation

Southern Watch, flying combat missions while enforcing the no-fly zone over Iraq. The JFK/CVW-1 team set new records in bombing accuracy while employing the most lethal combination of precision weaponry ever put to sea, amassing 10,302 arrested landings along the way.

On Jan. 1st, JFK became the Carrier of the New Millennium by being the only carrier underway as the year 2000 arrived. Her triumphant return to Mayport on March 19, 2000, marked the completion of yet another successful forward deployment as one of our nation's most visible guarantors of support for our allies and freedom of the seas.

Kennedy returned to Mayport on March 19, 2000, and after a few weeks in port, Kennedy returned to the sea, heading north for New York where Big John participated in the 2000 International Naval Review over the July 4 holiday. After Independence Day, JFK went even further north to Boston for Sail Boston 2000.

Upon returning to Mayport, Kennedy underwent a brief but extensive availability period, installing components of the most recent technology. As a test bed for Cooperative Engagement Capability (CEC), Big John is flagship to the most technologically advanced battle group in history. CEC enables battle group ships and aircraft to share sensor data and provide a single, integrated picture to all. With CEC, Kennedy can see and respond, with fire-control accuracy, to air contacts further from the ship than was previously possible. Secretary of the Navy, Gordon England, recognized Kennedy Battle Group members for their participation in the test and evaluation of CEC in January 2002 with a Meritorious Unit Citation.

Kennedy's 17th deployment was accelerated by three weeks in response to terrorist attacks on America; Kennedy deployed

on Feb. 7, 2002, to the North Arabian Sea in support of Operation Enduring Freedom. During Kennedy's four months in the North Arabian Sea, their air wing, CVW 7, dropped more than 64,000 pounds of ordnance on Taliban and al Qaeda targets. Kennedy's air wing flew day- and night-missions over Afghanistan, supporting American and coalition forces on the ground with close air support. In early April, Kennedy became the only aircraft carrier operating in support of Operation Enduring Freedom.

Big John returned in mid-August, after more than six months away from homeport. The ship continues to support, serving as the east coast platform for carrier qualifications and will begin an extensive maintenance period in 2003.

In July 2004, USS John F. Kennedy (CV-67) collided with a dhow in the Persian Gulf, resulting in the tragic loss of all individuals on the traditional Arab sailing boat. Following the incident, the Navy relieved the commander of the Kennedy, Capt. Stephen B. Squires. Although the carrier itself remained unscathed, two jet fighters on the deck were damaged when one slid into the other as the ship executed a hard turn to avoid the small vessel. Capt. Squires made the decision to turn at the last possible moment to allow for the recovery of aircraft returning from airstrikes that were critically low on fuel.

Budget cutbacks and evolving naval tactics, coupled with the fact that the Kennedy was the most expensive carrier in the fleet to maintain and was due for a costly overhaul, prompted the U.S. Navy to retire the Kennedy. On April 1, 2005, the Navy formally announced the cancellation of the carrier's scheduled 15-month overhaul.

Before decommissioning, the Kennedy made several stops to allow the public to "say farewell" to her, including a stop at her "homeport" Boston Harbor. She was officially decommissioned in Mayport, Florida, on March 23, 2007.

After being decommissioned in March 2007, USS John F. Kennedy (CV-67) was initially placed in the Reserve Fleet, berthed at the Naval Inactive Ship Maintenance Facility in Philadelphia. Despite interest from several cities to convert the Kennedy into a museum ship, these plans did not materialize due to high costs and logistical challenges. Officially stricken from the Naval Vessel Register in October 2009, the Kennedy remained on the Navy's donation hold list until 2017, when the decision was made to scrap the ship. In early 2019, the Kennedy was towed to Brownsville, Texas, where the dismantling and recycling process began, marking the end of its storied history and significant contributions to naval aviation.

CVN-68 • USS NIMITZ

(Propulsion: Two nuclear reactors, four shafts. Length: 1,092 feet (332.85 meters). Beam: 134 feet (40.84 meters); Flight Deck Width: 252 feet (76.8 meters); Displacement: Approximately 97,000 tons (87,996.9 metric tons) full load. Speed: 30+ knots (34.5+ miles per hour). Crew: Ship's Company: 3,200 - Air Wing: 2,480. Armament: Two or three (depending on modification) NATO Sea Sparrow launchers, 20mm Phalanx CIWS mounts: (3 on Nimitz and Dwight D. Eisenhower and 4 on Vinson and later ships of the class.). Aircraft: 85.)

USS Nimitz has responded to numerous regional and international crises, earning a significant place in history, much like her namesake, Fleet Admiral Chester W. Nimitz.

The keel of USS Nimitz (CVN 68) was laid on June 22, 1968, at Newport News Shipbuilding and Dry Dock Company in Newport News, Virginia. The ship was commissioned on May 3, 1975, at Pier 12, Naval Station Norfolk, Va., by President Gerald R. Ford. Notable guests included Secretary of Defense James R. Schlesinger, Secretary of the Navy J. William Middendorf, III, Chief of Naval Operations Admiral James L. Holloway, III, and Mrs. James T. Nimitz-Lay, the ship's sponsor.

In his remarks to a crowd of over 20,000, President Ford stated: "Wherever the United States Ship Nimitz shows her flag, she will be seen as we see her now - a solid symbol of United States strength; United States resolve. Made in America and manned by Americans. Whether its mission is one of defense, diplomacy or humanity, Nimitz will command awe and admiration from some, caution and circumspection from others, and respect from all."

The Nimitz's first deployment began on July 7, 1976, when it departed Norfolk for the Mediterranean. The task force included the nuclear-powered cruisers USS South Carolina and USS California. This deployment marked the first time in 10 years that nuclear-powered ships had deployed to the Mediterranean. In November 1976, Nimitz was awarded the Battle "E" from Commander, Naval Air Forces Atlantic Fleet, for being the most efficient and foremost aircraft carrier in the Atlantic Fleet. The ship returned to Norfolk on February 7, 1977, after a seven-month deployment.

Nimitz embarked on its second Mediterranean deployment on December 1, 1977, and returned to Norfolk on July 20, 1978. During its third Mediterranean cruise, starting on September

10, 1979, Nimitz was dispatched to strengthen the U.S. Naval presence in the Indian Ocean as tensions rose over Iran's hostage-taking of 52 Americans. Operation "Evening Light," an attempt to rescue the hostages, was launched from Nimitz but was aborted when the number of operational helicopters fell below the required minimum. During this deployment, the ship operated 144 continuous days at sea. Nimitz's homecoming on May 26, 1980, was the largest given to any carrier battle group returning to the U.S. since World War II, with President and Mrs. Carter, members of Congress, military leaders, and thousands of families and friends in attendance.

On May 15, 1981, Nimitz departed Norfolk for the final phases of its workup schedule for an upcoming Mediterranean cruise. On May 25, an EA-6B Prowler crash-landed on the flight deck, killing 14 crewmen and injuring 45 others. The carrier returned to port to repair damaged catapults and went back to sea less than 48 hours later to complete its training schedule. During its fourth deployment in August 1981, Nimitz and USS Forrestal conducted an open ocean missile exercise in the Gulf of Sidra. On August 19, Nimitz aircraft from VF-41 engaged and shot down two Libyan MiGs.

In June 1985, Nimitz was dispatched to the Eastern Mediterranean in response to the hijacking of TWA Flight 847. The ship remained off the coast of Lebanon until August. After an extended deployment, Nimitz left the Mediterranean in May 1987 and sailed to its new homeport in Bremerton, Washington, arriving on July 2, 1987.

In September 1988, Nimitz operated off South Korea to provide security for the Olympic Games in Seoul. It then participated in Operation "Earnest Will" in the North Arabian Sea, protecting shipping lanes and escorting U.S.-registered Kuwaiti tankers. Nimitz was deployed to the Western Pacific and the Arabian Gulf in February 1991 during Operation Desert Storm, returning to Bremerton on August 24, 1991. Another deployment to the Arabian Gulf followed in February 1993, and the ship returned in August 1993.

In November 1995, Nimitz was deployed to the Western Pacific, Indian Ocean, and Arabian Gulf, where its presence helped calm tensions between mainland China and Taiwan. In September 1997, Nimitz embarked on an around-the-world cruise, returning to the East Coast for a multi-year overhaul in Newport News.

On September 21, 2001, after sea trials in the Virginia Capes, Nimitz began its transit around South America to its new homeport of NAS North Island in San Diego, California, arriving on November 13, 2001. The ship underwent a four-month post-shakedown maintenance availability at North Island in early 2002. Nimitz's eleventh operational deployment began on March 3, 2003, in support of Operation Iraqi Freedom (OIF) and Operation Enduring Freedom (OEF). It returned to San Diego on November 5, 2003.

Nimitz deployed again to the Persian Gulf on May 7, 2005, marking three decades of service. This deployment was depicted in the Emmy award-winning 2008 PBS documentary series "Carrier." In April 2007, Nimitz was deployed to the Arabian Sea, supporting OIF and anchoring off Chennai, India, to expand bilateral defense cooperation between India and the United States. The ship returned to North Island on September 30, 2007.

In January 2008, Nimitz deployed to the Pacific for a "surge" deployment. In February 2008, Russian Tu-95 'Bear' bombers overflew the carrier in the Western Pacific, intercepted by F/A-18C Hornets. Nimitz was awarded the Navy Battle "E" for battle efficiency for 2007. The ship returned to San Diego on June 3, 2008.

Nimitz Strike Group, including CVW-11, departed for a Western Pacific deployment on July 31, 2009, and began flying combat missions in support of Operation Enduring Freedom on September 21. In January 2010, while in the Persian Gulf, Nimitz was awarded the Meritorious Unit Commendation for its deployments in support of the wars in Iraq and Afghanistan in 2007 and 2008. The award was presented by Admiral Gary Roughead in a ceremony on the ship on January 6, 2010.

USS Nimitz (CVN-68), after 2010, continued its significant role in U.S. Navy operations. The carrier undertook several notable deployments and operations, highlighting its versatility and enduring capability.

In 2013, the Nimitz was deployed to the Persian Gulf and the Mediterranean Sea, where it played a crucial role in Operation Enduring Freedom and supported U.S. military operations in the Middle East. The carrier returned to its home port in Bremerton, Washington, after completing these missions.

From 2015 to 2017, the Nimitz underwent a Planned Incremental Availability (PIA) at the Puget Sound Naval Shipyard for maintenance and upgrades. This period was essential for ensuring the ship remained in optimal condition for future deployments.

In 2017, the Nimitz Carrier Strike Group (CSG) was deployed to the Western Pacific and the Indian Ocean, where it participated in various joint exercises and operations, including operations against ISIS in the Middle East. The deployment concluded in December 2017.

In April 2020, the coronavirus was reported to have spread to Nimitz when the first case was reported on 7 April. One sailor had received a positive result the previous week after exhibiting symptoms and was subsequently placed in isolation and removed from the ship. Another crew member also tested positive but was reported to have not been working on the ship. On 27 April, Nimitz completed a 27-day quarantine and began COMPTUEX training.

On 5 July 2020, the ship was deployed in the South China Sea along with USS Ronald Reagan.

On 31 December 2020, acting Secretary of Defense Chris Miller ordered Nimitz to return directly to her home port following a nearly ten-month deployment in the Fifth Fleet area of operation. The carrier was at the time supporting the withdrawal of U.S. troops in Somalia along with USS Makin Island and her amphibious ready group.

On 3 January 2021, in an abrupt reversal, acting Defense Secretary Miller ordered Nimitz to redeploy due to "recent

threats issued by Iranian leaders against President Trump and other U.S. government officials."

In May 2022, Nimitz led Carrier Strike Group 11 in the Eastern Pacific Ocean.

On 28 November 2022, Nimitz departed Kitsap and sailed to San Diego to pick up the carrier strike group leadership. She left San Diego on 3 December 2022 for her Pacific deployment leading Carrier Strike Group 11 and embarking Carrier Air Wing 17. On 21 January 2023, she made a port call to Singapore, departing on 26 January 2023. On 26 February 2023, Nimitz made a port call in Guam. She arrived in Busan for a scheduled port visit on 28 March 2023. Nimitz logged her 350,000th arrested landing on 22 April 2023. On 24 April 2023, Nimitz arrived in Laem Chabang, Thailand for a scheduled port visit. She returned to her homeport on 2 July 2023.

CVN-69 • USS DWIGHT D. EISENHOWER

(Propulsion: Two nuclear reactors, four shafts. Length: 1,092 feet (332.85 meters). Beam: 134 feet (40.84 meters); Flight Deck Width: 252 feet (76.8 meters). Displacement: Approximately 97,000 tons (87,996.9 metric tons) full load. Speed: 30+ knots (34.5+ miles per hour). Crew: Ship's Company: 3,200 - Air Wing: 2,480. Armament: Two or three (depending on modification) NATO Sea Sparrow launchers, 20mm Phalanx CIWS mounts: (3 on Nimitz and Dwight D. Eisenhower and 4 on Vinson and later ships of the class.). Aircraft: 85. Construction of the third nuclear-powered and second Nimitz-class aircraft carrier, USS Dwight D. Eisenhower (CVN-69), was authorized by Congress in fiscal year 1970).

The carrier, named after the nation's 34th president, was christened at 11:11 a.m. on October 11, 1975, by Mrs. Mamie Doud-Eisenhower, the ship's sponsor and widow of the late president. IKE was commissioned on October 18, 1977, under the command of Capt. William E. Ramsey, USN, and assigned to the Atlantic Fleet. After 14 months of fleet training, IKE set sail for her first Mediterranean deployment.

Since that time, IKE has completed eight Mediterranean deployments. In 1980, IKE's second extended deployment tallied 254 days at sea with only a five-day stopover in Singapore.

After its fourth deployment, IKE sailed into Newport News and drydock in October 1985 for a complex overhaul. The 18-month yard period included the addition of the Close-in Weapons System, NATO Seasparrow Missile System, Navy Tactical Data System, Anti-Submarine Warfare module, communications upgrades, and the rehabilitation of 1,831 berths in 25 compartments. IKE re-entered the fleet in April 1987.

In 1990, IKE completed its sixth Mediterranean Sea deployment. The deployment became a commemorative event in the worldwide 'Dwight D. Eisenhower Centennial,' celebrating the 100th anniversary of the late president's birth. During D-Day anniversary ceremonies off the coast of Normandy, IKE's son, John Eisenhower, and D-Day veterans embarked on the ship while Carrier Air Wing Seven conducted a memorial flyover of the American cemetery at Omaha Beach.

In response to Iraq's invasion of Kuwait, IKE became the first carrier to conduct sustained operations in the Red Sea. IKE was the second nuclear-powered aircraft carrier to transit the Suez Canal. IKE served as a ready striking force in the event Iraq invaded Saudi Arabia and participated in maritime interception operations in support of a United Nations embargo against Iraq.

After completion of an extensive shipyard period and workups, IKE deployed on September 26, 1991, to the Arabian Gulf to continue multinational operations with coalition forces in support of Operation Desert Storm. IKE returned to Norfolk on April 2, 1992. On January 6, 1993, IKE entered Norfolk Naval Shipyard for overhaul and conversion, and returned to the fleet on November 12, 1993.

In September 1994, IKE and the U.S. Army 10th Mountain Division again made history with the concept of "adaptive force packaging." The division's soldiers and equipment were loaded on board, and the IKE Army and Navy team headed for Port-au-Prince to spearhead Operation Uphold Democracy, the U.S.-led effort to restore the democratically elected government of Haiti.

One month later, in October 1994, IKE departed for a six-month deployment, which included flying missions in support of Operations Southern Watch and Deny Flight. This deployment marked the first time that women had deployed as crewmembers of a U.S. Navy combatant. The IKE, Carrier Wing Three, and COMCRUDESGRU Eight team included more than 400 women.

IKE returned to Newport News Shipbuilding on July 17, 1995, for an 18-month complex overhaul which was completed on January 27, 1997. The ship departed on its 10th deployment on June 10, 1998, and returned in December. In February 1999, IKE returned to the Norfolk Navy Shipyard for a six-month refitting and returned to the fleet in June. Upon completion in June 1999, she returned to full duty in the fleet.

In September 1994, Eisenhower and elements of the U.S. 10th Mountain Division first tested the concept of adaptive force packaging. The division's soldiers and equipment were loaded on board, and the ship's Army/Navy team headed for Port-au-Prince to lead Operation Uphold Democracy, the U.S.-led effort to restore the elected government of Haiti. One month later, in October 1994, Eisenhower departed for a six-month deployment, which included flying missions in support of Operation Southern Watch and Operation Deny Flight. This deployment marked the first time that women had deployed as crew members of a U.S. Navy combatant. Eisenhower, Carrier Air Wing Three, and COMCRUDESGRU Eight team included more than 400 women. The integration of women caused some negative headlines for the Navy. During the deployment, 15 women serving aboard had to be reassigned ashore due to pregnancy, earning the ship the nickname "The Love Boat." There was also a case of a sailor who filmed himself having sex with a female. Eisenhower returned to Newport News Shipbuilding on July 17, 1995, for an 18-month complex overhaul, which was completed on January 27, 1997. During this refitting, the popular game show Wheel of Fortune taped several weeks' worth of shows upon the boat for its twelfth season. The ship departed on her 10th deployment on June 10, 1998, and returned in December. In February 1999, she returned to the Norfolk Navy Shipyard for a six-month refitting and returned to the fleet in June. Upon completion in June 1999, she returned to full duty in the fleet.

2000s: Deploying in February 2000 and returning that August on the "Millennium Cruise," for the first time IKE's embarked aircraft dropped ordnance in combat while supporting Operation Southern Watch's No-Fly Zone over Iraq. On May 21, 2001, the ship returned to Newport News Shipbuilding and

Drydock Company (NNS&DDC) to begin a 4-year, $2.5 billion refueling and complex overhaul. In addition to replacing the cores in the two nuclear reactors, Eisenhower underwent complex renovations and major technological upgrades during this scheduled half-life overhaul. On January 25, 2005, IKE left NNS&DDC under tow, and returned to Norfolk Naval Station and the fleet. The modernizations are expected to extend the ship's service life well beyond 2025.

Persian Gulf / Indian Ocean Deployment: On October 3, 2006, with Carrier Air Wing 7 (CVW-7), Eisenhower returned to sea as the flagship of Rear Admiral Allen G. Myers, commanding Carrier Strike Group 8. CSG-8 also included the guided-missile cruiser USS Anzio (CG-68), guided-missile destroyers Ramage (DDG-61) and Mason (DDG-87), and the fast-attack submarine Newport News (SSN-750). She visited Naples, Italy, and then Limassol, Cyprus, for three days in October 2006 before departing to the east. She entered the Persian Gulf in December 2006. On January 8, 2007, a US AC-130 gunship based out of Djibouti was dispatched to target Al-Qaeda operatives located in Somalia. Eisenhower "was deployed in the Indian Ocean to provide air cover for the operation and, if needed, to evacuate downed airmen and other casualties." It joined other US and allied vessels from Bahrain-based Combined Task Force 150. The US spokesperson did not say what particular ships comprised the cordon, but the task force includes vessels from Canada, France, Germany, Pakistan, the UK, and the US. US ships of Combined Task Force 150 include the Arleigh Burke-class destroyer USS Ramage (DDG-61) and the Ticonderoga-class cruiser USS Bunker Hill (CG-52). The aim of the patrols is to "stop SICC leaders or foreign militant supporters escaping" Somalia. In March 2007, following the Iranian seizure of Royal Navy personnel, Eisenhower began battle group exercises off the Iranian coastline. In April 2007, the ship was relieved by the USS Nimitz.

On October 4, 2008, Eisenhower sailor Robert Lemar Robinson was killed on the ship during training exercises off the coast of North Carolina. The sailor was killed after being struck by an airplane at 8:15 p.m. on the carrier's flight deck. On February 21, 2009, Eisenhower deployed for the Arabian Sea and environs, rotating into the forward-deployed forces there. She served as the flagship of Carrier Strike Group 8 commanded by Rear Admiral Kurt W. Tidd. Also embarked were Carrier Air Wing 7 and the staff of Commander, Destroyer Squadron 28.

Other ships of Strike Group 8 were USS Bainbridge (DDG-96), USS Halyburton (FFG-40), USS Scranton (SSN-756), USS Vicksburg (CG-69), and USS Gettysburg (CG-64). In addition to supporting Operations Iraqi Freedom and Enduring Freedom, the strike group conducted maritime security operations, including anti-piracy operations. On 16 May, Eisenhower became the first Nimitz-class carrier to dock pierside in Manama, Bahrain. The last carrier to moor pierside in Bahrain was USS Rendova (CVE-114) in 1948. On 30 July 2009, Eisenhower returned to Naval Station Norfolk after almost a six-month deployment.

On 2 January 2010, Eisenhower again deployed to the Middle East, the U.S. 5th and 6th Fleet areas of operations. She again served as the flagship of the Dwight D. Eisenhower Carrier Strike Group commanded by Rear Admiral Philip S. Davidson. While in theater, the strike group provided security cooperation,

forward naval presence, maritime security, and crisis response. In addition to the "Ike," the strike group was made up of Carrier Air Wing 7; Commander, Destroyer Squadron 28; the guided-missile cruiser USS Hue City; and the guided-missile destroyers USS McFaul, USS Carney, and USS Farragut (DDG-99).

In June 2012, Eisenhower deployed to the Middle East for a nine-month deployment, supporting Operation Enduring Freedom. During this deployment, the ship participated in maritime security operations and theater security cooperation efforts in the U.S. 5th Fleet area of operations.

In 2013, Eisenhower underwent an extended maintenance period, known as a Drydocking Planned Incremental Availability (DPIA), at Norfolk Naval Shipyard. The ship received upgrades to its combat systems, hull, mechanical and electrical systems, and flight deck.

In 2016, Eisenhower again deployed to the Middle East, supporting Operation Inherent Resolve. The ship conducted airstrikes against ISIS targets in Iraq and Syria and supported coalition forces with maritime security operations.

In 2018, Eisenhower completed another DPIA at Norfolk Naval Shipyard, which included maintenance and modernization to extend the ship's operational capabilities.

In 2020, Eisenhower deployed to the U.S. 5th and 6th Fleet areas of operations. The deployment was notable for the extensive precautions taken due to the COVID-19 pandemic, with the ship conducting operations without port visits to minimize the risk of infection. During this deployment, Eisenhower set a new record for consecutive days at sea for a U.S. Navy aircraft carrier, surpassing the previous record set in 1973.

In 2021, Eisenhower participated in a Composite Training Unit Exercise (COMPTUEX) before deploying to the Middle East, supporting Operation Freedom's Sentinel and Operation Inherent Resolve. The ship continued to demonstrate its capabilities in maritime security and power projection.

In 2023, Eisenhower completed its most recent deployment, participating in various multinational exercises and providing a forward naval presence in key regions. The ship continues to serve as a critical asset in the U.S. Navy's carrier fleet, projecting power and ensuring maritime security worldwide.

CVN-70 • USS CARL VINSON

(Propulsion: Two nuclear reactors, four shafts. Length: 1,092 feet (332.85 meters). Beam: 134 feet (40.84 meters); Flight Deck Width: 252 feet (76.8 meters). Displacement: Approximately 97,000 tons (87,996.9 metric tons) full load. Speed: 30+ knots (34.5+ miles per hour). Crew: Ship's Company: 3,200 - Air Wing: 2,480. Armament: Two or three (depending on modification) NATO Sea Sparrow launchers, 20mm Phalanx CIWS mounts: (3 on Nimitz and Dwight D. Eisenhower and 4 on Vinson and later ships of the class.). Aircraft: 85.)

Commissioned on March 13, 1982, the Carl Vinson embarked on its maiden deployment in 1983, an eight-month global voyage that set the stage for its future operational legacy. This deployment saw the carrier navigate diverse waters, including the Caribbean Sea, Atlantic Ocean, Mediterranean Sea, South Atlantic and Indian Oceans, South China Sea, East Sea/Sea of Japan, and the Pacific Ocean, ultimately arriving at its new homeport of Naval Air Station Alameda, California.

In 1984, the Carl Vinson achieved the highest marks ever awarded to an aircraft carrier during an operational readiness examination. It then participated in RIMPAC '84, a multinational exercise involving naval forces from around the Pacific Rim. The ship's operational tempo continued to escalate with deployments to the Western Pacific and Indian Oceans, receiving accolades such as the Meritorious Unit Commendation and the Admiral Flatley Memorial Award for operational readiness and aviation safety.

The late 1980s and early 1990s were marked by a series of significant deployments and achievements. Notably, in 1989, the Carl Vinson participated in PACEX '89, the largest peacetime naval exercise since World War II, and conducted operations in the icy waters of the Bering Sea and the Western Pacific. In 1990, it embarked on its fifth deployment to the Western Pacific and Indian Ocean, earning the COMNAVAIRPAC Battle "E" award.

After a complex overhaul at Puget Sound Naval Shipyard from 1990 to 1993, the Carl Vinson resumed its operational duties, participating in operations such as Southern Watch in the Arabian Gulf. The mid-1990s saw continued excellence, with the carrier receiving multiple Admiral Flatley Awards and participating in commemorations and high-profile exercises.

The dawn of the 21st century brought new challenges and missions. In 2001, following the September 11 attacks, the Carl Vinson was redirected to the North Arabian Sea, launching the first strikes in support of Operation Enduring Freedom. This deployment underscored the carrier's pivotal role in the War on Terrorism, conducting over 4,200 combat sorties.

Post-2001, the Carl Vinson underwent numerous upgrades and maintenance periods to enhance its operational capabilities. The 2000s saw the carrier participating in significant operations and exercises, including RIMPAC '98, Operation Desert Fox, and Operation Southern Watch. In 2003, the carrier responded to global security needs by ensuring theater security in cooperation with allies in the Western Pacific.

The carrier's legacy of excellence continued through the 2010s. On January 2, 2010, the Carl Vinson deployed to the Middle East, serving as the flagship of Carrier Strike Group 1. The strike group, including Carrier Air Wing 17 and several guided-missile destroyers, provided security cooperation, forward naval presence, maritime security, and crisis response. This deployment was notable for its comprehensive engagement in the U.S. 5th and 6th Fleet areas of operations.

USS Carl Vinson's operations and activities after 2010: After 2010, the USS Carl Vinson (CVN-70) continued to play a pivotal role in U.S. naval operations, reflecting its robust capability and strategic importance.

In early 2011, the USS Carl Vinson gained widespread attention for hosting the burial at sea of Osama bin Laden after the successful raid by U.S. Navy SEALs in Pakistan. The carrier was deployed in the North Arabian Sea at the time, supporting operations in Afghanistan as part of Operation Enduring Freedom.

In November 2012, Carl Vinson departed from its homeport in San Diego for a deployment to the Western Pacific and the Middle East. During this deployment, the ship supported maritime security operations and conducted various exercises with allied navies. The carrier's air wing, Carrier Air Wing (CVW) 17, conducted numerous sorties, enhancing regional stability and providing critical support to ongoing operations in Afghanistan.

Carl Vinson returned to San Diego in May 2013 after completing its deployment. The carrier underwent a

maintenance period to ensure its readiness for future operations. This Planned Incremental Availability (PIA) included upgrades to its combat systems, hull, mechanical and electrical systems, and flight deck.

In August 2014, Carl Vinson departed San Diego for another deployment to the Western Pacific and the Middle East. This deployment was marked by the ship's participation in Operation Inherent Resolve, where it played a crucial role in conducting airstrikes against ISIS targets in Iraq and Syria. The carrier and its air wing launched numerous sorties, delivering precision strikes and providing close air support to coalition forces on the ground.

The carrier continued its operations in the Middle East well into 2015, contributing significantly to the fight against ISIS. The ship returned to San Diego in June 2015 after a successful deployment. Following this, Carl Vinson entered another maintenance period, during which it received further upgrades to enhance its operational capabilities.

In January 2017, Carl Vinson deployed again, this time to the Western Pacific. This deployment was notable for the heightened tensions in the region, particularly concerning North Korea's missile tests. The carrier conducted various freedom of navigation operations and exercises with allied navies, reinforcing the U.S. commitment to regional security and stability. In April 2017, the ship was redirected towards the Korean Peninsula as a show of force amidst rising tensions with North Korea.

In February 2018, Carl Vinson made a historic port visit to Vietnam, marking the first time a U.S. aircraft carrier had visited the country since the end of the Vietnam War. This visit was seen as a significant step in strengthening U.S.-Vietnam relations and demonstrating the U.S. commitment to a free and open Indo-Pacific. The carrier continued its deployment in the Western Pacific, participating in various exercises and operations with regional allies.

The ship underwent another maintenance period in 2019, during which it received significant upgrades and overhauls

to extend its service life. This maintenance was crucial in ensuring the carrier's continued readiness to meet future operational demands.

In January 2020, Carl Vinson deployed once again, initially operating in the Western Pacific before transitioning to the Middle East. This deployment was marked by the ongoing COVID-19 pandemic, which required the ship to implement strict health and safety protocols to prevent the spread of the virus among the crew. Despite these challenges, the carrier successfully conducted numerous operations, including airstrikes against terrorist targets and maritime security missions.

The carrier returned to its homeport in San Diego in May 2021. Following its return, Carl Vinson underwent another maintenance period to address wear and tear from its extensive deployment. The ship also received further upgrades to its combat and communication systems, ensuring it remained at the cutting edge of naval technology.

In August 2022, Carl Vinson departed for another deployment, this time focusing on the Indo-Pacific region. The deployment aimed to strengthen alliances and partnerships, conduct freedom of navigation operations, and enhance regional security. The carrier participated in several joint exercises with allied navies, showcasing its capabilities and reinforcing the U.S. presence in the region.

In early 2023, Carl Vinson continued its operations in the Indo-Pacific, engaging in numerous training exercises and port visits to enhance cooperation with regional partners. The ship's air wing conducted various sorties, ensuring readiness for any potential conflicts. The carrier returned to San Diego in July 2023, concluding another successful deployment.

CVN-71 • USS THEODORE ROOSEVELT

(Propulsion: Two nuclear reactors, four shafts. Length: 1,092 feet (332.85 meters). Beam: 134 feet (40.84 meters); Flight Deck Width: 252 feet (76.8 meters). Displacement: Approximately 97,000 tons (87,996.9 metric tons) full load. Speed: 30+ knots (34.5+ miles per hour). Crew: Ship's Company: 3,200 - Air Wing: 2,480. Armament: Two or three (depending on modification) NATO Sea Sparrow launchers, 20mm Phalanx CIWS mounts: (3 on Nimitz and Dwight D. Eisenhower and 4 on Vinson and later ships of the class.). Aircraft: 85.)

USS Theodore Roosevelt (CVN-71) is the fourth Nimitz-class carrier. Her history began on September 30, 1980, when a contract was awarded to Newport News Shipbuilding.

Construction began on October 31, 1981, when Secretary of Defense Casper Weinberger authenticated the keel laying of TR by initiating the first weld. Capt. Paul W. Parcells was named Prospective Commanding Officer in February 1984, and that October, the ship was officially christened. On October 25, 1986, TR was placed in active service.

Capt. Dayton W. Ritt became TR's second Commanding Officer on October 3, 1987, and on December 30, 1988, TR started her maiden deployment, which was also the maiden deployment of the first 10-squadron air wing, Carrier Air Wing Eight. USS Theodore Roosevelt was awarded the Battle "E" from Commander, Naval Air Force, U.S. Atlantic Fleet, on March 20, 1990.

On June 9, 1990, Capt. Charles S. Abbot became the ship's third Commanding Officer, and on December 28, TR and CVW-8 deployed for Operation Desert Shield. TR entered the war on January 9, 1991, eventually flying over 4,200 sorties (more than any other carrier) and dropping more than 4,800,000 pounds of ordnance before the cease-fire on February 28.

When Iraqi forces turned on the Kurds, TR and CVW-8 were among the first coalition forces in Operation Provide Comfort, flying patrols over northern Iraq. After a 189-day deployment, with 169 days at sea, TR returned to Norfolk on June 28, 1991. On February 14, 1992, the ship won its second Battle "E." This was followed by the award of the Battenburg Cup for 1991 as the Atlantic Fleet's premier ship.

Capt. Stanley W. Bryant became TR's fourth Commanding Officer on August 27, 1992.

TR and CVW-8 began their third deployment on March 11, 1993, teamed with the Special Purpose Marine Air Ground Task

Force (SPMAGTF) to test the concept of embarking a multi-purpose Marine force in a carrier. TR hosted President Bill Clinton's first visit to a U.S. Navy ship, then sailed to the Adriatic as CVW-8 planes enforced Operation Deny Flight in the U.S. no-fly zone over Bosnia. In June, on the way to only her second port visit, TR was ordered to turn around and transit the Suez Canal en route to the Red Sea to participate in Operation Southern Watch, enforcing the no-fly zone over Iraq.

Deployed for 184 days, TR spent 169 days underway. Her flight deck logged over 16,000 hours, and CVW-8 flew more sorties than during the Persian Gulf War. For its accomplishments, the ship received its second Meritorious Unit Commendation.

In November 1993, TR entered Norfolk Naval Shipyard (NNSY) for a Selected Restricted Availability (SRA). Heading back to sea on April 14, 1994, TR became the first nuclear carrier to complete an SRA ahead of schedule at NNSY.

Awards for 1993 continued. TR received the CINCLANTFLT Golden Anchor Award for the best retention in an Atlantic Fleet carrier. On March 10, 1994, TR received its third Battle "E." Then on June 3, TR was awarded its second Battenburg Cup as the best ship in the Atlantic Fleet.

On July 8, 1994, Capt. Ronald L. Christenson became TR's fifth Commanding Officer.

TR and CVW-8 began their fourth deployment in March 1995, operating in the Red Sea in support of Operation Southern Watch. TR also provided a "Forward... from the Sea" presence, conducting flight operations in support of Operations Deny Flight and Sharp Guard over the skies of Bosnia and in the Adriatic operating areas. Deny Flight evolved into Operation Deliberate Force, as CVW-8 aircraft led NATO strikes against strategic Bosnian Serb targets in Sarajevo-Herzegovina. During TR's transit home, Secretary of the Navy John Dalton came

aboard and awarded the Theodore Roosevelt Battle Group the Navy Unit Commendation for its Bosnia operations.

In 1996, TR received its third consecutive Golden Anchor Award and Commander, Naval Air Force, U.S. Atlantic Fleet's first Security Excellence Award. CVW-3 joined TR in May 1996 prior to her port visit to Halifax, Nova Scotia. On November 1, 1996, Capt. David Architzel became TR's sixth Commanding Officer. TR deployed for her fifth deployment on November 25, 1996, conducting operations in the Mediterranean and Arabian Gulf in support of Operation Southern Watch.

On July 8, 1997, TR entered the Newport News Shipbuilding yard for a one-year Extended Drydock and Selected Restricted Availability (EDSRA), her first major overhaul since commissioning. In February 1998, TR received her fifth Golden Anchor Award while in the shipyard. One year later, TR returned to her homeport at the Norfolk Naval Station.

Capt. David R. Bryant became TR's seventh Commanding Officer on September 22, 1998.

TR began her sixth deployment on March 26, 1999, with CVW-8. They were immediately called to duty in the Ionian Sea to support NATO's Operation Allied Force. TR and CVW-8 aircraft conducted air strikes for two months over the skies of Kosovo against the Serbians. TR and CVW-8 were then dispatched to support Operation Southern Watch, enforcing the "no-fly" zone over Southern Iraq. TR returned to her homeport of Norfolk, Va., on September 24, 1999.

On January 7, 2000, TR entered a Planned Incremental Availability at the Norfolk Naval Shipyard, Portsmouth, Virginia. This six-month maintenance period ended on June 30, 2000, when the ship departed on sea trials.

TR successfully completed her Inter-Deployment Training Cycle in August 2001.

TR began her seventh deployment on September 19, 2001, with Carrier Air Wing One. After a routine transit of the Atlantic Ocean, TR and CVW-1 were called to support Operation Enduring Freedom. Receiving support from around the country, TR and CVW-1 led coalition forces in creating a more stable government in Afghanistan and hunting down members of al-Qaeda. After breaking the record for the longest period underway since WWII, TR's crew only hit two ports after 159 days at sea. TR was again awarded the Battenberg Cup, Naval Unit Commendation, and Battle E. TR returned on March 17, 2002.

On February 19, 2004, TR made its way up the Elizabeth River to start a Docked Planned Incremental Availability (DPIA) at Norfolk Naval Shipyard (NNSY) in Portsmouth, VA. TR entered its slot at Dry Dock 8 and the aft dock wall closed behind it. As the river slowly began draining the massive concrete void, Sailors observing must have wondered exactly how such a seemingly insurmountable task would be possible to complete. Several extensive evolutions awaited the several thousand people making up the TR project team, consisting of its own Sailors, shipyard workers, and contractors. The mission of transforming an 18-year-old, 1,092-foot-long, 97,000-ton aircraft carrier into what was essentially a brand new ship was no small one. She entered the shipyard with the most extensive work package ever attempted within the allotted time period. The purpose of the DPIA was to ensure TR maintained its material condition as well as kept up with technological advances to maximize mission capability.

USS Theodore Roosevelt conducted a "fast cruise" from December 6-10, 2004, marking the end of the ship's Docked Planned Incremental Availability (DPIA) at Norfolk Naval Shipyard (NNSY) in Portsmouth, VA. Although sea trials do not officially begin until TR leaves NNSY and gets underway, the fast cruise was meant to showcase more of a change from one evolution to the other. It also showed that TR was ready for

tasking. The ship was next expected to return home to Naval Station Norfolk upon completion of sea trials before the ship's first holiday leave period.

USS Theodore Roosevelt (TR) (CVN-71) met its goal of completing its Docked Planned Incremental Availability (DPIA) at Norfolk Naval Shipyard (NNSY) in Portsmouth, VA, and sea trials, as well as delivering the ship back to the fleet on December 17, 2004. During the ship's DPIA, everything from converters for hot water heaters and improved head facilities, to upgrades for the elevators and major work on two of the four catapults, was overhauled or replaced. A major project while in dry dock was the removal, repair, and reinstallation of the ship's propellers (or screws). Two of the four were replaced altogether.

By mid-January 2005, USS Theodore Roosevelt (CVN-71) and Carrier Air Wing Eight (CVW-8) were underway again,

and the flight deck was back in business. TR conducted carrier flight deck certifications, which took a tremendous amount of manpower to make this evolution happen. Certifications became a necessity for aircraft carriers coming out of dry-docking planned incremental availability (DPIA). Every job performed by personnel from the hangar bay and flight deck to those in Supply and Medical was observed and evaluated.

By April 2005, the nuclear-powered aircraft carrier and Carrier Air Wing Eight (CAG-8) were underway for Tailored Ship Training Availability and Final Evaluation Problem (TSTA/FEP) in the Atlantic Ocean.

A ceremony held aboard USS Theodore Roosevelt (CVN-71) on May 4, 2005, recognized the ongoing friendship between Spain and the United States with the joining of a Spanish ship to the TR Carrier Strike Group. Alvaro de Bazan (F-101), a medium-sized Spanish frigate, officially became a part of Carrier Strike Group 2.

By May 19, 2005, USS Theodore Roosevelt (CVN-71) was underway off the coast of Virginia, conducting carrier qualifications in the Atlantic Ocean.

In September 2005, USS Theodore Roosevelt deployed to the Persian Gulf in support of Operation Iraqi Freedom and maritime security operations. During this deployment, the carrier's air wing flew numerous sorties, providing close air support to coalition forces and conducting surveillance missions. The deployment lasted until March 2006, when TR returned to its homeport in Norfolk, Virginia.

USS Theodore Roosevelt underwent another maintenance period in 2007, which included upgrades to its systems and infrastructure to ensure continued operational readiness. In September 2008, TR deployed again to the Persian Gulf, participating in Operation Enduring Freedom and conducting maritime security operations. The carrier's air wing played a crucial role in providing air support to ground forces in Afghanistan.

In March 2009, TR returned to Norfolk after completing its deployment. Later that year, the carrier underwent another Planned Incremental Availability (PIA) to address maintenance needs and enhance its capabilities.

In May 2011, USS Theodore Roosevelt entered Norfolk Naval Shipyard for a major mid-life Refueling and Complex Overhaul (RCOH). This extensive maintenance period, which lasted over three years, included refueling the ship's reactors, upgrading combat systems, and refurbishing living quarters and other essential areas. The RCOH was completed in August 2013, significantly extending the carrier's service life.

After completing the RCOH, USS Theodore Roosevelt began sea trials and post-overhaul certifications. In March 2015, TR deployed to the Middle East to support Operation Inherent Resolve, conducting airstrikes against ISIS targets in Iraq and Syria. The deployment demonstrated the carrier's enhanced capabilities following its overhaul.

In January 2017, TR deployed again to the Western Pacific and the Middle East. The carrier conducted freedom of navigation operations, participated in joint exercises with allied navies, and provided a forward naval presence to deter regional threats. The deployment reinforced the U.S. commitment to maintaining stability and security in key regions worldwide.

In October 2018, USS Theodore Roosevelt returned to its homeport in San Diego, California, after a successful deployment. The carrier continued to participate in various training exercises and underwent routine maintenance to maintain its readiness.

In January 2020, USS Theodore Roosevelt deployed to the Indo-Pacific region. However, the deployment was interrupted by a significant outbreak of COVID-19 among the crew, leading

to a port visit to Guam for quarantine and medical treatment. The situation highlighted the challenges of maintaining operational readiness during a global health crisis.

In February 2021, TR returned to San Diego after completing its deployment. The carrier underwent further maintenance and upgrades to prepare for future missions.

In April 2022, USS Theodore Roosevelt participated in a large-scale naval exercise in the Pacific, demonstrating its capabilities in joint operations with allied forces. The exercise aimed to enhance interoperability and strengthen regional security partnerships.

By mid-2023, USS Theodore Roosevelt was again deployed to the Western Pacific, where it continued to play a vital role in maintaining maritime security and stability. The carrier conducted numerous exercises with partner navies and provided a forward presence to deter potential adversaries.

CVN-72 • USS LINCOLN

(Propulsion: Two nuclear reactors, four shafts. Length: 1,092 feet (332.85 meters). Beam: 134 feet (40.84 meters); Flight Deck Width: 252 feet (76.8 meters). Displacement: Approximately 97,000 tons (87,996.9 metric tons) full load. Speed: 30+ knots (34.5+ miles per hour). Crew: Ship's Company: 3,200 - Air Wing: 2,480. Armament: Two or three (depending on modification) NATO Sea Sparrow launchers, 20mm Phalanx CIWS mounts: (3 on Nimitz and Dwight D. Eisenhower and 4 on Vinson and later ships of the class.). Aircraft: 85.)

Lincoln's contract was awarded to Newport News Shipbuilding on December 27, 1982; her keel was laid on November 3, 1984, at Newport News, Virginia. The ship was launched on February 13, 1988, and commissioned on November 11, 1989. She cost $4.5 billion in 2007 dollars.

Abraham Lincoln was transferred to the Pacific in September 1990, performing Gringo-Gaucho with the Argentine Naval Aviation during her transit. Her maiden Western Pacific deployment came unexpectedly on May 28, 1991, in response to Operation Desert Shield/Desert Storm. While heading towards the Indian Ocean, the ship was diverted to support evacuation operations after Mount Pinatubo erupted on Luzon Island in the Philippines. In support of Operation Fiery Vigil, Lincoln led a 23-ship armada that moved over 45,000 people from the Subic Bay Naval Station to the port of Cebu in the Visayas. It was the largest peacetime evacuation of active military personnel and their families in history. After Fiery Vigil, Lincoln steamed toward the Persian Gulf to run reconnaissance and combat air patrols in Iraq and Kuwait, assisting allied and U.S. troops involved with Desert Storm.

In early 1992, the ship was at Naval Air Station Alameda on Ship's Restricted Availability (SRA) for minor maintenance and refitting. In October 1993, the carrier was ordered to the coast of Somalia to assist UN humanitarian operations. For four weeks, Abraham Lincoln flew air patrols over Mogadishu in support of Operation Restore Hope. Abraham Lincoln was the first carrier to integrate female aviators into the crew after the Combat Exclusion Laws were lifted on April 28, 1993. The ship left San Diego on October 24, 1994, to begin refresher training. The next day, Lieutenant Kara Spears Hultgreen, the first female

F-14 Tomcat pilot, died when her plane crashed into the sea on final approach due to a combination of engine malfunction and pilot error. Abraham Lincoln's third deployment began in April 1995 when Lincoln was sent to the Persian Gulf, where the ship assisted in Southern Watch and Operation Vigilant Sentinel. During an underway replenishment, Lincoln was hit by USS Sacramento due to Sacramento's steering difficulties caused by a split rudder. The collision damaged the port side of USS Sacramento, partially crushed a female crew berthing area, and punched a large hole in Sacramento's superstructure. Lincoln was able to continue its mission while Sacramento had to dock at Jebel Ali, UAE, for several weeks for repair.

Abraham Lincoln began a fourth deployment in June 1998. Once again, the ship headed for the Persian Gulf in support of Southern Watch. The ship spent three months in the gulf during one of the hottest summers in recent years, with temperatures on the flight deck reported at 150 degrees Fahrenheit (66 °C). In 1999, the ship participated in several internal Navy exercises and underwent upkeep at Puget Sound Naval Shipyard in Bremerton, Washington.

The carrier's fifth deployment commenced in August 2000 when Abraham Lincoln again traveled to the Persian Gulf in support of Southern Watch. On this deployment, the carrier, air wing, and battle group ships earned the Navy Meritorious Unit Commendation. Additionally, the ship earned the prestigious Arleigh Burke Award as the most improved command in the Pacific Fleet. Abraham Lincoln was in port on September 11, 2001. She was put to sea on July 20, 2002, to support Operation Enduring Freedom. She took up station once more in support of Operation Southern Watch before making a port visit to Perth, Western Australia. During this time, Lincoln was ordered to the Persian Gulf to take part in Operation Iraqi Freedom, which forced the Navy to extend Lincoln's stay from January 20, 2003, to May 6, 2003. The news of this extension was delivered to the ship's crew on New Year's morning by the then Battle Group Commander, RADM Kelly, with the famous phrase, "We don't need to be home holding our loved ones; we need to be here holding the line. Get over it!"

The USS Abraham Lincoln returned to port carrying its "Mission Accomplished" banner on May 2, 2003. Abraham Lincoln and the carrier battle group and air wing helped deliver the opening salvos and airstrikes in Operation Iraqi Freedom. During her deployment, some 16,500 sorties were flown, and 1.6 million pounds of ordnance were used. Sea Control Squadron 35 (VS-35), the "Bluewolves," was instrumental in delivering over 1 million pounds of fuel to these strike aircraft, one of the largest aerial refueling undertakings by a carrier aviation squadron in history. The carrier returned home in May 2003, receiving a visit from President George W. Bush before officially ending Lincoln's deployment by docking at San Diego and then returning to her homeport in Everett, WA. Bush delivered a speech that day announcing the end of major combat operations in the War on Iraq. A large sign in the background for his speech read "Mission Accomplished." The White House said their services constructed the banner. As explained by Cmdr. Conrad Chun, a Navy spokesman, "The banner was a Navy idea, the ship's idea. The idea popped up in one of the meetings aboard the ship preparing for its homecoming and thought it would be good to have a banner, 'Mission Accomplished.' The sailors then asked if the White House could get the sign made. The banner signified the successful completion of the ship's deployment," Cmdr. Chun continued, noting that Abraham Lincoln was deployed for 290 days, longer than any other nuclear-powered aircraft carrier in history.

Abraham Lincoln departed for her next voyage on October 15, 2004. The carrier was on a port call in Hong Kong when a 9.0-magnitude earthquake struck southern Asia on December 26, 2004. To help with the international relief effort and assist with search and rescue efforts already underway, Lincoln deployed to the hard-hit western coast of Sumatra to provide humanitarian assistance for Operation Unified Assistance. In mid-January 2005, the carrier left Indonesia after the Indonesian government refused to allow fighter pilots assigned to Lincoln to conduct air patrols and training flights. By law, U.S. carrier-based pilots must practice at least once every two to three weeks to remain "fit"; otherwise, they are grounded. Despite the move into international waters, Lincoln continued to provide support to the region until February 4. During the carrier's 33 days on station, she and her strike group delivered 5.7 million pounds of relief supplies. The 17 helicopters assigned to HSL-47 Saberhawks and HS-2 "Golden Falcons," attached to CVW-2, flew 1,747 relief missions along the western coast of Sumatra. The carrier's departure coincided with the arrival of the hospital ship Mercy.

At the end of February 2006, Abraham Lincoln went underway as part of a scheduled "WestPac" deployment. The carrier returned to Naval Station Everett on August 8, 2006, and on August 27, relocated to Puget Sound Naval Shipyard for an upkeep/overhaul period. On December 18, 2006, the USS Abraham Lincoln left dry dock at the shipyard ahead of schedule and under budget. The Puget Sound Naval Shipyard and Intermediate Maintenance Facility (PSNS & IMF) completed ship tank maintenance in less than half the scheduled time. In 89 days, 18 tanks were completed. The Tank Value Stream Team achieved this by partnering with Ship's Force and the Lincoln Project Team. While in dry dock, the whole ship was painted by the crew at nights and on weekends rather than waiting for contractors to do the job. The refit was completed on March 26, 2007, when Rear Adm. Scott Van Buskirk assumed command of Carrier Strike Group Nine (CSG 9) from Rear Adm. Bill Goodwin.

On March 19, 2008, USS Abraham Lincoln departed Naval Station Everett for a seven-month deployment to the Persian Gulf. The ship returned to Naval Station Everett on October 12, 2008.

In January 2010, Abraham Lincoln departed for another deployment to the Persian Gulf, where she supported operations in Iraq and Afghanistan. This deployment also saw the carrier participate in anti-piracy operations off the coast of Somalia.

USS Abraham Lincoln deployed again in December 2011 to support maritime security operations and coalition efforts in the Middle East. The carrier's air wing conducted missions over Afghanistan and supported counter-piracy efforts in the region.

In March 2013, Abraham Lincoln entered Newport News Shipbuilding for a four-year Refueling and Complex Overhaul (RCOH). This extensive maintenance period included refueling the ship's reactors, upgrading combat systems, and refurbishing living quarters and other essential areas. The RCOH was completed in May 2017, significantly extending the carrier's service life.

In April 2019, Abraham Lincoln deployed from Norfolk, Virginia, to the Middle East amid escalating tensions with Iran. The carrier conducted air operations and provided a visible presence in the region, reinforcing U.S. commitments to its allies.

During the COVID-19 pandemic, Abraham Lincoln was involved in various exercises and operations while implementing health protocols to safeguard the crew. The carrier continued to play a crucial role in ensuring maritime security and stability in the Indo-Pacific region.

2022: In January 2022, USS Abraham Lincoln (CVN-72) made history by deploying with the first female to command a nuclear-powered aircraft carrier, Capt. Amy Bauernschmidt. The deployment took the carrier to the Indo-Pacific region, where it participated in various exercises and operations aimed at strengthening alliances and maintaining regional security.

During the deployment, USS Abraham Lincoln participated in the multinational exercise RIMPAC 2022 (Rim of the Pacific Exercise), which is the world's largest international maritime warfare exercise. The exercise involved navies from 26 countries and included complex warfare scenarios to improve interoperability among the participating forces.

The carrier also conducted freedom of navigation operations in the South China Sea, demonstrating the United States' commitment to maintaining free and open sea lanes in contested waters. This was particularly significant given the rising tensions in the region due to territorial disputes.

Additionally, USS Abraham Lincoln was involved in joint exercises with the Japanese Maritime Self-Defense Force and the Royal Australian Navy, enhancing collaboration and readiness among allied forces.

In early 2023, USS Abraham Lincoln continued its operational activities in the Indo-Pacific region. The carrier conducted numerous training exercises to maintain the proficiency and readiness of its crew and embarked air wing. These exercises included flight operations, damage control drills, and joint training with other U.S. Navy vessels and allied ships.

In March 2023, the carrier made a port visit to Yokosuka, Japan, home to the U.S. Navy's 7th Fleet. This visit underscored the strong partnership between the United States and Japan and provided an opportunity for the crew to engage with the local community and conduct necessary maintenance.

During the mid-year, USS Abraham Lincoln participated in additional freedom of navigation operations, reaffirming the

U.S. commitment to uphold international law and the rights of all nations to access global commons.

In August 2023, the carrier took part in another significant exercise, Talisman Sabre, a biennial exercise between the United States and Australia. This exercise focused on high-end warfighting scenarios, involving complex air, land, and sea operations to enhance the combat readiness and interoperability of U.S. and Australian forces.

Throughout 2023, USS Abraham Lincoln remained a critical component of the U.S. Navy's forward-deployed presence in the Indo-Pacific, serving as a deterrent to potential adversaries and reassuring allies of the U.S. commitment to regional stability and security.

By the end of 2023, the carrier had completed its deployment and returned to its homeport, where it began preparations for its next series of training cycles and maintenance periods to ensure continued readiness for future missions.

Overall, USS Abraham Lincoln's activities in 2022 and 2023 demonstrated its versatility and importance in maintaining maritime security and supporting U.S. strategic interests in the Indo-Pacific region.

CVN-73 • USS GEORGE WASHINGTON

(Propulsion: Two nuclear reactors, four shafts. Length: 1,092 feet (332.85 meters). Beam: 134 feet (40.84 meters); Flight Deck Width: 252 feet (76.8 meters). Displacement: Approximately 97,000 tons (87,996.9 metric tons) full load. Speed: 30+ knots (34.5+ miles per hour). Crew: Ship's Company: 3,200 - Air Wing: 2,480. Armament: Two or three (depending on modification) NATO Sea Sparrow launchers, 20mm Phalanx CIWS mounts: (3 on Nimitz and Dwight D. Eisenhower and 4 on Vinson and later ships of the class.). Aircraft: 85.)

In 1994, during George Washington's maiden deployment, the ship served as the backdrop for the 50th anniversary of D-Day. The USS George Washington was also critical in responding to Iraq's threatened second invasion of Kuwait, being called to the Persian Gulf for the second time in the deployment. When the ship entered the Red Sea, Iraq began its pullback of troops from the border.

In 1996, during its second deployment, George Washington hosted a meeting of the Joint Military Commission, composed of the military leaders of the former warring factions in Bosnia and Herzegovina. The ship also played a peacekeeping role in Operation Decisive Endeavor in Bosnia and Herzegovina and enforced the "No-Fly Zone" over Southern Iraq as part of Operation Southern Watch (OSW).

On its third deployment from October 1997 to April 1998, George Washington spent most of her six-month cruise in the Persian Gulf as the cornerstone of the U.S. military force.

On its fourth deployment from June 21, 2000, to December 2000, George Washington again spent a large portion of the six-month deployment in the Persian Gulf as the centerpiece of the U.S. military presence there. The deployment included operations in the Mediterranean Sea, Indian Ocean, and Persian Gulf. While in the Persian Gulf, the battle group supported OSW by flying more than 800 sorties over Iraq. Surface forces supported UN sanctions against Iraq by conducting Maritime Interception Operations and diverting more than 20,000 metric tons of oil smuggled out of Iraq in violation of UN sanctions. In the Adriatic, the battle group was a stabilizing presence when tensions rose in the region after presidential elections in Yugoslavia. During the deployment, battle group ships

steamed more than 400,000 nautical miles (740,000 km) and spent a combined 1,800 days underway. The aircraft of Carrier Air Wing 17 (CVW-17) flew more than 9,000 sorties and made 9,000 arrested landings aboard George Washington. George Washington returned to homeport on December 19, 2000.

On February 13, George Washington began a six-month Planned Incremental Availability (PIA) at Norfolk Naval Shipyard. The ship received upgrades to many ship systems, including berthing spaces, ventilation systems, and computer networking. On July 31, the ship began four days of sea trials before returning to homeport in preparation for workups for a planned 2002 deployment.

On September 6, George Washington was presented with the 2000 Battenberg Cup, awarded annually to the Atlantic Fleet ship or submarine with the greatest crew accomplishments during the previous calendar year. It was the second time GW had won the award (the ship was also the winner of the 1997 award). The crew also accepted the Flatley Award for aircraft carriers with the best aviation safety record. On the morning of September 11, George Washington was operating off the coast of Virginia conducting routine carrier qualifications when the September 11 attacks took place. She was diverted north and arrived in New York City the following day. For the next three days, the ship and her air wing (most of which was hastily transferred from USS John F. Kennedy (CV-67)) provided airspace defense for the city and surrounding area in coordination with NORAD. The ship then returned home and resumed the Inter-Deployment Training Cycle. From November 2 to 29, George Washington participated in Tailored Ship's Training Availability (TSTA) I/II. For the remainder of the year and into 2002, the ship hosted more carrier qualifications.

Washington transited the Atlantic Ocean as she and her embarked Carrier Air Wing 17 (CVW-17) returned after conducting missions in support of Operation Enduring Freedom and Operation Southern Watch. The George Washington Battle Group deployed on June 20, 2002, and headed for the North Arabian Sea, where it relieved the USS John F. Kennedy (CV-67) Battle Group on July 19.

On September 11, 2002, George Washington was relieved by the USS Abraham Lincoln (CVN-72) Battle Group in the Persian Gulf, where George Washington had been supporting Operation Southern Watch and Operation Enduring Freedom. On October 2, the crew was treated to a flight deck concert by the band 3 Doors Down while the ship was making a port visit to Lisbon, Portugal. Footage from this concert and the band's tour of the ship was used in the video for the band's single "When I'm Gone". The George Washington Battle Group returned to Norfolk on December 20, 2002. During the six-month

deployment, George Washington launched approximately 10,000 sorties.

Washington personnel carried injured crew members across the ship's flight deck after four personnel were rescued from the burning merchant fishing vessel Diamond Shoal. On June 23, George Washington was presented with the 2002 Battenberg Cup. It was the third time the ship had won the prestigious award. On September 11, while the ship was operating off the coast of Virginia, an arresting wire parted while an F/A-18 was landing. As the wire parted, it snapped back violently across the deck, injuring eleven crewmen, two critically. The sailors, who were part of the ship's company, VAW-120, and VFA-106, had to be medically evacuated from the ship, but all survived the mishap. The aircraft was lost over the side, but the pilot was able to eject safely. Footage of the mishap was later used in a National Geographic special about the ship that was being filmed at the time.

GW returned to Norfolk, Va., on December 19, 2003, after 40 days at sea, where she successfully completed her Composite Training Unit Exercise (COMPTUEX). The GWSTRKGRU was composed of Destroyer Squadron 28, Carrier Air Wing 7 (CVW-7), USS Vella Gulf (CG-72), USS Ramage (DDG-61), USS Ross (DDG-71), USS Bulkeley (DDG-84), USS Elrod (FFG-55), and USS Supply (T-AOE-6). The exercise involved more than 7,600 sailors operating off the coasts of Virginia, North Carolina, Georgia, and Florida. Having completed this final phase of the training cycle, the strike group assumed surge status and was fully qualified to deploy.

The Navy announced on January 13, 2004, that the George Washington Carrier Strike Group would depart for a "surge" deployment a week later. On January 20, GW, with CVW-3 embarked, deployed in support of the global war on terrorism. George Washington made a port visit at Souda Bay, Crete, from February 6 to 10, 2004. On February 16, George Washington transited the Suez Canal and entered the Red Sea on February 17.

On February 20, George Washington entered the Gulf of Aden and a week later was conducting operations in the Persian Gulf. On March 13, the ship made the first of three port visits to Jebel Ali, United Arab Emirates. On April 8, F/A-18 Hornets from Carrier Air Wing Seven participated in Operation Vigilant Resolve. One of the Naval Air Station Oceana-based "Wildcats" from Strike Fighter Squadron 131 (VFA-131) conducted a 20 mm strafing run against an enemy position. Another VFA-131 Hornet dropped two 500-pound GBU-12 laser-guided bombs on another enemy position in Fallujah, Iraq, on April 9. This was the first live ordnance dropped by CVW-7 aircraft since George Washington deployed in support of Operations Iraqi Freedom and Enduring Freedom. After being relieved by the USS John F. Kennedy (CV-67), the ship began her transit home, making a final port visit at Naples, Italy, from July 14 to 18. After traveling more than 51,000 nautical miles (94,000 km) and spending six months at sea, George Washington completed its sixth Mediterranean and Persian Gulf deployment and returned to Norfolk on July 26.

On January 28, 2005, the ship entered dry dock for Planned Incremental Availability (PIA). Many of the ship's systems were upgraded, and maintenance was done to the hull. The ship's four jet blast deflectors were removed and upgraded to handle the increased heat generated by the F/A-18E/F Super Hornet. The ship's defensive weapons configuration was also altered, as one Phalanx CIWS mount and one Sea Sparrow launcher were removed and replaced with two Rolling Airframe Missile launchers. During the 11 months the ship was dry docked, the

crew contributed 20,000 hours of volunteer community service to the city of Newport News. The availability was completed on schedule, and George Washington returned to her Norfolk homeport on December 17.

On December 1, 2005, the United States Navy announced that George Washington would replace USS Kitty Hawk (CV-63) as the forward-deployed carrier at Yokosuka Naval Base in Yokosuka, Japan, becoming the first nuclear-powered surface warship permanently stationed outside the continental US. In an attempt to explain the carrier's mission to the Japanese public, the U.S. Navy printed a manga about life aboard GW, titled "CVN-73".

GW and CVW-17 left Norfolk on April 4 for a scheduled two-month deployment to operate as part of SOUTHCOM's "Partnership of the Americas." This deployment included counter-drug operations in the Caribbean Sea, crew exchanges, and exercises with Latin American and South American navies, and port visits for the carrier and strike group, which consisted of USS Monterey (CG-61), USS Stout (DDG-55), and USS Underwood (FFG-36). The first of these port visits took place from April 14-17 in St. Maarten and from May 15-18 in Antigua. GW returned to Norfolk on May 24. In a ceremony held on September 1, Commanding Officer Garry R. White was promoted to Rear Admiral, marking a rare occasion when a Flag Officer commanded a ship. She entered Norfolk Naval Shipyard for a Planned Incremental Availability in September to prepare for her upcoming homeport transfer to Yokosuka Naval Base in Yokosuka, Japan. Work included the removal and replacement of the ship's radar mast, propeller screws, and re-alignment of the Aviation Intermediate Maintenance Department. On December 14, Rear Admiral White was relieved by Captain David Dykhoff in a ceremony held at Naval Station Norfolk.

In April 2008, USS George Washington embarked on a significant journey, departing Norfolk, Virginia, for a transit around South America. This voyage included the Gringo-Gaucho exercise with the Argentine Naval Aviation, enhancing interoperability between the two navies. The carrier's ultimate destination was Yokosuka, Japan, where it would replace USS Kitty Hawk (CV-63) as the forward-deployed carrier. This transition marked a historic moment as George Washington became the first nuclear-powered surface warship to be permanently stationed outside the continental United States. The carrier conducted a planned turnover with Kitty Hawk at Naval Station Pearl Harbor, Hawaii, before proceeding to its new homeport.

In its initial years in Japan, George Washington quickly integrated into the U.S. Seventh Fleet's operations, participating in numerous exercises and missions. The carrier's presence was crucial in maintaining stability and responding to regional threats in the Indo-Pacific. Throughout 2009 and 2010, George Washington conducted several high-profile exercises, including the Annual Exercise (ANNUALEX) with the Japan Maritime Self-Defense Force and joint operations with other regional allies.

In 2011, USS George Washington played a pivotal role in humanitarian assistance following the devastating earthquake and tsunami that struck Japan in March. As part of Operation Tomodachi, the carrier provided essential support to relief efforts, delivering supplies, and assisting in search and rescue operations. This mission underscored the carrier's versatility and its importance in fostering international cooperation during times of crisis.

The carrier continued to participate in key military exercises throughout 2012 and 2013, enhancing its capabilities and strengthening ties with allied navies. Exercises such as Talisman Sabre with the Australian Navy and RIMPAC (Rim of the Pacific Exercise) highlighted George Washington's role in promoting regional security and interoperability.

In August 2014, George Washington returned to the United States to undergo a mid-life Refueling and Complex Overhaul (RCOH) at Newport News Shipbuilding. This extensive maintenance period, expected to last approximately four years, involved refueling the ship's nuclear reactors, upgrading combat systems, and refurbishing living quarters and other essential areas. The RCOH was crucial in extending the carrier's operational life and ensuring it remained at the forefront of naval technology.

During the RCOH, George Washington underwent significant upgrades and overhauls. These included modernizing its flight deck, installing new radar systems, and improving its defensive capabilities. The carrier's crew also underwent training and preparations to ensure they were ready to resume full operations once the overhaul was complete.

In mid-2018, USS George Washington completed its RCOH and began sea trials to test its new systems and ensure operational readiness. The carrier's return to service was marked by rigorous training exercises and a renewed focus on integrating new technologies and capabilities acquired

After rejoining the fleet, George Washington resumed its role as a vital component of the U.S. Navy's forward-deployed forces. The carrier participated in various exercises and missions, reinforcing its position as a key player in the Indo-Pacific region. The ship continued to foster cooperation with allies and enhance maritime security through joint operations and training.

The COVID-19 pandemic presented unique challenges for the crew of USS George Washington. The carrier implemented strict health protocols to safeguard the crew while maintaining operational readiness. Despite these challenges, George Washington continued to participate in essential missions and exercises, demonstrating resilience and adaptability.

In 2022, USS George Washington deployed to the Indo-Pacific region under the command of Capt. Amy Bauernschmidt, the first female to command a nuclear-powered aircraft carrier. The carrier participated in RIMPAC 2022, the world's largest international maritime warfare exercise, and conducted freedom of navigation operations in the South China Sea. These operations reinforced the United States' commitment to maintaining free and open sea lanes and supporting regional allies.

In 2023, USS George Washington continued its operational activities, participating in numerous training exercises to maintain proficiency and readiness. The carrier made a port visit to Yokosuka, Japan, and engaged in joint exercises with regional allies. George Washington's activities underscored its critical role in promoting regional stability and security.

CVN-74 • USS JOHN C STENNIS

(Propulsion: Two nuclear reactors, four shafts. Length: 1,092 feet (332.85 meters). Beam: 134 feet (40.84 meters); Flight Deck Width: 252 feet (76.8 meters). Displacement: Approximately 97,000 tons (87,996.9 metric tons) full load. Speed: 30+ knots (34.5+ miles per hour). Crew: Ship's Company: 3,200 - Air Wing: 2,480. Armament: Two or three (depending on modification) NATO Sea Sparrow launchers, 20mm Phalanx CIWS mounts: (3 on Nimitz and Dwight D. Eisenhower and 4 on Vinson and later ships of the class.). Aircraft: 85).

The nuclear-powered USS John C. Stennis (CVN-74) was contracted on March 29, 1988, and the keel was laid on March 13, 1991, at Newport News Shipbuilding Co., Newport News, VA.

The ship was christened on November 11, 1993, in honor of Senator John Cornelius Stennis (D-Mississippi), who served in the Senate from 1947 to 1989. The daughter of the ship's namesake, Mrs. Margaret Stennis-Womble, was the ship's sponsor. Stennis was commissioned on December 9, 1995, at Naval Station Norfolk, VA, and conducted flight deck certification in January 1996. The first arrested landing was by a VX-23 F-14B. The ship conducted numerous Carrier Qualifications and Independent Steaming Exercises off the East Coast throughout the next two years. Among these events was the first carrier landing of an F/A-18E/F Super Hornet on January 18, 1997.

USS John C. Stennis and the smaller British Invincible-class HMS Illustrious on a joint patrol, April 1998. On February 26, 1998, with Carrier Air Wing Seven embarked, Stennis left Norfolk for her maiden deployment, transiting the Suez Canal on March 7 and arriving in the Persian Gulf on March 11, 1998. The ship traveled 8,020 nautical miles in 274 hours, an average speed of 29.4 knots (54.4 km/h), to relieve USS George Washington (CVN-73) in conducting Operation Southern Watch missions. Stennis departed the Persian Gulf on July 19, 1998, for her new homeport of Naval Air Station North Island in San Diego, California, arriving on August 26, 1998. In October 1998, Stennis entered a six-month Phased Incremental Availability for maintenance and upgrades at North Island, returning to sea in April 1999. During the maintenance period, a jet blast deflector collapsed, severely injuring two sailors. On November 30, 1999, Stennis ran aground in a shallow area adjacent to the turning

basin near North Island. Silt clogged the intake pipes to the steam condensing systems for the nuclear reactor plants, causing the carrier's two nuclear reactors to be shut down (one reactor by crew, the other automatically) for a period of 45 minutes. Stennis was towed back to her pier for maintenance and observation for the next two days. The cleanup cost was about $2 million.

On January 7, 2000, Stennis deployed to the Persian Gulf to relieve USS John F. Kennedy (CV-67) in Operation Southern Watch. During the deployment, the ship made port visits to South Korea, Hong Kong, Malaysia, Bahrain, the United Arab Emirates, and Australia, before returning to San Diego on July 3, 2000. Following the September 11, 2001, terrorist attacks, Stennis conducted Noble Eagle missions off the U.S. West Coast.

On November 12, 2001, two months earlier than scheduled, the ship left on her third deployment to the U.S. Fifth Fleet area of responsibility in support of Operation Enduring Freedom, returning to San Diego on May 28, 2002. From June 2002 to January 2003, JCS underwent a seven-month Planned Incremental Availability (PIA).

From May 24 to November 1, 2004, Stennis conducted her fourth major overseas deployment, participating in Exercise Northern Edge 2004 in the Gulf of Alaska, Rim of the Pacific (RimPac) Exercise off Hawaii, exercises with Kitty Hawk off Japan, and goodwill visits to Japan, Malaysia, and Western Australia. Shortly after returning from deployment to San Diego, JCS changed her homeport to Naval Station Bremerton, Washington, on January 19, 2005. Once at Bremerton, Stennis underwent an 11-month Docking Planned Incremental Availability (DPIA), the first time she had been dry-docked since commissioning. Upgrades included a new mast.

On January 20, 2007, Stennis set sail for the Persian Gulf as part of an increase in U.S. military presence within the Middle East. Stennis arrived in the area on February 19, 2007, joining USS Dwight D. Eisenhower in the United States Fifth Fleet area of operations. This marked the first time since 2003 that there were two aircraft carrier battle groups in the region simultaneously. On May 23, 2007, Stennis, along with eight other warships including the aircraft carrier Nimitz and amphibious assault ship Bonhomme Richard, passed through the Strait of

Hormuz. U.S. Navy officials said it was the largest such move since 2003. On August 31, 2007, Stennis returned to Bremerton.

Stennis departed Bremerton for a six-month deployment to the Western Pacific on January 13, 2009. On April 24, the ship arrived in Singapore. That same day, one of the ship's sailors was crushed and killed while working from a small harbor boat to secure a drain that discharges oily water from Stennis' aircraft catapults. On April 29, the ship's executive officer, Commander David L. Burnham, was relieved by Rear Admiral Mark A. Vance over unspecified personal conduct. Burnham was reassigned to a base in San Diego, pending an investigation. After participating in operations in the Persian Gulf, exercises with the Japan Maritime Self Defense Force and the Republic of Korea, as well as joint exercise Northern Edge 2009, USS Stennis returned from deployment in early July 2009. Carrier Air Wing 9 debarked on July 6 at NAS North Island, prior to the ship's arrival at her homeport of Bremerton, WA, on July 10.

In 2010, USS John C. Stennis underwent a series of maintenance and training exercises to ensure continued readiness for future deployments. This period included carrier qualifications, flight deck certifications, and other essential drills to maintain the crew's proficiency and the ship's operational capabilities.

In July 2011, Stennis departed for a scheduled deployment to the Western Pacific and the Middle East. The carrier joined the Fifth Fleet area of responsibility, where it supported Operations Enduring Freedom and New Dawn. During this deployment, Stennis conducted numerous sorties in support of ground operations in Afghanistan and provided maritime security in the region.

After a successful deployment, Stennis returned to her homeport in Bremerton, Washington, in early 2012. The ship then underwent routine maintenance and preparations for her next deployment. The crew conducted various training exercises to ensure readiness for future missions.

In late 2012, Stennis was called upon for a quick turnaround deployment due to escalating tensions in the Middle East. The carrier departed in August 2012 and once again operated in the Fifth Fleet area of responsibility. Stennis played a crucial role in maintaining regional stability and supporting coalition forces in Afghanistan. The ship returned to Bremerton in May 2013.

In 2014, USS John C. Stennis underwent a scheduled maintenance period, including upgrades to various systems and equipment to enhance the ship's operational capabilities. This period allowed the crew to perform necessary repairs and ensure the carrier was prepared for future deployments.

In January 2016, Stennis departed Bremerton for another deployment to the Western Pacific. The carrier participated in numerous exercises with allied navies, including the Republic of Korea Navy and the Japan Maritime Self-Defense Force. These joint exercises aimed to enhance interoperability and strengthen regional security alliances. During this deployment, Stennis also conducted freedom of navigation operations in the South China Sea, reaffirming the U.S. commitment to maintaining open sea lanes and challenging excessive maritime claims. The carrier returned to Bremerton in August 2016.

Throughout 2017, USS John C. Stennis focused on maintaining readiness through various training exercises and certifications. The crew conducted flight deck operations, damage control drills, and other essential training to ensure the ship's operational capabilities remained at their peak.

In early 2018, Stennis entered a Planned Incremental Availability (PIA) period at Puget Sound Naval Shipyard. This maintenance period included upgrades to the ship's systems, repairs to the hull, and improvements to the living quarters for the crew. The PIA was completed in late 2018, and Stennis returned to sea for post-maintenance sea trials and certifications.

In 2019, USS John C. Stennis embarked on a deployment to the Western Pacific and the Middle East. During this deployment, the carrier supported various operations, including maritime security and joint exercises with allied navies. In May

2019, it was announced that Stennis would undergo a homeport change to Norfolk, Virginia, following the deployment.

In early 2020, USS John C. Stennis arrived at Newport News Shipbuilding in Virginia to begin her Refueling and Complex Overhaul (RCOH). This extensive maintenance period, expected to last several years, includes refueling the ship's reactors, upgrading combat systems, and refurbishing living quarters and other essential areas. The RCOH will extend the carrier's service life by another 25 years, ensuring it remains a critical asset for the U.S. Navy.

Throughout 2021 and 2022, USS John C. Stennis continued to undergo significant upgrades and maintenance as part of the RCOH. This period involved extensive work on the ship's nuclear propulsion systems, combat systems, and other critical components. The crew also participated in various training programs to prepare for the ship's return to active service.

As of 2023, the RCOH for USS John C. Stennis is progressing on schedule. The carrier is expected to complete the overhaul and return to the fleet in the coming years, ready to resume its role as a key element of the U.S. Navy's carrier strike capabilities.

CVN-75 • USS HARRY S TRUMAN

(Propulsion: Two nuclear reactors, four shafts. Length: 1,092 feet (332.85 meters). Beam: 134 feet (40.84 meters); Flight Deck Width: 252 feet (76.8 meters). Displacement: Approximately 97,000 tons (87,996.9 metric tons) full load. Speed: 30+ knots (34.5+ miles per hour). Crew: Ship's Company: 3,200 - Air Wing: 2,480. Armament: Two or three (depending on modification) NATO Sea Sparrow launchers, 20mm Phalanx CIWS mounts: (3 on Nimitz and Dwight D. Eisenhower and 4 on Vinson and later ships of the class.). Aircraft: 85.)

USS Harry S. Truman (CVN-75) is the ninth nuclear-powered aircraft carrier and the eighth in the Nimitz class. The keel was laid by Newport News Shipbuilding on November 29, 1993, and the ship, named after the 33rd President of the United States, was christened on September 7, 1996. The ship was launched on September 13, 1996, and delivered to the United States Navy on June 30, 1998. Harry S. Truman was commissioned on July 25, 1998.

In August 1998, Harry S. Truman put to sea for the first time as a U.S. Navy carrier to conduct flight deck certifications, an evolution designed to test the ship's ability to successfully launch and recover aircraft.

After an extensive work-up cycle that tested the ship and challenged the crew, Harry S. Truman embarked on its maiden deployment on November 28, 2000. On February 16, 2001, the Truman team was called into action in support of Operation Southern Watch. In response to surface-to-air missile fire against coalition forces enforcing United Nations Security Council Resolutions, aircraft from Carrier Air Wing Three (CVW-3) took part in a sanctioned response and struck Iraqi integrated air defense system sites. The air wing flew 869 combat sorties, totaling more than 2,700 flight hours.

After nearly 180 days and traveling more than 44,000 nautical miles, Harry S. Truman completed its first deployment on May 23, 2001.

Harry S. Truman entered the Norfolk Naval Shipyard for its first Planned Incremental Availability on September 5, 2001. Six months later, Harry S. Truman began training for its second deployment in support of Operation Enduring Freedom, which began on December 5, 2002. Tensions with Iraq were high as the Truman team prepared for the possibility of conflict.

In early February 2003, Harry S. Truman took up station in the Eastern Mediterranean and waited for word to launch strikes against Saddam Hussein's regime in Iraq. When the order came in the late-night hours of March 19, Operation Iraqi Freedom was in full swing. Harry S. Truman successfully launched 1,280 sorties, dropped more than 700 tons of ordnance, and helped liberate the people of Iraq. Harry S. Truman was relieved of its Operation Iraqi Freedom support duties on April 18, 2003.

In August 2003, Harry S. Truman entered Norfolk Naval Shipyard for its second Planned Incremental Availability. The ship pulled out of the shipyard on February 13, 2004, four days early and under budget.

After leaving the shipyard, the Harry S. Truman team focused on preparing to re-enter the Global War on Terrorism. Under the Navy's new Fleet Response Plan, Harry S. Truman surged to the Mediterranean as one of seven carriers participating in Summer Pulse '04, while simultaneously completing the required training exercises that would ensure the ship and air wing team were ready to act when called upon.

After another successful yard period and an intense work-up cycle, Truman departed Norfolk on a seven-month combat deployment in November 2007. Throughout the deployment, Truman supported Maritime Security Operations in the Mediterranean Sea and the Persian Gulf and safely launched and recovered over 15,300 fixed and rotary-wing aircraft. Simultaneously, aircraft from CVW-3 flew more than 26,500 hours during 9,500 sorties. Of these, 2,459 were combat sorties directly supporting Coalition forces serving in Operation Iraqi Freedom. The air wing flew just under 14,000 combat hours and expended 77,500 pounds of ordnance.

Truman completed her third Planned Incremental Availability in February 2009 and has undergone numerous underways since that time, including an ammunition onload in March, Tailored Ship's Training Availability (TSTA) and

Final Evaluation Problem (FEP) in April and May, Board of Inspection and Survey (INSURV) and Combined Training Unit Exercise (COMTUEX) in June, squadron Carrier Qualifications in July and August, and Joint Task Force Exercise (JTFEX) in September. Planned Carrier Qualifications were scheduled up to her next deployment in 2010. USS Harry S. Truman (CVN-75) has won five Battle "E" awards: 2003, 2004, 2005, 2008, and 2009.

In May 2010, USS Harry S. Truman deployed for her fifth overseas deployment. The carrier was part of Operation Enduring Freedom in the Arabian Sea, where it launched sorties to support coalition forces in Afghanistan. The air wing flew over 2,000 combat sorties, providing crucial air support to ground operations. Truman returned to Norfolk in December 2010 after a seven-month deployment.

Following her return, Harry S. Truman underwent maintenance and upgrades during another Planned Incremental Availability (PIA) period. This included updates to the ship's combat systems and general maintenance to ensure operational readiness. The carrier also participated in various training exercises to maintain the crew's proficiency.

In July 2013, Truman deployed to the U.S. Fifth Fleet area of responsibility. During this deployment, the carrier supported Operation Enduring Freedom and participated in numerous maritime security operations. The deployment was initially scheduled to end in April 2014, but it was extended due to the ongoing instability in the region. Truman finally returned to Norfolk in November 2014, marking one of her longest deployments.

In 2015, USS Harry S. Truman entered another Planned Incremental Availability (PIA) at Norfolk Naval Shipyard. This period involved significant maintenance and upgrades, including improvements to the ship's radar systems, flight deck,

and living quarters for the crew. The carrier emerged from PIA in early 2016, ready for another operational cycle.

In November 2015, Harry S. Truman departed for another deployment, this time to support Operation Inherent Resolve against ISIS in Iraq and Syria. The carrier conducted numerous sorties, delivering airstrikes against ISIS targets and providing vital support to coalition forces. Truman returned to Norfolk in July 2016 after a successful deployment.

After her return, the carrier underwent routine maintenance and upgrades to ensure continued readiness. This period included modernization efforts to enhance the ship's capabilities and extend its operational life.

In April 2019, USS Harry S. Truman deployed once again, this time participating in high-tempo operations across multiple theaters, including the Mediterranean Sea and the Arabian Gulf. The carrier conducted freedom of navigation operations, participated in joint exercises with allied navies, and supported ongoing military operations in the region. Truman returned to Norfolk in December 2019.

The COVID-19 pandemic in 2020 brought unique challenges to the crew of USS Harry S. Truman. The carrier was deployed in the Atlantic Ocean and the Mediterranean Sea, where it conducted operations while implementing strict health protocols to protect the crew from the virus. Despite the pandemic, Truman successfully completed her mission and returned to Norfolk in June 2020.

Throughout 2021, USS Harry S. Truman focused on maintenance and training to ensure operational readiness. This included routine repairs, system upgrades, and various training exercises to keep the crew prepared for future deployments.

In December 2021, USS Harry S. Truman deployed to the Sixth Fleet area of operations, primarily in the Mediterranean Sea. The carrier supported NATO allies and participated in several high-profile exercises to enhance interoperability and regional security. Truman's presence was a deterrent against potential threats and a reassurance to allies in the region. The carrier returned to Norfolk in mid-2022.

In 2023, USS Harry S. Truman continued to maintain high readiness levels through scheduled maintenance, training exercises, and preparations for future deployments. The carrier participated in joint exercises with international partners and played a key role in demonstrating the U.S. Navy's commitment to global maritime security.

CVN-76 • USS RONALD REAGAN

(Propulsion: Two nuclear reactors, four shafts. Length: 1,092 feet (332.85 meters). Beam: 134 feet (40.84 meters); Flight Deck Width: 252 feet (76.8 meters). Displacement: Approximately 97,000 tons (87,996.9 metric tons) full load. Speed: 30+ knots (34.5+ miles per hour). Crew: Ship's Company: 3,200 - Air Wing: 2,480. Armament: Two or three (depending on modification) NATO Sea Sparrow launchers, 20mm Phalanx CIWS mounts: (3 on Nimitz and Dwight D. Eisenhower and 4 on Vinson and later ships of the class.). Aircraft: 85.)

The contract to build USS Ronald Reagan was awarded to Northrop Grumman Newport News and Dry Dock Company in Newport News, Virginia, on December 8, 1994. Her keel was laid down on February 12, 1998. The budget for the ship had to be increased several times, ultimately costing $4.5 billion. Reagan was christened by Reagan's wife Nancy on March 4, 2001, at Newport News Shipbuilding. The crew moved aboard on October 30, 2002, and the ship was commissioned on July 12, 2003, at Naval Station Norfolk, with Captain J. W. Goodwin in command. Vice President Dick Cheney and Lynne Cheney were present at the ceremony, as well as Nancy Reagan, who gave the ship's crew the traditional first order as an active unit of the Navy: "Man the ship and bring her to life." Ronald Reagan made her maiden voyage on July 21, 2003. President Reagan, who did not attend either the launch or the commissioning due to Alzheimer's disease, died eleven months later. At the end of the graveside services, the ship's commanding officer at that time, Captain James Symonds, presented the flag that draped the former president's casket to Mrs. Reagan at her request. This was also the flag that had flown over Capitol Hill on January 20, 1981, when the president was inaugurated. At a later date, Captain Symonds also presented Mrs. Reagan with the flag that had been flying over Ronald Reagan when the former president died.

Ronald Reagan conducted her maiden port visit in Ft. Lauderdale, Florida, from November 11-14, 2003. After a five-month post shakedown availability (PSA), Reagan received flight deck certification on May 8, 2004, and then began her transit from Naval Station Norfolk, Virginia, around Cape Horn, South America, to her new homeport of Naval Air Station North Island, San Diego, California. Carrier Air Wing Eleven, which is normally assigned to USS Nimitz (CVN-68), embarked only 25 percent of its total strength for the transit. The squadrons making the transit were VFA-14 and VFA-41 flying the F/A-18E/F Super Hornet, VAW-117 flying the E-2C Hawkeye 2000, HS-6 flying the SH-60F Seahawk, and VRC-30 flying the C-2A Greyhound. The ship transited the Strait of Magellan on June 20-21 and made port visits to Rio de Janeiro, Brazil; Valparaiso, Chile; and Callao, Peru before arriving in San Diego on July 23, 2004.

USS Ronald Reagan departed San Diego on January 4, 2006, on her maiden deployment to conduct naval operations in support of Operation Iraqi Freedom and Operation Enduring Freedom, as well as to conduct maritime security operations (MSO) in the Persian Gulf. The ship entered the Gulf on February 22, 2006, and returned from deployment on July 6, 2006.

USS Ronald Reagan and the Reagan Carrier Strike Group (CSG) departed North Island, Coronado in San Diego on January 27, 2007, on an unscheduled surge deployment to the Western Pacific, fulfilling the role of the forward-deployed carrier Kitty Hawk while it underwent maintenance in Japan. On April 20, 2007, USS Ronald Reagan and her CSG returned to Coronado. The "surge deployment" was part of the Navy's Fleet Response Plan (FRP), which provides the U.S. with the ability to respond to any global commitment with flexible and sustainable forces and the ability to rapidly respond to a range of situations on short notice. In January 2007, it was announced that USS Ronald Reagan had earned the 2006 Commander Naval Air Force, Pacific Carrier Battle Efficiency "E" award for the West Coast, the first Battle "E" ever for the carrier. On January 28, 2006, an F/A-18 Hornet strike fighter attempting a night landing aboard Reagan crashed into the ship's flight deck about 200 km (120 miles) southeast of Brisbane, Australia. The aircraft struck the ramp at a low angle, caught fire, and skidded overboard. The pilot ejected safely, but the aircraft was lost. Reagan returned to Naval Air Station North Island on April 20, 2007, following the three-month deployment in support of operations in the Western Pacific.

On December 15, 2007, the carrier answered a distress call from a cruise ship off the coast of Baja California. An Illinois teenager whose appendix had ruptured while on a Mexican cruise was airlifted by an SH-60 helicopter to Ronald Reagan, where an emergency appendectomy was performed by the ship's surgeon. In 2008, USS Ronald Reagan, with CVW-14 embarked, departed San Diego on May 19 for a scheduled 7th Fleet and 5th Fleet deployment. The Reagan Carrier Strike Group performed humanitarian assistance and disaster relief operations in the Philippines on June 24, 2008, after the country was devastated by Typhoon Fengshen, killing hundreds from the central island regions and the main island of Luzon. The typhoon also capsized the passenger ferry MV Princess of the Stars. Working in support of the Armed Forces of the Philippines, Reagan and her escorts of Carrier Strike Group 7 focused their efforts on the island of Panay in the Central Visayas. For eight days, SH-60 Seahawk helicopters and C-2A Greyhound aircraft of the Ronald Reagan Strike Group helped deliver more than 519,000 lbs. of rice, fresh water, and other supplies to areas of Panay, which were not reachable via truck due

to flooded roads. The mission in Panay would earn the entire strike group the Navy's Humanitarian Service Medal. The Strike Group arrived in the U.S. Fifth Fleet area on August 28, 2008, where she launched more than 1,150 sorties into Afghanistan in support of Operation Enduring Freedom. Reagan returned to San Diego on November 25, 2008.

In February 2009, USS Ronald Reagan received word that the ship had won its second Battle Efficiency Award. On May 28, 2009, Reagan deployed with Carrier Air Wing 14 to the 7th and 5th Fleet Areas of Responsibility. Reagan relieved the Eisenhower CSG and launched its first sorties in support of OEF on July 6. Reagan returned to homeport on October 21, 2009, after a five-month deployment.

In 2010, USS Ronald Reagan underwent a scheduled maintenance period and various training exercises to maintain operational readiness. The carrier continued to support Operation Enduring Freedom with deployments in 2011 and 2012, conducting numerous sorties and providing crucial support to coalition forces in Afghanistan. In 2013, Ronald Reagan entered dry dock for a comprehensive maintenance period, including upgrades to various systems and general maintenance.

After completing the maintenance period, USS Ronald Reagan returned to operational status and participated in various training exercises and operations. The carrier continued to demonstrate its capability to project power and support global missions.

In 2015, the USS Ronald Reagan replaced the USS George Washington (CVN-73) as the forward-deployed carrier in Yokosuka, Japan. This marked a significant change in the carrier's operational focus, with an increased emphasis on supporting U.S. interests in the Indo-Pacific region. Throughout 2016 and 2017, Ronald Reagan participated in numerous joint exercises with regional allies, including the Republic of Korea and Japan Maritime Self-Defense Force, enhancing interoperability and regional security.

USS Ronald Reagan continued its forward-deployed operations in the Indo-Pacific region, participating in high-profile exercises such as RIMPAC and Talisman Sabre. The carrier conducted freedom of navigation operations and supported various missions to ensure regional stability. In 2020, despite the challenges posed by the COVID-19 pandemic, Ronald Reagan maintained its operational readiness and continued to project power in the region.

In 2021, USS Ronald Reagan continued its deployment in the Indo-Pacific, participating in joint exercises and supporting regional security efforts. The carrier played a crucial role in maintaining a forward U.S. presence and enhancing partnerships with regional allies. Throughout 2022 and 2023, Ronald Reagan remained a key asset in the U.S. Navy's strategy, participating in numerous training exercises, operations, and missions to ensure global maritime security.

Conclusion: From her commissioning in 2003 to the present day, USS Ronald Reagan has demonstrated exceptional versatility and capability. Whether conducting combat operations, humanitarian missions, or supporting regional security, the carrier has consistently fulfilled its mission. As USS Ronald Reagan continues to sail the world's oceans, she remains a symbol of the U.S. Navy's enduring strength and commitment to global security.

CVN-77 • USS GEORGE HW BUSH

(Propulsion: Two nuclear reactors, four shafts. Length: 1,092 feet (332.85 meters). Beam: 134 feet (40.84 meters); Flight Deck Width: 252 feet (76.8 meters). Displacement: Approximately 97,000 tons (87,996.9 metric tons) full load. Speed: 30+ knots (34.5+ miles per hour). Crew: Ship's Company: 3,200 - Air Wing: 2,480. Armament: Two or three (depending on modification) NATO Sea Sparrow launchers, 20mm Phalanx CIWS mounts: (3 on Nimitz and Dwight D. Eisenhower and 4 on Vinson and later ships of the class.). Aircraft: 85.)

Construction and Commissioning:
The contract to build USS George H.W. Bush (CVN-77) was awarded on January 26, 2001. The ship's construction spanned seven years, during which significant milestones were achieved: • The keel was laid on September 6, 2003. • The ship was christened on October 7, 2006. • USS George H.W. Bush was commissioned on January 10, 2009.

The construction of the ship involved the use of 47,000 tons of structural steel and about a million pounds of aluminum. A modular construction process was employed, forming large individual units of the ship much like interlocking building blocks. These units were welded together to form a module or superlift, with each superlift weighing up to 900 tons.

Following its commissioning in January 2009, USS George H.W. Bush began initial operations and sea trials. These trials tested the ship's systems, capabilities, and crew readiness. The carrier successfully completed flight deck certifications, carrier qualifications, and other essential evaluations, ensuring that the ship was ready for operational deployment.

In May 2010, USS George H.W. Bush embarked on its maiden deployment. The carrier, along with Carrier Air Wing Eight (CVW-8), conducted operations in the Mediterranean Sea and the Arabian Sea. The deployment included participation in various multinational exercises, maritime security operations, and support for Operation Enduring Freedom. The carrier returned to its homeport in Norfolk, Virginia, in December 2010.

From 2011 to 2013, USS George H.W. Bush continued to conduct operations and participate in numerous exercises. In 2011, the carrier was involved in operations in the Arabian

Sea, providing air support for coalition forces in Afghanistan. The ship also participated in joint exercises with allied navies, enhancing interoperability and regional security.

In 2012, USS George H.W. Bush underwent maintenance and upgrades to ensure continued operational readiness. The carrier participated in various training exercises and certifications, preparing the crew for future deployments.

In February 2014, USS George H.W. Bush deployed to the U.S. Fifth Fleet area of responsibility, conducting operations in the Arabian Gulf. The carrier supported Operation Enduring Freedom and later played a crucial role in Operation Inherent Resolve, launching airstrikes against ISIS targets in Iraq and Syria. The deployment demonstrated the carrier's ability to project power and provide sustained air support for ground operations. The ship returned to Norfolk in November 2014.

Throughout 2015 and 2016, USS George H.W. Bush underwent a scheduled maintenance period, including a Planned Incremental Availability (PIA) at Norfolk Naval Shipyard. This maintenance period involved overhauls of critical systems, upgrades to combat capabilities, and general upkeep to ensure the carrier's long-term operational readiness. The crew also participated in various training exercises and certifications to maintain their proficiency.

In January 2017, USS George H.W. Bush deployed to the Mediterranean Sea and the Arabian Gulf. The carrier conducted combat operations against ISIS as part of Operation Inherent Resolve, launching numerous sorties and delivering precision airstrikes. The deployment also included participation in multinational exercises with NATO allies, enhancing cooperation and interoperability. The carrier returned to Norfolk in August 2017 after a successful and impactful deployment.

Following the 2017 deployment, USS George H.W. Bush focused on maintaining operational readiness through routine maintenance, training exercises, and certifications. In 2018, the carrier underwent a Drydocking Planned Incremental Availability (DPIA) at Norfolk Naval Shipyard, ensuring the ship remained in peak condition. The crew continued to participate in training exercises, preparing for future deployments and operational requirements.

The COVID-19 pandemic in 2020 brought unique challenges to the crew of USS George H.W. Bush. Despite the global health crisis, the carrier maintained its readiness and continued to support U.S. Navy operations. Strict health protocols were implemented to protect the crew, ensuring the carrier could continue its mission. In 2021, USS George H.W. Bush participated in various exercises and training operations, demonstrating resilience and adaptability.

In 2022, USS George H.W. Bush continued to maintain high readiness levels through scheduled maintenance, training exercises, and preparations for future deployments. The carrier participated in joint exercises with international partners, enhancing interoperability and strengthening alliances. The ship's activities underscored its critical role in supporting U.S. Navy operations and maintaining global maritime security.

*An artist's conception (2004) of CVN-78

CVN-78 • USS GERALD R FORD

GERALD R. FORD CLASS:
BUILDER: Huntington Ingalls Industries – Newport News Shipbuilding
COMMISSIONED: July 22, 2017
PROPULSION: Two nuclear reactors, four shafts
LENGTH: 1,092 feet
BEAM: 134 feet, Flight Deck Width: 256 feet
DISPLACEMENT: Approximately 100,000 long tons full load
SPEED: 30+ knots (34.5+ miles per hour)
CREW: Approximately 4,550 (ship, air wing, and staff)
AIRCRAFT: 75+
ARMAMENT: Evolved Sea Sparrow Missile, Rolling Airframe Missile, CIWS

Overview The USS Gerald R. Ford (CVN-78) is the lead ship of the Ford-class aircraft carriers of the United States Navy. It is the most technologically advanced aircraft carrier ever built and is designed to replace the Nimitz-class carriers.

Early History—Construction and Launch: The contract for the construction of USS Gerald R. Ford was awarded to Huntington Ingalls Industries – Newport News Shipbuilding on January 10, 2007. The keel was laid on November 14, 2009. The ship was launched on October 11, 2013, in a ceremony attended by shipbuilders, Navy officials, and the Ford family. The ship was christened by Susan Ford Bales, daughter of the ship's namesake, President Gerald R. Ford.

Commissioning: USS Gerald R. Ford was commissioned on July 22, 2017, at Naval Station Norfolk. The commissioning ceremony was a significant event, marking the introduction of the most advanced aircraft carrier in the world into the U.S. Navy fleet. President Donald Trump delivered the principal address, highlighting the ship's advanced capabilities and its role in strengthening national security.

Technological Advancements: Electromagnetic Aircraft Launch System (EMALS). One of the most notable features of the USS Gerald R. Ford is the Electromagnetic Aircraft Launch System

(EMALS). This system replaces traditional steam catapults with electromagnetic energy to launch aircraft. EMALS provides a smoother acceleration, which reduces stress on the airframes of the aircraft and allows for the launch of a wider range of aircraft types.

Advanced Arresting Gear (AAG): The Advanced Arresting Gear (AAG) is another innovative feature. It uses a new system of cables and winches to catch aircraft landing on the carrier. This system is designed to handle a broader range of aircraft, including lighter unmanned aerial vehicles and heavier manned aircraft, providing greater flexibility and safety.

New Features: The USS Gerald R. Ford incorporates numerous other advanced technologies, including:

- New Radar System: Enhances the carrier's ability to detect and track targets.
- Advanced Weapons Elevators: Designed to move ordnance more efficiently and safely.
- Flight Deck Design: Redesigned to improve aircraft handling, sortie generation rates, and operational efficiency.

Deployment and Service: The USS Gerald R. Ford is currently deployed in the Atlantic Ocean. Its advanced systems and technologies are continually tested and refined to ensure operational readiness. The carrier is expected to remain in service for at least 50 years, playing a crucial role in projecting U.S. naval power and ensuring maritime security.

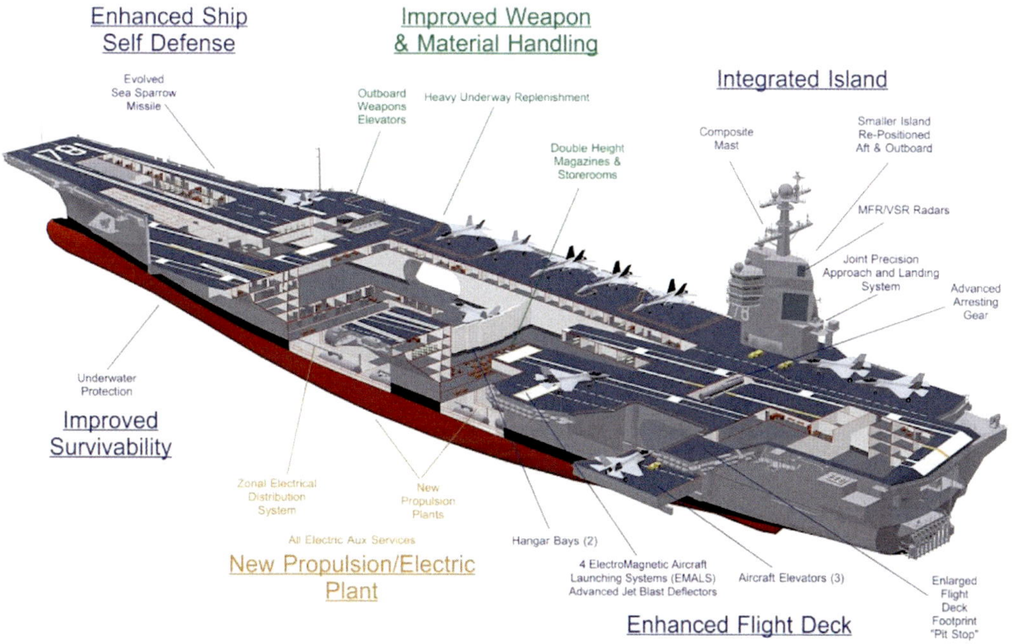

CVN-79 • USS JOHN F KENNEDY

*An artist's conception (2004) of CVN-79

**An artist's conception of CVN-80*

CVN-80 • USS ENTERPRISE

CVN-81 • USS DORIS MILLER

CANCELLED CARRIERS

CV-35 .... USS REPRISAL
CVB-44 .... UNNAMED • NEVER LAID DOWN
CV-46 .... IWO JIMA • NEVER COMPLETED
CV-50 .... UNNAMED • NEVER COMPLETED
CV-51 .... UNNAMED • NEVER COMPLETED
CV-52 .... UNNAMED • NEVER COMPLETED
CV-53 .... UNNAMED • NEVER COMPLETED
CV-54 .... UNNAMED • NEVER COMPLETED
CV-55 .... UNNAMED • NEVER COMPLETED
CVB-56 .... UNNAMED • NEVER COMPLETED
CVB-57 .... UNNAMED • NEVER COMPLETED
CVA-58 .... UNITED STATES

Made in the USA
Las Vegas, NV
07 November 2026

2f30248e-7d42-436d-8c0e-3463f90f1cd8R01